Alex Bellos is the *Guardian*'s maths blogger, and has [illegible] paper in London and Rio de Janeiro, whe[illegible] foreign correspondent. He is a curat[illegible] Museum and has a degree in Mathematics and Philosophy from the University of Oxford. In 2002 he wrote a critically acclaimed book about Brazilian football, *Futebol: The Brazilian Way of Life* and in 2006 ghostwrote Pelé's autobiography, which was a number one bestseller. *Alex Through the Looking-Glass: How Life Reflects Numbers and Numbers Reflect Life*, a *Sunday Times* bestseller, was published in April 2014. He lives in London.
www.alexbellos.com

Praise for *Alex's Adventures in Numberland*:

An Amazon.com Top Ten Science Book of the Year

'A mathematical wonder that will leave you hooked on numbers . . . It's hard not to get swept away by Bellos's enthusiasm' *Daily Telegraph*

'[Bellos] is like an intrepid cosmic explorer, floating in an airship over a strange planet, and describing the fascinating things he sees' *Observer*

'[Bellos] renders the world of numbers accessible and captivating'
Daily Express

'An entertaining and accessible account of how maths underpins our lives . . . covers everything from the Rubik's cube to the probability of winning on slot machines. Maths will never appear so forbidding again' *TES*

'It is to be hoped that the uncountable delights of Bellos's book, its verve and feeling for mathematics, convey its enchantments to a new generation'
Times Literary Supplement

'I would urge all those who are innumerate and proud of it to read this brilliant book and think again – maths can be human, quirky and fun'
Roger Highfield, editor of *New Scientist*

'There is no shortage of good popular maths books. It's only so often, though, that one comes along that is outstanding – and *Alex's Adventures in Numberland* is one of them. The style is laced with humour, but at all times, the star of the show is the mathematics. It's a beautiful book and deserves to be a huge success' *Prospect*

'What Bellos calls "the wow factor" of mathematics leaps out at the reader from every page of this remarkable foray into the realm of numbers. The stories prove so engaging, the personalities so colourful, that readers may forget they are mastering some powerful mathematical concepts. It is, indeed, these concepts – not the piquant personalities – that deliver the most wows. And in the wonderland of multiple infinities, Bellos saves the biggest wow for last. Intellectual entertainment of the first order'

Booklist

'A smorgasbord for math fans of all abilities' *Kirkus Reviews*

'An unforgettable journey of intellectual discovery, a true transcultural Magical Mystery Tour' Apostolos Doxiadis, author of *Logicomix*

Praise for *Alex Through the Looking-Glass*:

'Another sparkling romp through the world of numbers, with the inimitable Alex Bellos as your friendly, informed, and crystal-clear guide. A brilliant successor to *Alex's Adventures in Numberland*'

Ian Stewart, author of *The Great Mathematical Problems*

'Fresh, fascinating and endlessly charming.'

Tim Harford, author of *The Undercover Economist*

'To read *Alex Through The Looking-Glass* is to have one's mind quietly but continually blown with the knowledge that the world, so seemingly complex, is constantly conforming to patterns' *Sunday Express*

'If anything, *Looking-Glass* is a better work than *Numberland* – it feels more immediate, more relevant and fun. ****' *Daily Telegraph*

'Bellos [is] on fine form, guiding readers through the mathematical landscape on a series of adventures that range from jaunts in the foothills to clambering up some quite challenging peaks'

Simon Singh, *Observer*

Alex's Adventures in Numberland

ALEX BELLOS

Illustrations by Andy Riley

BLOOMSBURY

LONDON • NEW DELHI • NEW YORK • SYDNEY

For my mother and father

First published in Great Britain in 2010
This paperback edition published 2011

Bloomsbury Publishing Plc, 50 Bedford Square, London WC1B 3DP

Bloomsbury is a trademark of Bloomsbury Publishing Plc

Bloomsbury Publishing, London, New Delhi, New York and Sydney

A CIP catalogue record for this book is available from the British Library

ISBN 978 1 4088 0959 4

17

Designed by Grade Design Consultants www.gradedesign.com
Mathematical diagrams by Oxford Designers and Illustrators

Printed and bound in Great Britain by CPI Group (UK) Ltd, Croydon CR0 4YY

www.bloomsbury.com/alexbellos
www.alexbellos.com

Plate Section Picture Credits:
p. 1 (top), p. 1 (bottom), p. 6 (top), p. 6 (bottom), p. 7 (top), p. 7 (bottom), p. 12 (top), p. 15 (top), p. 15 (bottom), p. 16 (top) © Alex Bellos; pp. 2–3 SR Euclid Collection, UCL Library Services, Special Collections; p. 4 (top), p. 4 (bottom) © Robert Lang; p. 5 (top) © Eva Madrazo, 2009. Used under license from Shutterstock.com; p. 5 (bottom) © Neil Mason; p. 8 Le Casse-tête en portraits, Gandais, Paris, 1818, from the Slocum Puzzle Collection, Lilly Library; p. 9 Thanks to Jerry Slocum; p. 10 (top left), p. 10 (top right), p. 10 (bottom left), p. 10 (bottom right), p. 11 © Christopher Lane; p. 12 (bottom), p. 13 Thanks to Eddy Levin; p. 14 © FLC/ ADAGP, Paris and DACS, London 2009; p. 16 (bottom) © Daina Taimina.

Contents

Introduction

In the summer of 1992 I was working as a cub reporter at the *Evening Argus* in Brighton. My days were spent watching recidivist teenagers appear at the local magistrates court, interviewing shop-keepers about the recession and, twice a week, updating the opening hours of the Bluebell Railway for the paper's listings page. It wasn't a great time if you were a petty thief, or a shopkeeper, but for me it was a happy period in my life.

John Major had recently been re-elected as prime minister and, flush from victory, he delivered one of his most remembered (and ridiculed) policy initiatives. With presidential seriousness, he announced the creation of a telephone hotline for information about traffic cones – a banal proposal dressed up as if the future of the world depended on it.

In Brighton, however, cones were big news. You couldn't drive into town without getting stuck in roadworks. The main route from London – the A23 (M) – was a corridor of striped orange cones all the way from Crawley to Preston Park. With its tongue firmly in its cheek, the *Argus* challenged its readers to guess the number of cones that lined the many miles of the A23 (M). Senior staff congratulated themselves on such a brilliant idea. The village fête-style challenge explained the story while also poking fun at central government: perfect local-paper stuff.

Yet only a few hours after the competition was launched, the first entry was received, and in it the reader had guessed the correct number of cones. I remember the senior editors sitting in dejected silence in the newsroom, as if an important local councillor had just died. They had aimed to parody the prime minister, but it was they who had been made to look like fools.

The editors had assumed that guessing how many cones there were on 20 or so miles of motorway was an impossible task. It self-evidently wasn't and I think I was the only person in the building who could see why. Assuming that cones are positioned at identical intervals, all you need to do is make one calculation:

Number of cones = length of road ÷ distance between cones

The length of road can be measured by driving down it or by reading a map. To calculate the distance between cones you just need a tape measure. Even though the space between cones may vary a little, and the estimated length of road may also be subject to error, over large distances the accuracy of this calculation is good enough for the purposes of winning competitions in local papers (and was presumably exactly how the traffic police had counted the cones in the first place when they supplied the *Argus* with the right answer).

I remember this incident very clearly because it was the first moment in my career as a journalist that I realized the value of having a mathematical mind. It was also disquieting to realize just how innumerate most journalists are. There was nothing very complicated about finding out how many cones were lined alongside a road, yet for my colleagues the calculation was a step too far.

Two years previously I had graduated in mathematics and philosophy, a degree with one foot in science and the other in the liberal arts. Entering journalism was a decision, at least superficially, to abandon the former and embrace the latter. I left the *Argus* shortly after the cones fiasco, moving to work on papers in London. Eventually, I became a foreign correspondent in Rio de Janeiro. Occasionally my heightened aptitude for numbers was helpful, such as when finding the European country whose area was closest to the most recently deforested swathe of Amazon jungle, or when calculating exchange rates during various currency crises. But essentially, it felt very much as if I had left maths behind.

Then, a few years ago, I came back to the UK not knowing what I wanted to do next. I sold T-shirts of Brazilian footballers, I started a blog, I toyed with the idea of importing tropical fruit. Nothing worked out. During this process of reassessment, I looked again at the subject that had consumed me for so much of my youth, and it was there that I found the spark of inspiration that led me to write this book.

Entering the world of maths as an adult was very different from entering it as a child, where the requirement to pass exams means that often the really engrossing stuff is passed over. Now I was free to wander down avenues just because they sounded curious and interesting. I learned about 'ethnomathematics', the study of how different cultures approach maths, and about how maths was shaped by religion. I became intrigued by recent work in behavioural psychology and neuroscience that is piecing together exactly why and how the brain thinks of numbers.

I realized that I was behaving just like a foreign correspondent on assignment, except the country I was visiting was an abstract one – 'Numberland'.

My journey soon became geographical, since I wanted to experience mathematics in the real world. So, I flew to India to learn how the country invented 'zero', one of the greatest intellectual breakthroughs in human history. I booked myself into a mega-casino in Reno to see probability in action. And in Japan, I met the world's most numerate chimpanzee.

As my research progressed, I found myself being in the strange position of being both an expert and a non-specialist at the same time. Relearning school maths was like reacquainting myself with old friends, but there were many friends of friends I had never met back then and there are also a lot of new kids on the block. Before I wrote this book, for example, I was unaware that for hundreds of years there have been campaigns to introduce two new numbers to our ten-number system. I didn't know why Britain was the first nation to mint a heptagonal coin. And I had no idea of the maths behind Sudoku (because it hadn't been invented).

I was led to unexpected places, such as Braintree, Essex, and Scottsdale, Arizona, and to unexpected shelves on the library. I spent a memorable day reading a book on the history of rituals surrounding plants to understand why Pythagoras was a notoriously fussy eater.

The book starts at Chapter Zero, since I wanted to emphasize that the subject discussed here is pre-mathematics. This chapter is about how numbers emerged. At the beginning of Chapter One numbers have indeed emerged and we can get down to business. Between that point and the end of Chapter Eleven the book covers arithmetic, algebra, geometry, statistics and as many other fields as I could squeeze into 400-ish pages. I have tried to keep the technical material to a minimum, although sometimes there was no way out and I had to spell out equations and proofs. If you feel your brain hurting, skip to the beginning of the next section and it will get easier again. Each chapter is self-contained, meaning that to understand it one does not have to have read the previous chapters. You can read the chapters in any order, although I hope you read them from the first to the last since they follow a rough chronology of ideas and I occasionally refer back to points made earlier. I have aimed the book at the reader with no mathematical knowledge, and it covers material from primary school level to concepts that are taught only at the end of an undergraduate degree.

I have included a fair bit of historical material, since maths is the history of maths. Unlike the humanities, which are in a permanent state of reinvention, as new ideas or fashions replace old ones, and unlike applied science, where theories are undergoing continual refinement, mathematics does not age. The theorems of Pythagoras and Euclid are as valid now as they always were – which is why Pythagoras and Euclid are the oldest names we study at school. The GCSE syllabus contains almost no maths beyond what was already known in the mid seventeenth century, and likewise A-level with the mid eighteenth century. (In my degree the most modern maths I studied was from the 1920s.)

When writing this book, my motivation was at all times to communicate the excitement and wonder of mathematical discovery. (And to show that mathematicians are funny. We are the kings of logic, which gives us an extremely discriminating sense of the illogical.) Maths suffers from a reputation that it is dry and difficult. Often it is. Yet maths can also be inspiring, accessible and, above all, brilliantly creative. Abstract mathematical thought is one of the great achievements of the human race, and arguably the foundation of all human progress.

Numberland is a remarkable place. I would recommend a visit.

Alex Bellos
January 2010

>2

CHAPTER ZERO

A Head for Numbers

When I walked into Pierre Pica's cramped Paris apartment I was overwhelmed by the stench of mosquito repellent. Pica had just returned from spending five months with a community of Indians in the Amazon rainforest, and he was disinfecting the gifts he had brought back. The walls of his study were decorated with tribal masks, feathered headdresses and woven baskets. Academic books overloaded the shelves. A lone Rubik's Cube lay unsolved on a ledge.

I asked Pica how the trip had been.

'Difficult,' he replied.

Pica is a linguist and, perhaps because of this, speaks slowly and carefully, with painstaking attention to individual words. He is in his fifties, but looks boyish – with bright blue eyes, a reddish complexion and soft, dishevelled silvery hair. His voice is quiet; his manner intense.

Pica was a student of the eminent American linguist Noam Chomsky and is now employed by France's National Centre for Scientific Research. For the last ten years the focus of his work has been the Munduruku, an indigenous group of about 7000 people in the Brazilian Amazon. The Munduruku are hunter-gatherers who live in small villages spread across an area of rainforest twice the size of Wales. Pica's interest is the Munduruku language: it has no tenses, no plurals and no words for numbers beyond five.

To undertake his fieldwork, Pica embarks on a journey worthy of the great adventurers. The nearest large airport to the Indians is Santarém, a town 500 miles up the Amazon from the Atlantic Ocean. From there, a 15-hour ferry ride takes him almost 200 miles along the Tapajós River to Itaituba, a former gold-rush town and the last stop to stock up on food and fuel. On his most recent trip Pica hired a jeep in Itaituba and loaded it up with his equipment, which included computers, solar panels, batteries, books and 120 gallons of petrol. Then he set off down the Trans-Amazon Highway, a 1970s folly of nationalistic infrastructure that has deteriorated into a precarious and often impassable muddy track.

Pica's destination was Jacareacanga, a small settlement a further 200 miles southwest of Itaituba. I asked him how long it took to drive there. 'Depends,' he shrugged. 'It can take a lifetime. It can take two days.'

How long did it take *this* time, I repeated.

'You know, you never know how long it will take because it never takes the same time. It takes between ten and twelve hours during the rainy season. If everything goes well.'

Jacareacanga is on the edge of the Munduruku's demarcated territory. To get inside the area, Pica had to wait for some Indians to arrive so he could negotiate with them to take him there by canoe.

'How long did you wait?' I enquired.

'I waited quite a lot. But, again, don't ask me how many days.'

'So, it was a couple of days?' I suggested tentatively.

A few seconds passed as he furrowed his brow. 'It was about two weeks.'

More than a month after he left Paris, Pica was finally approaching his destination. Inevitably, I wanted to know how long it took to get from Jacareacanga to the villages.

But by now Pica was demonstrably impatient with my line of questioning: 'Same answer to everything – *it depends!*'

I stood my ground. How long did it take *this* time?

He stuttered: 'I don't know. I think … perhaps … two days … a day and a night …'

The more I pushed Pica for facts and figures, the more reluctant he was to provide them. I became exasperated. It was unclear if underlying his responses was French intransigence, academic pedantry or simply a general contrariness. I stopped my line of questioning and we moved on to other subjects. It was only when, a few hours later, we talked about what it was like to come home after so long in the middle of nowhere that he opened up. 'When I come back from Amazonia I lose sense of time and sense of number, and perhaps sense of space,' he said. He forgets appointments. He is disoriented by simple directions. 'I have extreme difficulty adjusting to Paris again, with its angles and straight lines.' Pica's inability to give me quantitative data was part of his culture shock. He had spent so long with people who can barely count that he had lost the ability to describe the world in terms of numbers.

No one knows for certain, but numbers are probably no more than about 10,000 years old. By this I mean a working system of words and symbols for numbers. One theory is that such a practice emerged together with agriculture and trade, as numbers were an indispensable tool for taking stock and making sure you were not ripped off. The Munduruku are only subsistence farmers and money has only recently begun to circulate in their villages, so they never evolved counting skills. In the case of the indigenous tribes of Papua New Guinea, it has been argued that the appearance of numbers was triggered by elaborate customs of gift exchange. The Amazon, by contrast, has no such traditions.

Tens of thousands of years ago, well before the arrival of numbers, however, our ancestors must have had certain sensibilities about amounts. They would have been able to distinguish one mammoth from two mammoths, and to recognize that one night is different from two nights. The intellectual leap from the concrete idea of two things to the invention of a symbol or word for the abstract idea of 'two', however, will have taken many ages to come about. This occurrence, in fact, is as far as some communities in the Amazon have come. There are tribes whose only number words are 'one', 'two' and 'many'. The Munduruku, who go all the way up to five, are a relatively sophisticated bunch.

Numbers are so prevalent in our lives that it is hard to imagine how people survive without them. Yet while Pierre Pica stayed with the Munduruku he easily slipped into a numberless existence. He slept in a hammock. He went hunting and ate tapir, armadillo and wild boar. He told the time from the position of the sun. If it rained, he stayed in; if it was sunny, he went out. There was never any need to count.

Still, I thought it odd that numbers larger than five did not crop up at all in Amazonian daily life. I asked Pica how an Indian would say 'six fish'. For example, just say that he or she was preparing a meal for six people and he wanted to make sure everyone had a fish each.

'It is impossible,' he said. 'The sentence "I want fish for six people" does not exist.'

What if you asked a Munduruku who had six children: 'How many kids do you have?'

Pica gave the same response: 'He will say "I don't know". It is impossible to express.'

However, added Pica, the issue was a cultural one. It was not the case that the Munduruku counted his first child, his second, his third, his fourth, his fifth and then scratched his head because he could go no further. For the Munduruku, the whole idea of counting children was ludicrous. The whole idea, in fact, of counting anything was ludicrous.

Why would a Munduruku adult want to count his children, asked Pica? The children are looked after by all the adults in the community, he said, and no one is counting who belongs to whom. He compared the situation to the French expression '*j'ai une grande famille*', or 'I'm from a big family'. 'When I say that I have a big family I am telling you that I don't know [how many members it has]. Where does my family stop and where does the others' family begin? I don't know. Nobody ever told me that.' Similarly, if you asked an adult Munduruku how many children he is responsible for, there is no correct answer. 'He will answer "I don't know", which really is the case.'

The Munduruku are not alone in the sweep of history in not counting members of their community. When King David counted his own people he was punished with three days of pestilence and 77,000 deaths. Jews are meant to count Jews only indirectly, which is why in synagogues the way of making sure there are ten men present, a *minyan*, or sufficient community for prayers, is to say a ten-word prayer pointing at each person per word. Counting people with numbers is considered a way of singling people out, which makes them more vulnerable to malign influences. Ask an Orthodox rabbi to count his kids and you have as much chance of an answer as if you asked a Munduruku.

I once spoke to a Brazilian teacher who had spent a lot of time working in indigenous communities. She said that Indians thought that the constant questioning by outsiders of how many children they had was a peculiar compulsion, even though the visitors were simply asking the question to be polite. What is the purpose of counting children? It made the Indians very suspicious, she said.

The first written mention of the Munduruku dates from 1768, when a settler spotted some of them on the bank of a river. A century later, Franciscan missionaries set up a base on Munduruku land and more contact was made during the rubber boom of the late nineteenth century when rubber-tappers penetrated the region. Most Munduruku still live in relative isolation, but like many other Indian groups with a long history of contact, they tend to wear Western clothes like T-shirts and shorts. Inevitably, other features of modern life will eventually enter their world, such as electricity and television. And numbers. In fact, some Munduruku who live at the fringes of their territory have learned Portuguese, the national language of Brazil, and can count in Portuguese. 'They can count *um, dois, três,* up until the hundreds,' said Pica. 'Then you ask them, "By the way, how much is five minus three?"' He parodied a Gallic shrug. They have no idea.

In the rainforest Pica conducts his research using laptops powered by solar-charged batteries. Maintaining the hardware is a logistical nightmare because of the heat and the damp, although sometimes the trickiest challenge is assembling the participants. On one occasion the leader of a village demanded that Pica eat a large, red *sauba* ant in order to gain permission to interview a child. The ever-diligent linguist grimaced as he crunched the insect and swallowed it down.

The purpose of researching the mathematical abilities of people who have the capacity to count only on one hand is to discover the nature of our basic numerical intuitions. Pica wants to know what is universal to all humans, and what is shaped by culture. In one of his most fascinating experiments he examined the Indians' spatial understanding of numbers. How did they visualize numbers when spread out on a line? In the modern world we do this all the time – on tape measures, rulers, graphs and houses along a street. Since the Munduruku don't have numbers, Pica tested them using sets of dots on a screen. Each volunteer was presented with the figure overleaf, of an unmarked line. To the left side of the line was one dot; to the right, ten dots. Each volunteer was then shown random sets of between one and ten dots. For each set the subject had to point at where on the line he or she thought the number of dots should be located. Pica moved the cursor to this point and clicked. Through repeated clicks, he could see exactly how the Munduruku spaced numbers between one and ten.

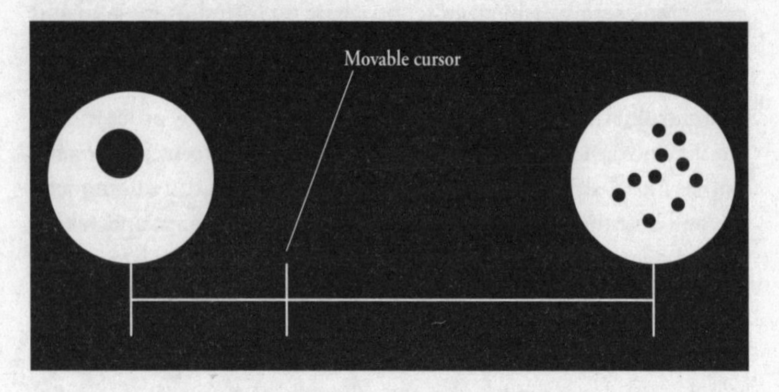

When American adults were given this test, they placed the numbers at equal intervals along the line. They recreated the number line we learn at school, in which adjacent digits are the same distance apart as if measured by a ruler. The Munduruku, however, responded quite differently. They thought that intervals between the numbers started large and became progressively smaller as the numbers increased. For example, the distance between the marks for one dot and two dots, and two dots and three dots were much larger than the distance between seven and eight dots, or eight and nine dots, as the following two graphs make clear.

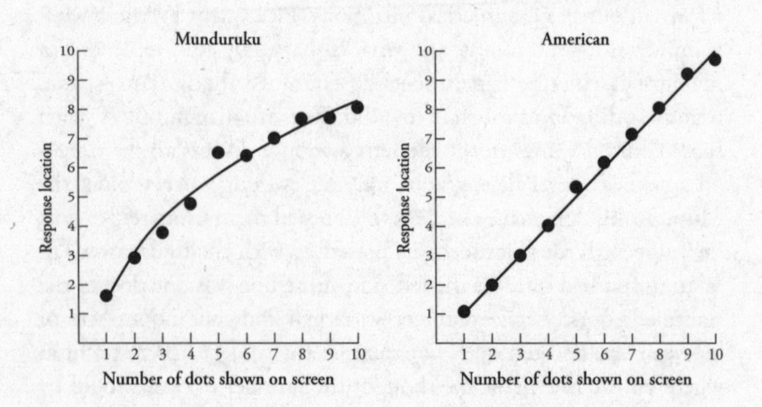

The results were striking. It is generally considered a self-evident truth that numbers are evenly spaced. We are taught this at school and we accept it easily. It is the basis of all measurement and science. Yet the Munduruku do not see the world like this. Stripped of a language of counting and number words, they visualize magnitudes in a completely different way.

When numbers are spread out evenly on a ruler, the scale is called *linear*. When numbers get closer as they get larger, the scale is called *logarithmic*.* It turns out that the logarithmic approach is not exclusive to Amazonian Indians. We are all born conceiving numbers this way. In 2004, Robert Siegler and Julie Booth at Carnegie Mellon University in Pennsylvania presented a similar version of the number-line experiment to a group of kindergarten pupils (with an average age of 5.8 years), first-graders (6.9) and second-graders (7.8). The results showed in slow motion how familiarity with counting moulds our intuitions. The kindergarten pupil, with no formal maths education, maps out numbers logarithmically. By the first year at school, when the pupils are being introduced to number words and symbols, the curve is straightening. And by the second year at school, the numbers are at last evenly laid out along the line.

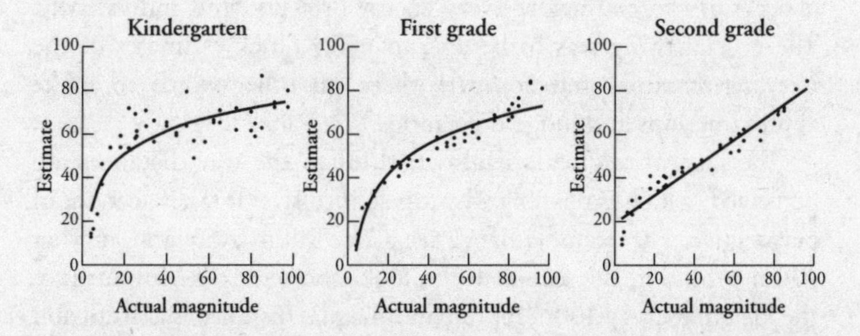

Why do Indians and children think that higher numbers are closer together than lower numbers? There is a simple explanation. In the experiments, the volunteers were presented with a set of dots and asked where this set should be located in relation to a line with one dot on the left and ten dots on the right. (Or, in the children's case,

* In fact, numbers need to get closer in a certain way for the scale to be logarithmic. For a fuller discussion of the scale, see p. 190.

100 dots). Imagine a Munduruku is presented with five dots. He will study it closely and see that five dots are *five times bigger* than one dot, but ten dots are only *twice as big* as five dots. The Munduruku and the children seem to be making their decisions about where numbers lie based on estimating the ratios between amounts. When considering ratios, it is logical that the distance between five and one is much greater than the distance between ten and five. And if you judge amounts using ratios, you will always produce a logarithmic scale.

It is Pica's belief that understanding quantities approximately in terms of estimating ratios is a universal human intuition. In fact, humans who do not have numbers – like Indians and young children – have no alternative but to see the world in this way. By contrast, understanding quantities in terms of exact numbers is not a universal intuition; it is a product of culture. The precedence of approximations and ratios over exact numbers, Pica suggests, is due to the fact that ratios are much more important for survival in the wild than the ability to count. Faced with a group of spear-wielding adversaries, we needed to know instantly whether there were more of them than us. When we saw two trees we needed to know instantly which had more fruit hanging from it. In neither case was it necessary to enumerate every enemy or every fruit individually. The crucial thing was to be able to make quick estimates of the relevant amounts and compare them, in other words to make approximations and judge their ratios.

The logarithmic scale is also faithful to the way distances are perceived, which is possibly why it is so intuitive. It takes account of perspective. For example, if we see a tree 100m away and another 100m behind it, the second 100m looks shorter. To a Munduruku, the idea that every 100m represents an equal distance is a distortion of how he perceives the environment.

Exact numbers provide us with a linear framework that contradicts our logarithmic intuition. Indeed, our proficiency with exact numbers means that the logarithmic intuition is overruled in most situations. But it is not eliminated altogether. We live with both a linear and a logarithmic understanding of quantity. For example, our understanding of the passing of time tends to be logarithmic. We often feel that time passes faster the older we get. Yet it works in the other direction too: yesterday seems a lot longer than the whole of last week. Our deep-seated logarithmic instinct surfaces most clearly when it comes to thinking about very large numbers. For example, we can all understand the difference between one and ten. It is unlikely we would confuse one pint of beer and ten pints of beer. Yet what about the difference between a billion gallons of water and ten billion gallons of water? Even though the difference is enormous, we tend to see both quantities as quite similar – as very large amounts of water. Likewise, the terms millionaire and billionaire are thrown around almost as synonyms – as if there is not so much difference between being very rich and very, very rich. Yet a billionaire is a thousand times richer than a millionaire. The higher numbers are, the closer together they feel.

The fact that Pica temporarily forgot how to use numbers after only a few months in the jungle indicates that our linear understanding of numbers is not as deeply rooted in our brains as our logarithmic one. Our understanding of numbers is surprisingly fragile, which is why without regular use we lose our ability to manipulate exact numbers and default to our intuitions judging amounts with approximations and ratios.

Pica said that his and others' research on our mathematical intuitions may have serious consequences for maths education – both in the Amazon and in the West. We require understanding of the linear number line to function in modern society – it is the basis of measuring, and facilitates calculations. Yet maybe in our dependence on linearity we have gone too far in stifling our own logarithmic intuition. Perhaps, said Pica, this is a reason why so many people find maths difficult. Perhaps we should pay more attention to judging ratios rather than manipulating exact numbers. Likewise, maybe it would be wrong to teach the Munduruku to count like we do since this might deprive them of the mathematical intuitions or knowledge that are necessary for their own survival.

Interest in the mathematical abilities of those who have no words or symbols for numbers has traditionally focused on animals. One of the best-known research subjects was a trotting stallion called Clever Hans. In the early 1900s, crowds gathered regularly in a Berlin courtyard to watch Hans's owner, Wilhelm von Osten, a retired maths instructor, set the horse simple arithmetical sums. Hans answered by stamping the ground with his hoof the correct number of times. His repertoire included addition and subtraction as well as fractions, square roots and factorization. Public fascination, and suspicion that the horse's supposed intelligence was some kind of trick, led to an investigation of his abilities by a committee of eminent scientists. They concluded that, *jawohl!*, Hans really was doing the math.

It took a less eminent but more rigorous psychologist to debunk the equine Einstein. Oscar Pfungst noticed that Hans was reacting to cues in von Osten's body language. Hans would start stamping his hoof on the ground and stopped only when he could sense a build-up or release of tension in von Osten's face, indicating the answer had been reached. The horse was sensitive to the tiniest visual signals, such as the leaning of the head, the raising of the eyebrows and even the dilation of the nostrils. Von Osten was not even aware he was making these gestures. Hans was clever at reading people, certainly, but was no arithmetician.

Many further attempts were made in the last century to teach animals to count, not all for the purposes of circus-like entertainment. In 1943 the German scientist Otto Koehler trained his pet raven Jakob to select a pot with a specified number of spots on its lid from a selection of pots with a variety of numbers of spots on their lids. The bird could perform this task when the number of spots on any one lid was between one and seven spots. In recent years, avian intelligence has reached more impressive heights. Irene Pepperberg of Harvard University taught an African grey parrot called Alex numerals from 1 to 6. When shown an assortment of coloured blocks he could reply, for example, how many blue blocks there were by squawking the English number word. So renowned had Alex become among scientists and birdlovers that when he died unexpectedly in 2007, his obituary appeared in *The Economist*.

The lesson of Clever Hans was that when teaching animals to count, supreme care must be taken to eliminate involuntary human prompting. For the maths education of Ai, a chimpanzee brought to Japan from West Africa in the late 1970s, the chances of human cues were eliminated because she learned using a touch-screen computer.

Ai is now 31 and lives at the Primate Research Institute in Inuyama, a small tourist town in central Japan. Her forehead is high and balding, the hair on her chin is white and she has the dark sunken eyes of ape middle age. She is known there as a 'student', never a 'research subject'. Every day Ai attends classes where she is given tasks. She turns up at 9 a.m. on the dot after spending the night outdoors with a group of other chimps on a giant tree-like construction of wood, metal and rope. On the day I saw her she sat with her head close to a computer, tapping sequences of digits on the screen when they appeared. When she completed a task correctly an 8mm cube of apple whizzed down a tube to her right. Ai caught it in her hand and scoffed it instantly. Her mindless gaze, the nonchalant tapping of a flashing, beeping computer and the mundanity of continual reward reminded me of an old lady doing the slots.

When Ai was a child she became a great ape in both senses of the word by becoming the first non-human to count with Arabic numerals. (These are the symbols 1, 2, 3 and so on, that are used in almost all countries except, ironically, in parts of the Arab world.) In order for her to do this satisfactorily, Tetsuro Matsuzawa, director of the Primate Research Institute, needed to teach her the two elements that comprise human understanding of number: quantity and order.

Numbers express an amount, and they also express a position. These two concepts are linked, but different. For example, when I refer to 'five carrots' I mean that the quantity of carrots in the group is five. Mathematicians call this aspect of number 'cardinality'. On the other hand, when I count from 1 to 20 I am using the convenient feature that numbers can be ordered in succession. I am not referring to 20 objects, I am simply reciting a sequence. Mathematicians call this aspect of number 'ordinality'. At school we are taught notions of cardinality and ordinality together and we slip effortlessly between them. To chimpanzees, however, the interconnection is not obvious at all.

Matsuzawa first taught Ai that one red pencil refers to the symbol '1' and two red pencils to '2'. After 1 and 2, she learned 3 and then all the other digits up to 9. When shown, say, the number 5 she could tap a square with five objects, and when shown a square with five objects she could tap the digit 5. Her education was reward-driven: whenever she got a computer task correct, a tube by the computer dispensed a piece of food.

Once Ai had mastered the cardinality of the digits from 1 to 9, Matsuzawa introduced tasks to teach her how they were ordered. His tests flashed digits up on the screen and Ai had to tap them in ascending order. If the screen showed 4 and 2, she had to touch 2 and then 4 to win her cube of apple. She grasped this pretty quickly. Ai's competence in both the cardinality and ordinality tasks meant that Matsuzawa could reasonably say that his student had learned to count. The achievement made her a national hero in Japan and a global icon for her species.

Matsuzawa then introduced the concept of zero. Ai picked up the cardinality of the symbol 0 easily. Whenever a square appeared on the screen with nothing in it, she would tap the digit. Then Matsuzawa wanted to see if she was able to infer an understanding of the ordinality of zero. Ai was shown a random sequence of screens with two digits, just like when she was learning the ordinality of 1 to 9, although now sometimes one of the digits was a 0. Where did she think zero's place was in the ordering of numbers?

In the first session Ai placed 0 between 6 and 7. Matsuzawa calculated this by averaging out which numbers she thought 0 came after and which numbers she thought it came before. In subsequent sessions Ai's positioning of 0 went under 6, then under 5, 4 and after a few hundred trials 0 was down to around 1. She remained confused, however, if 0 was more or less than 1. Even though Ai had learned to manipulate numbers perfectly well, she lacked the depth of human numerical understanding.

A habit she did learn, however, was showmanship. She is now a total pro, tending to perform better at her computer tasks in front of visitors, especially camera crews.

Investigating animals' mastery of numbers is an active academic pursuit. Experiments have revealed an unexpected capacity for 'quantity discrimination' in animals as varied as salamanders, rats and dolphins. Even though horses may still be incapable of calculating square roots, scientists now believe that the numerical capacities of animals are much more sophisticated than previously thought. All creatures seem to be born with brains that have a predisposition for maths.

After all, numerical competence is crucial to survival in the wild. A chimpanzee is less likely to go hungry if he can look up a tree and quantify the amount of ripe fruit he will have for his lunch. Karen McComb at the University of Sussex monitored a pride of lions in the Serengeti in order to show that lions use a sense of number when deciding whether to attack other lions. In one experiment a solitary lioness was walking back to the pride at dusk. McComb had installed a loudspeaker hidden in the bushes and played a recording of a single roar. The lioness heard it and continued walking home. In a second experiment five lionesses were together. McComb played the roars of three lionesses through her hidden loudspeaker. The group of five heard the roars of three and peered in the direction of the noise. One lioness started to roar and soon all five were charging into the bushes to attack.

McComb's conclusion was that the lionesses were comparing quantities in their heads. One vs one meant it was too risky to attack, but with five-to-three advantage, the attack was on.

Not all animal number research is as glamorous as camping in the Serengeti or bonding with a celebrity chimpanzee. At the University of Ulm, in Germany, academics put some Saharan desert ants at the end of a tunnel and sent them down it foraging for food. Once they reached the food, however, some of the ants had the bottom of their legs clipped off and other ants were given stilts made from pig bristles. (Apparently this is not as cruel as it sounds, since the legs of desert ants are routinely frazzled off in the Saharan sun.) The ants with shorter legs undershot the journey home, while the ones with longer legs overshot it, suggesting that instead of using their eyes, the ants judged distance with an internal pedometer. Ants' great skill in wandering for hours and then always navigating their way back to the nest may just be due to a proficiency at counting strides.

Research into the numerical competence of animals has taken some unexpected turns. Chimpanzees may have limits to their mathematical proficiency, yet, while studying this, Matsuzawa discovered that they have other cognitive abilities that are vastly superior to ours.

In 2000 Ai gave birth to a son, Ayumu. On the day I visited the Primate Research Institute, Ayumu was in class right next to his mum. He is smaller, with pinker skin on his face and hands and blacker hair. Ayumu was sitting in front of his own computer, tapping away at the screen when numbers flashed up and avidly

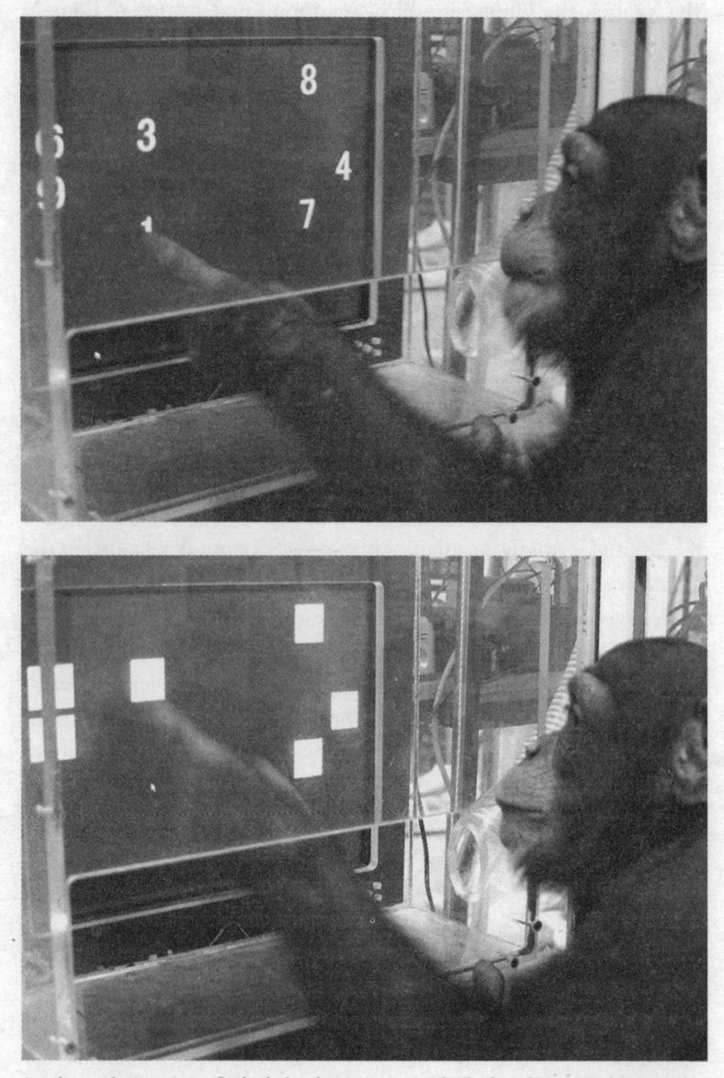

In this task Ayumu is flashed the digits 1 to 7, which then become white squares. He must remember the positions of the numbers so that he can then tap the squares in order to win the food reward.

scoffing the apple cubes when he won them. He is a self-confident lad, living up to his privileged status as son and heir of the dominant female in the group.

Ayumu was never taught how to use the touch-screen displays, although as a baby he would sit by his mother as she attended class every day. One day Matsuzawa opened the classroom door only halfway, just enough for Ayumu to come in but too narrow for Ai to join him. Ayumu went straight up to the computer monitor. The staff watched him eagerly to see what he had learned. He pressed the screen to start, and the digits 1 and 2 appeared. This was a simple ordering task. Ayumu clicked on 2. Wrong. He kept on pressing 2. Wrong again. Then he tried to press 1 and 2 at the same time. Wrong. Eventually he got it right: he pressed 1, then 2 and an apple cube shot down into his palm. Before long, Ayumu was better at all the computer tasks than his mum.

A couple of years ago Matsuzawa introduced a new type of number task. On pressing the start button, the numbers 1 to 5 were displayed in a random pattern on the screen. After 0.65 seconds the numbers turned into small white squares. The task was to tap the white squares in the correct order, remembering which square had been which number.

Ayumu completed this task correctly about 80 per cent of the time, which was about the same amount as a sample group of Japanese children. Matsuzawa then reduced the time that the numbers were visible, to 0.43 seconds, and while Ayumu barely noticed the difference, the children's performances dropped significantly, to a success rate of about 60 per cent. When Matsuzawa reduced the time that the numbers were visible again – to only 0.21 seconds, Ayumu was still registering 80 per cent, but the kids dropped to 40.

This experiment revealed that Ayumu had an extraordinary photographic memory, as do the other chimps in Inuyama, although none is as good as he is. Matsuzawa has increased the number of digits in further experiments and now Ayumu can remember the positioning of eight digits made visible for only 0.21 seconds. Matsuzawa also reduced the time interval and Ayumu can now remember the positioning of five digits visible for only 0.09 seconds – which is barely enough time for a human to register the numbers, let alone remember them. This astonishing talent for instant memorization may well be because making snap decisions, for example, about numbers of foes, is vital in the wild.

Studies into the limits of animals' numerical capabilities bring us naturally to the question of innate human abilities. Scientists wanting to investigate minds as uncontaminated as possible by acquired knowledge require subjects who are as young as possible. As a result, infants only a few months old are now routinely tested on their maths skills. Since at this age babies cannot talk or properly control their limbs, testing them for signs of numerical prowess relies on their eyes. The theory is that they will stare for longer at pictures they find interesting. In 1980 Prentice Starkey at the University of Pennsylvania showed babies between 16 and 30 weeks old a screen with two dots, and then showed another screen with two dots. The babies looked at the second screen for 1.9 seconds. But when Starkey repeated the test, showing a screen with three dots after the screen with two dots, the babies looked at it for 2.5 seconds: almost a third longer. Starkey argued that this extra stare-time meant the babies had noticed something different about three dots compared to two dots, and therefore had a rudimentary understanding of number. This method of judging numerical cognition through the length of attention span is now standard. Elizabeth Spelke at Harvard showed in 2000 that six-month-old babies can tell the difference between 8 and 16 dots, and in 2005 that they can distinguish between 16 and 32.

A related experiment showed that babies had a grasp of arithmetic. In 1992, Karen Wynn, at the University of Arizona, sat a five-month-old baby in front of a small stage. An adult placed a Mickey Mouse doll on the stage and then put up a screen to hide it. The adult then placed a second Mickey Mouse doll behind the screen, and the screen was then pulled away to reveal two dolls. Wynn then repeated the experiment, this time with the screen pulling away to reveal a wrong number of dolls: just one doll or three of them. When there were one or three dolls, the baby stared at the stage for longer than when the answer was two, indicating that the infant was surprised when the arithmetic was wrong. Babies understood, argued Wynn, that one doll plus one doll equals two dolls.

The Mickey experiment was later performed with the *Sesame Street* puppets Elmo and Ernie. Elmo was placed on the stage. The screen came down. Then another Elmo was placed behind the screen. The screen was taken away. Sometimes two Elmos were revealed, sometimes an Elmo and an Ernie together and sometimes only one Elmo or only one Ernie. The babies stared for longer when just one puppet was revealed, rather than when two of the *wrong* puppets were revealed. In other words, the arithmetical impossibility of

Character put on stage

Screen appears and hides character

Second character put behind screen

Screen hides characters

Screen removed to reveal one of the above scenarios

In Karen Wynn's experiment, babies were tested on their ability to distinguish the correct number of dolls behind a screen.

1 + 1 = 1 was much more disturbing than the metamorphosis of Elmos into Ernies. Babies' knowledge of mathematical laws seems much more deeply rooted than their knowledge of physical ones.

The Swiss psychologist Jean Piaget (1896–1980) argued that babies build up an understanding of numbers slowly, through experience, so there was no point in teaching arithmetic to children younger than six or seven. This influenced generations of educators, who often preferred to let primary-age pupils play around with blocks in lessons rather than introduce them to formal mathematics. Now Piaget's views are considered outdated. Pupils come face to face with Arabic numerals and sums as soon as they get to school.

Dot experiments are also the cornerstone of research into adult numerical cognition. A classic experiment is to show a person dots on a screen and ask how many dots he or she sees. When there are one, two or three dots, the response comes almost instantly. When there are four dots, the response is significantly slower, and with five slower still.

So what! you might say. Well, this probably explains why in several cultures the numerals for 1, 2 and 3 have been one, two and three lines, while the number for 4 is *not* four lines. When there are three lines or fewer we can tell the number of lines straight away, but when there are four of them our brain has to work too hard and a different symbol is necessary. The Chinese characters for one to four are 一, 二, 三 and 四. Ancient Indian numerals were —, =, ≡ and +. (If you join the lines, you can see how they turned into the modern 1, 2, 3 and 4.)

In fact, there is some debate about whether the limit of the number of lines we can grasp instantly is three or four. The Romans actually had the alternatives IIII and IV for four. The IV is much more instantly recognizable, but clock faces – perhaps for aesthetic reasons – tended to use the IIII. Certainly, the number of lines, dots, or sabre-toothed tigers that we can enumerate rapidly, confidently and accurately is no more than four. While we have an *exact* sense of 1, 2 and 3, beyond 4 our exact sense wanes and our judgements about numbers become *approximate*. For example, try to guess quickly how many dots are at the top of the page opposite.

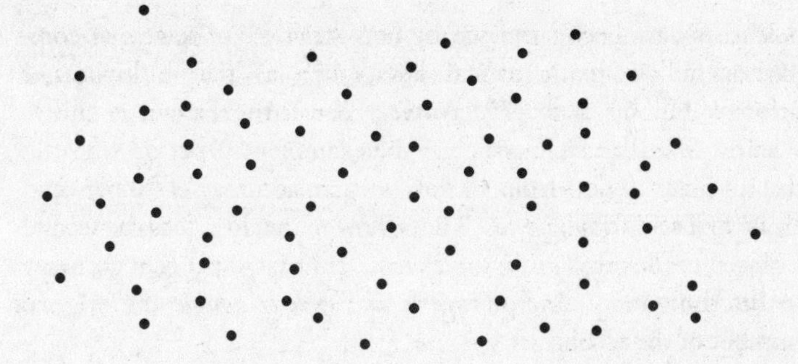

It's impossible. (Unless you are an autistic savant, like the character played by Dustin Hoffman in *Rain Man*, who would be able to grunt in a split second 'Seventy-five'.) Our only strategy is to estimate, and we'd probably be far off the mark.

Researchers have tested the extent of our intuition of amounts by showing volunteers images of different numbers of dots and asking which set is larger, and our proficiency at discriminating dots, it turns out, follows regular patterns. It is easier, for example, to tell the difference between a group of 80 dots and a group of 100 dots than it is between two groups of 81 and 82 dots. Similarly, it is easier to discriminate between 20 and 40 dots than it is between 80 and 100 dots. In both A and B below, the left set of dots is larger than the right set of dots, yet the length of time it takes us to process the information is noticeably longer in case B.

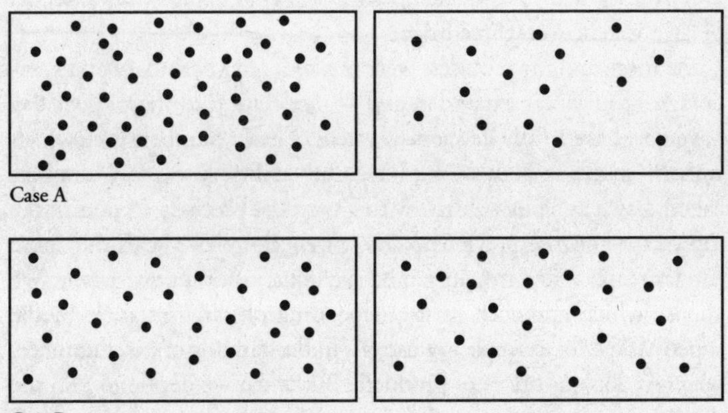

Case A

Case B

Scientists have been surprised by how strictly our powers of comparison follow mathematical laws, such as the multiplicative principle. In his book *The Number Sense*, the French cognitive scientist Stanislas Dehaene gives the example of a person who can discriminate 10 dots from 13 dots with an accuracy of 90 per cent. If the first set is doubled to 20 dots, how many dots does the second set need to include so that this person still has 90 per cent accuracy in discrimination? The answer is 26, *exactly double* the original number of the second set.

Animals are also able to compare sets of dots. While they do not score as highly as we do, the same mathematical laws also seem to govern their skills. This is pretty remarkable. Humans are unique in having a wonderfully elaborate system of counting. Our life is filled with numbers. Yet, for all our mathematical talent, when it comes to perceiving and estimating large numbers our brains function just like those of our feathered and furry friends.

Human intuitions about quantities led, over millions of years, to the creation of numbers. It is impossible to know exactly how this happened, but it is reasonable to speculate that it arose from our desire to track things – such as moons, mountains, predators or drum beats. At first we may have used visual symbols, such as our fingers, or notches on wood, in a one-to-one correspondence with the object we were tracking – two notches or two fingers means two mammoths, three notches or three fingers means three, and so on. Later on we will have come up with words to express the concepts of 'two notches' or 'three fingers'.

As more and more objects were tracked, our vocabulary and symbology of numbers expanded and – accelerating to the present day – we now have a fully developed system of exact numbers with which we can count as high as we like. Our ability to express numbers exactly, such as being able to say that there are precisely 75 dots in the top picture on the previous page, sits cheek-by-jowl with our more fundamental ability to understand such quantities approximately. We choose which approach to use depending on circumstance: in the supermarket, for example, we use our understanding of exact numbers when we look at prices of products. But when we decide to join the shortest checkout queue we are using our instinctive, approximate sense. We do not count every person in every queue. We look at the queues and estimate which one has the fewest people in it.

In fact, we use our imprecise approach to numbers constantly, even when using precise terminology. Ask someone how long it takes them to get to work and most often the answer will be a range, say, 'Thirty-five, forty minutes.' In fact, I have noticed that I am incapable of giving single-number answers to questions involving quantity. How many people were at the party? 'Twenty, thirty ...' How long did you stay? 'Three and a half, four hours ...' How many drinks did you have? 'Four, five ... *ten* ...' I used to think that I was just being indecisive. Now I'm not so sure. I prefer to think that I was drawing on my inner number sense, an intuitive, animal propensity to deal in approximations.

Since the approximate number sense is essential for survival, it might be thought that all humans would have comparable abilities. In a 2008 paper, psychologists at Johns Hopkins University and the Kennedy Krieger Institute investigated whether or not this was the case among a group of 14-year-olds. The teenagers were shown varying numbers of yellow and blue dots together on a screen for 0.2 seconds, and asked only whether there were more blue or yellow dots. The results astonished the researchers, since the scores showed an unexpectedly wide variation in performance. Some pupils could easily tell the difference between 9 blue dots and 10 yellow, but others had abilities comparable to those of infants – hardly even able to say if 5 yellow dots beat 3 blue.

An even more startling finding became apparent when the teenagers' dot-comparing scores were then compared to their maths scores since kindergarten. Researchers had previously assumed that the intuitive ability to discriminate amounts does not contribute much to how good a student is at tasks such as solving equations and drawing triangles. Yet this study found a strong correlation between a talent at reckoning and success in formal maths. The better one's approximate number sense, it seems, the higher one's chance of getting good grades. This might have serious consequences for education. If a flair for estimation fosters mathematical aptitude, maybe maths classes should be less about times tables and more about honing skills at comparing sets of dots.

Stanislas Dehaene is perhaps the leading figure in the cross-disciplinary field of numerical cognition. He started off as a mathematician, and is now a neuroscientist, a professor at the Collège de France and one of the directors of NeuroSpin, a state-of-the-art research institute near Paris. Shortly after he published *The Number Sense* in 1997, he was having lunch in the canteen of Paris's Science Museum with the Harvard development psychologist Elizabeth Spelke. There they sat down by chance next to Pierre Pica. Pica brought up his experiences with the Munduruku and, after excited discussions, the three decided to collaborate. The chance to study a community that doesn't have counting was a wonderful opportunity for new research.

Dehaene devised experiments for Pica to take to the Amazon, one of which was very simple: he wanted to know just what they understood by their number words. Back in the rainforest Pica assembled a group of volunteers and showed them varying numbers of dots on a screen, asking them to say aloud the number of dots they saw.

The Munduruku numbers are:

one *pũg*
two *xep xep*
three *ebapug*
four *ebadipdip*
five *pũg pogbi*

When there was one dot on the screen, the Munduruku said *pũg*. When there were two, they said *xep xep*. But beyond two they were not precise. When three dots showed up, *ebapug* was said only about 80 per cent of the time. The reaction to four dots was *ebadipdip* in only 70 per cent of cases. When shown five dots, *pũg pogbi* was the answer managed only 28 per cent of the time, with *ebadipdip* being given instead in 15 per cent of answers. In other words, for three and above the Munduruku's number words were really just estimates. They were counting 'one', 'two', 'threeish', 'fourish', 'fiveish'. Pica started to wonder whether *pũg pogbi*, which literally means 'handful', even really qualified as a number. Maybe they could not count up to five, but only to fourish?

Pica also noticed an interesting linguistic feature of their number words. He pointed out to me that from one to four, the number of syllables of each word is equal to the number itself. This observation really excited him. 'It is as if the syllables are an aural way of counting,' he said. In the same way that the Romans counted I, II, III and IIII but switched to V at five, the Munduruku started with one syllable for one, added another for two, another for three, another for four but did not use five syllables for five. Even though the words for three and four were not used precisely, they contained precise numbers of syllables. When the number of syllables was no longer important, the word was maybe not a number word at all. 'This is amazing since it seems to corroborate the idea that humans possess a number system that can only track up to four exact objects at a time,' Pica said.

Pica also tested the Munduruku's abilities to estimate large numbers. In one test, illustrated overleaf, the subjects were shown a computer animation of two sets of several dots falling into a can. They were then asked to say if these two sets added together in the can – no longer visible for comparison – amounted to more than a third set of dots that then appeared on the screen. This tested whether they could calculate additions in an approximate way. They could, performing just as well as a group of French adults given the same task.

In a related experiment, also illustrated overleaf, Pica's computer screen showed an animation of six dots falling into a can and then four dots falling out. The Munduruku were then asked to point at one of three choices for how many dots were left in the can. In other words, what is 6 minus 4? This test was designed to see if the Munduruku understood exact numbers for which they had no words. They could not do the task. When shown the animation of a subtraction that contained either 6, 7 or 8 dots, the solution always eluded them. 'They could not calculate even in simple cases,' said Pica.

The results of these dot experiments showed that the Munduruku were very proficient in dealing with rough amounts, but were abysmal in exact numbers above five. Pica was fascinated by the similarities this revealed between the Munduruku and Westerners: both had a fully functioning, exact system for tracking small numbers and an approximate system for large numbers. The significant difference was that the Munduruku had failed to

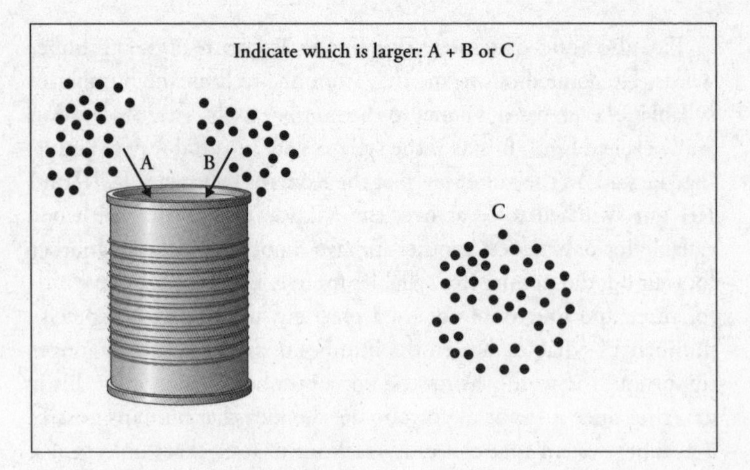

Approximate addition and comparison.

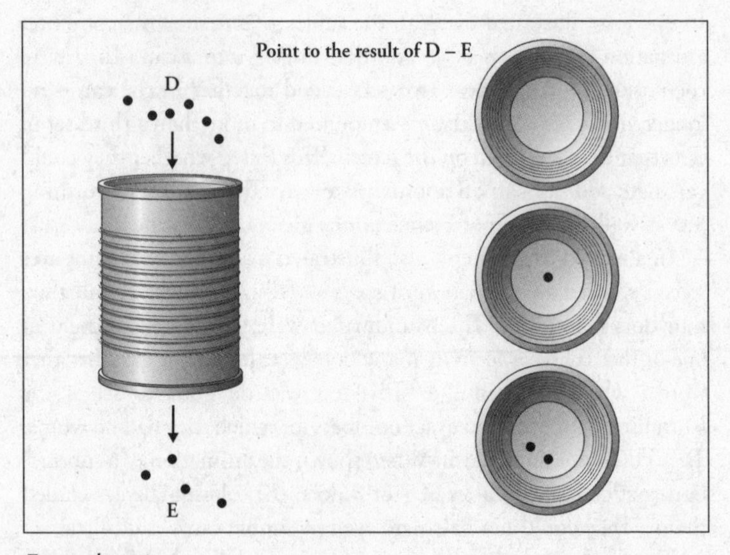

Exact subtraction.

combine these two independent systems together to reach numbers beyond five. Pica said that this must be because keeping the systems separate was more useful. He suggested that in the interests of cultural diversity it was important to try to protect the Munduruku way of counting, since it would surely become threatened by the inevitable increase in contact between the Indians and Brazilian settlers.

The fact, however, that there were some Munduruku who had learned to count in Portuguese but still failed to grasp basic arithmetic was an indication of just how powerful their own mathematical system was and how well suited it was to their needs. It also showed how difficult the conceptual leap must be to having a proper understanding of exact numbers above five.

Could it be that humans need words for numbers above four in order to have an exact understanding of them? Professor Brian Butterworth, of University College London, believes that we don't. He thinks that the brain contains a ready-built capacity to understand exact numbers, which he calls the 'exact number module'. According to his interpretation, humans understand the exact number of items in small collections, and by adding to these collections one by one we can learn to understand how bigger numbers behave. He has been conducting research in the only place outside the Amazon where there are indigenous groups with almost no number words: the Australian Outback.

The Warlpiri aboriginal community live near Alice Springs and have words only for one, two and many, and the Anindilyakwa of Groote Eylande in the Gulf of Carpentaria have words only for one, two, three (which sometimes means four) and many. In one experiment with children of both groups, a block of wood was tapped with a stick up to seven times and counters were placed on a mat. Sometimes the number of taps matched the number of counters, sometimes not. The children were perfectly able to say when the numbers matched and when they didn't. Butterworth argued that to get the answer right the children were producing a mental representation of exact number that was abstract enough to represent both auditory and visual enumeration. These children had no words for the numbers four, five, six and seven, yet were perfectly able to hold those amounts in their heads. Words were useful to understand exactness, Butterworth concluded, but not necessary.

Another important focus of Butterworth's work – and of Stanislas Dehaene's – is a condition called *dyscalculia*, or number blindness, in which one's number sense is defective. It occurs in an estimated 3–6 per cent of the population. Dyscalculics do not 'get' numbers the way most people do. For example, which of these two figures is biggest?

65 24

Easy, it's 65. Almost all of us will get the correct answer in less than half a second. If you have dyscalculia, however, it can take up to three seconds. The nature of the condition varies from person to person, but those diagnosed with it often have problems in correlating the symbol for a number, say 5, with the number of objects the symbol represents. They also find it hard to count. Dyscalculia does not mean you cannot count, but sufferers tend to lack basic intuitions about number and instead rely on alternative strategies to cope with numbers in everyday life, for instance by using their fingers more. Severe dyscalculics can barely read the time.

If you were smart in all your subjects at school but failed ever to pass an exam in maths, you may well be dyscalculic. (Although if you always failed at maths, you are probably not reading this book.) The condition is thought to be a principal cause of low numeracy. Understanding dyscalculia has a social urgency, since adults with low numeracy are much more likely to be unemployed or depressed than their peers. Yet dyscalculia is little understood. It can be thought of as the number version of dyslexia; the conditions are comparable in that they both affect roughly the same proportion of the population and they appear to have no bearing on overall intelligence. However, a lot more is known about dyslexia than about dyscalculia. It is estimated, in fact, that academic papers on dyslexia outnumber those on dyscalculia by about ten to one. Among the reasons why dyscalculia research is so far behind is that there are many *other* reasons why one might be bad at maths – the subject is often taught badly at school, and it is easy to fall behind if you miss lessons when crucial concepts are introduced. There is also less of a social taboo around being rubbish with numbers than there is around being rubbish at reading.

Brian Butterworth frequently writes references for people he has tested for dyscalculia, explaining to prospective employers that the failure to achieve school maths qualifications is not due to laziness or lack of intelligence. Dyscalculics can be high achievers in all other areas beyond numbers. It is even possible, says Butterworth, to be dyscalculic and very good at maths. There are several branches of mathematics, such as logic and geometry, that prioritize deductive reasoning or spatial awareness rather than dexterity with numbers or equations. Usually, however, dyscalculics are not at all good at maths.

Much of the research into dyscalculia is behavioural, such as the screening of tens of thousands of schoolchildren by giving them tests on a computer in which they must say which of two numbers is the biggest. Some is neurological, in which magnetic resonance scans of dyscalculic and non-dyscalculic brains are studied to see how their circuitry differs. In cognitive science, advances in understanding a mental faculty often come from studying cases where the faculty is faulty. Gradually, a clearer picture is emerging of what dyscalculia is – and of how the number sense works in the brain.

Neuroscience, in fact, is providing some of the most exciting new discoveries in the field of numerical cognition. It is now possible to see what happens to individual neurons in a monkey's brain when that monkey thinks of a precise number of dots.

Andreas Nieder at the University of Tübingen in southern Germany trained rhesus macaques to think of a number. He did this by showing them one set of dots on a computer, then, after a one-second interval, showing another set of dots. The monkeys were taught that if the second set was equal to the first set, then pressing a lever would earn them a reward of a sip of apple juice. If the second set was not equal to the first, then there was no apple juice. After about a year, the monkeys learned to press the lever only when the number of dots on the first and second screens was equal. Nieder and his colleagues reasoned that during the one-second interval between screens the monkeys were thinking about the number of dots they had just seen.

Nieder decided he wanted to see what was happening in the monkeys' brains when they were holding the number in their heads. So, he inserted an electrode two microns in diameter through a hole in their skulls and into the neural tissue. Don't worry, no monkeys were hurt. At that size, an electrode is tiny enough to slide through the brain without causing damage or pain. (The insertion of electrodes into human brains for research contravenes ethical guidelines, although it is allowed for therapeutic reasons such as the treatment of epilepsy.) Nieder positioned the electrode so that it faced a section of the monkeys' pre-frontal cortex, and then began the experiment.

The electrode was so sensitive that it could pick up electrical discharge in individual neurons. When the monkeys thought of numbers, Nieder saw that certain neurons became very active. A whole patch of their brains was lighting up.

On closer analysis, he made a fascinating discovery. The number-sensitive neurons reacted with varying charges depending on the number that the monkey was thinking of at the time. And each neuron had a 'preferred' number – a number that made it most active. There was, for example, a population of several thousand neurons that preferred the number one. These neurons shone brightly when a monkey thought of one, less brightly when he thought of two, even less brightly when he thought of three, and so on. There was another set of neurons that preferred the number two. These neurons shone brightest when a monkey thought of two, less brightly when he thought of one or three, dimmer still when the monkey thought of four. Another group of neurons preferred the number three, and another the number four. Nieder conducted experiments up to the number 30, and for each number found neurons that preferred that number.

The results offered an explanation for why our intuitions favour an approximate understanding of numbers. When a monkey is thinking 'four', the neurons that prefer four are the most active, of course. But the neurons that prefer three and the neurons that prefer five are also active, though less so, because its brain is also thinking of the numbers surrounding four. 'It is a noisy sense of number,' explained Nieder. 'The monkeys can only represent cardinalities in an approximate way.'

It is almost certain that the same thing happens in human brains. Which raises an interesting question. If our brains can represent numbers only approximately, then how were we able to 'invent' numbers in the first place? 'The "exact number sense" is a [uniquely] human property that probably stems from our ability to represent number very precisely with symbols,' concluded Nieder. Which reinforces the point that numbers are a cultural artefact, a man-made construct rather than something that we acquire innately.

11 blind mice
1010 green bottles
11000 blackbirds baked in a pie

CHAPTER ONE

The Counter Culture

In Lincolnshire during medieval times, a *pimp* plus a *dik* got you a *bumfit*. There was nothing dishonourable about this. The words were simply the numbers five, ten and fifteen in a jargon used by shepherds when counting their sheep. The full sequence ran:

1. Yan
2. Tan
3. Tethera
4. Pethera
5. Pimp
6. Sethera
7. Lethera
8. Hovera
9. Covera
10. Dik
11. Yan-a-dik
12. Tan-a-dik
13. Tethera-dik
14. Pethera-dik
15. Bumfit
16. Yan-a-bumfit
17. Tan-a-bumfit
18. Tethera-bumfit
19. Pethera-bumfit
20. Figgit

This is a different way from how we count now, and not just because all the words are unfamiliar. Lincolnshire shepherds organized their numbers in groups of twenty, starting counting with *yan* and ending with *figgit*. If a shepherd had more than twenty sheep – and provided he hadn't sent himself to sleep – he would make note of having completed one cycle by putting a pebble in his pocket, or making a mark on the ground, or scraping a line in his crook. He would then start from the beginning again: *'Yan, tan, tethera …'* If he had 80 sheep, he would have four pebbles in his pocket, or have marked four lines, at the end. The system is very efficient for the shepherd; he has four small items to represent 80 big ones.

In the modern world, of course, we group our numbers in tens, so our number system has ten digits – 0, 1, 2, 3, 4, 5, 6, 7, 8, 9. The number of the counting group, which is often also the number of symbols used, is called the base of a number system, so our decimal system is base ten, while the shepherds' base is 20.

Without a sensible base, numbers are unmanageable. Imagine that the shepherd had a base-one system, which would mean he had only one number word: *yan* for one. Two would be *yan yan*. Three would be *yan yan yan*. Eighty sheep would be *yan* said 80 times. This system is pretty useless for counting anything above about three. Alternatively, imagine that every number was a separate word so that being able to count up to 80 would require memory for 80 unique words. Now count to a thousand this way!

Many isolated communities still use unconventional bases. The Arara in the Amazon, for example, count in pairs, with the numbers from one to eight as follows: *anane, adak, adak anane, adak adak, adak adak anane, adak adak adak, adak adak adak anane, adak adak adak adak*. Counting in twos is not much of an improvement over counting in ones. Expressing 100 requires repeating *adak* 50 times in succession – which would make haggling at the market rather time-consuming. Systems in which numbers are grouped in threes and fours are also found in the Amazon.

The trick of a good base system is that the base number needs to be large enough to be able to express numbers like 100 without running out of breath, but not so large that we need to overexercise our memories. The most common bases throughout history have been five, ten and twenty, and there is an obvious reason why. These numbers are derived from the human body. We have five fingers on one hand, so five is the first obvious place to take a breath when counting upwards from one. The next natural pause comes at two hands, or ten fingers, and after that at hands and feet, or twenty

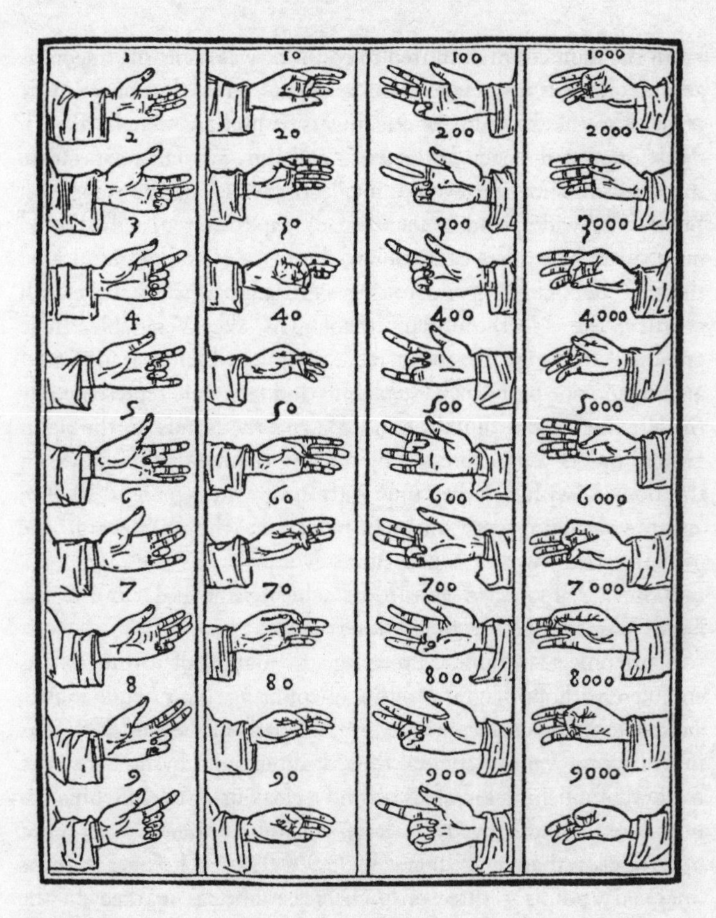

Finger counting from Luca Pacioli's Summa de arithmetica, geometria, proportioni et proportionalita *(1494).*

fingers and toes. (Some systems are composite. The Lincolnshire sheep-counting lexicon, for example, contains base five and ten as well as base 20: the first ten numbers are unique, and the next ten are grouped in fives.) The role that fingers have played in counting is reflected in much number vocabulary, not least the double meaning of digit. For example, five in Russian is *piat*, and the word for outstretched hand is *piast*. Similarly, Sanskrit for the word five, *pantcha*, is related to the Persian *pentcha*, hand.

From the moment man started to count he was using his fingers as an aid, and it is no exaggeration to credit a great deal of scientific progress to the versatility of our fingers. If humans were born with flat stumps at the ends of our arms and legs, it is fair to speculate that we would not have evolved intellectually beyond the Stone Age. Before the widespread availability of paper and pencil allowed numbers to be easily written down, they were often communicated through elaborate finger-counting sign languages. In the eighth century the Northumbrian theologian the Venerable Bede presented a system to count to a million, which was one part arithmetic, one part jazz hands. Units and tens were represented by the left fingers and thumb; hundreds and thousands on the right. Higher orders were expressed by moving the hands up and down the body – with a rather unpriestly image to represent 90,000: 'grasp your loins with the left hand, the thumb towards the genitals', Bede wrote. Much more evocative was the sign for a million, a self-satisfied gesture of achievement and closure: the hands clasped together, fingers intertwined.

Until only a few hundred years ago, no manual of arithmetic was complete without diagrams of finger-counting. Now, while mostly a lost art, the practice continues in some parts of the world. Traders in India who want to conceal their dealings from bystanders use a method of touching knuckles behind a cloak or cloth. In China, an ingenious – if rather overly intricate – technique allows you to count up to one less than ten billion – 9,999,999,999. Each finger has nine imaginary points – three on each crease line, as marked on the diagram opposite. These points on the right little finger represent the digits 1 to 9. The points on the right fourth finger take us from 10 to 90. The right middle finger goes from 100 to 900, and so on, with each new finger representing the next power of ten. It is therefore possible to count every single person on Earth with only your fingers, which is one way to have the whole world in your hands.

Some cultures count using more of their bodies than just fingers and toes. At the end of the nineteenth century an expedition of British anthropologists reached the islands of the Torres Strait, the stretch of water that separates Australia from Papua New Guinea. There they discovered a community that started with 'right hand little finger' for 1, 'right hand ring finger' for 2 and this continued through

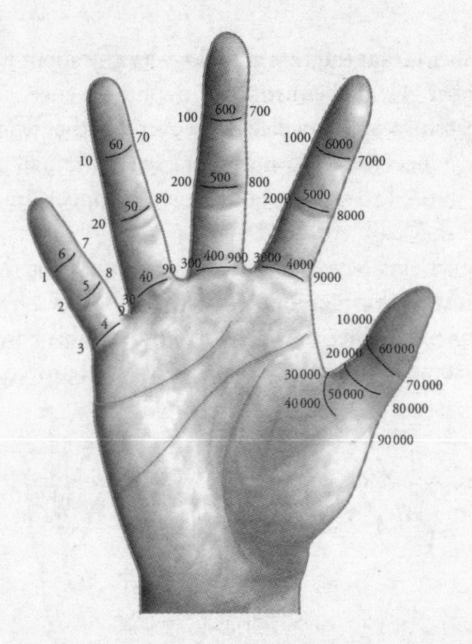

In this Chinese system, each finger has nine points, representing the digits 1 to 9 for each order of magnitude, so the right hand can express any number up to $10^5 - 1$ when the other hand touches the relevant points. Swapping hands, the numbers continue to $10^{10} - 1$. A 'zero' point is not needed on any finger, since when there are no values relating to that finger it is simply left alone by the other hand.

the fingers to 'right wrist' for 6, 'right elbow' for 7 and on through the shoulders, sternum, left arm and hand, feet and legs, ending at 'right foot little toe' for 33. Subsequent expeditions and research uncovered many communities in the region with similar 'body-tally' systems.

Perhaps the most curious is the Yupno, the only Papuan people for whom each individual owns a short melody that belongs to them like a name, or signature tune. They also have a counting system that enumerates the nostrils, eyes, nipples, belly button and climaxes in 31, for 'left testicle', 32, 'right testicle' and 33, 'penis'. While one can ponder the significance of 33 in the three great monotheistic religions (the age when Christ died, the length of King David's reign and the number of individual beads on a Muslim prayer string), what is particularly intriguing about the Yupno's phallic number is that they are actually very coy about it. They refer to the number 33 euphemistically in phrases such as 'the man thing'. Researchers were unable to discover whether women use the same terms, since they are not supposed to know the number system and refused to answer questions. The upper limit in Yupno is 34, which they call 'one dead man'.

Base-ten systems have been used in the West for thousands of years. Despite their harmoniousness with our bodies, however, many have questioned whether they are the most sensible base for counting. In fact, some have argued that their physical provenance makes

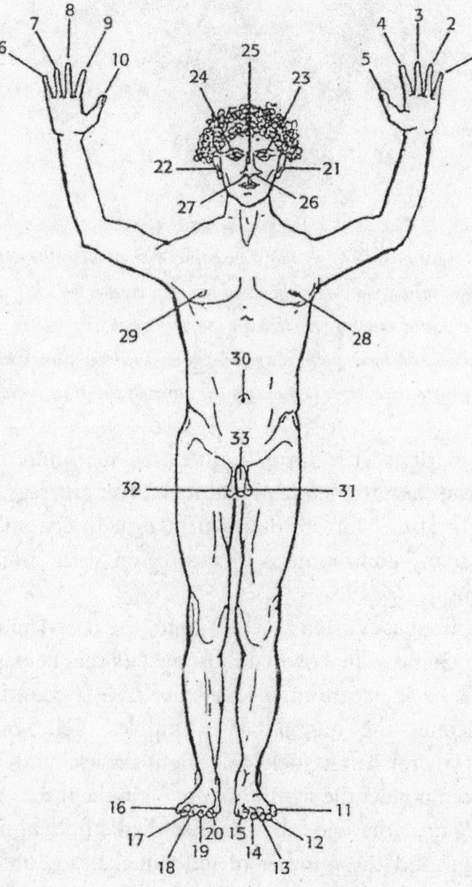

One dead Yupno.

them an actively *bad* choice. King Charles XII of Sweden dismissed base ten as the product of 'rustic and simple people' fumbling around with their fingers. In modern Scandinavia, he believed, a base was needed 'of more convenience and greater use'. So, in 1716, he ordered the scientist Emanuel Swedenborg to devise a new counting system with a base of 64. He arrived at this formidable number due to the fact that it was derived from a cube, $4 \times 4 \times 4$. Charles, who fought – and lost – the Great Northern War, believed that military calculations, such as measuring the volume of a box of gunpowder, would be made easier with a cube number as a base. Yet his brainwave, wrote Voltaire, 'could prove only that he loved the extraordinary and the difficult'. Base 64 requires 64 unique names (and symbols) for numbers – an absurdly inconvenient system. Swedenborg therefore simplified the system to base eight and came up with a new notation in which 0, 1, 2, 3, 4, 5, 6, 7 were renamed o, l, s, n, m, t, f, ŭ. In this system, therefore l + l = s, and m × m = so. (The words for the new numbers, however, were rather wonderful. The powers of 8, which would have been written lo, loo, looo, loooo and looooo, were to be pronounced, or yodelled, *lu, lo, li, le* and *la*.) In 1718, however, shortly before Swedenborg was due to present the system, a bullet shot the king – and his octonary dream – stone dead.

But Charles XII had a valid point. Why should we stick with the decimal system just because it was derived from the number of our fingers and toes? If humans were like Disney characters, for example, and had only three fingers and a thumb per hand, it is almost certain that we would live in a base-eight world: giving marks out of eight, compiling top eight charts and letting eight cents make a dime. Mathematics would not change by having an alternative way to group numbers. The bellicose Swede was correct to ask which base best suits our scientific needs – rather than opting for the one that suits our anatomy.

In late 1970s Chicago, Michael de Vlieger was watching the cartoons on Saturday morning TV. A short segment came on. The soundtrack was of disconcerting, off-key piano chords, wah-wah guitar and a menacing bass. Under a full moon and starry night sky a strange humanoid appeared. He had a blue and white striped top hat and tails, blond hair and a stick nose, rather in keeping with the glam-rock fashion of the era. If that wasn't creepy enough, he had five fingers and a thumb on each hand, and six toes on each foot. 'It was a little freaky, kind of spooky,' remembered Michael. The cartoon was *Little Twelvetoes*, an educational broadcast about base 12. 'I think the majority of the American population had no idea what was going on. But I thought it was so cool.'

Michael is now 38. I met him in his office, a business suite above some shops in a residential part of St Louis, Missouri. He has thick black hair with a few shoots of white, a round face, dark eyes and sallow skin. His mum is Filipino, while his father is white, and being a mixed-race kid made him the victim of taunts. A clever and sensitive child with an active imagination, he decided to invent his own language so that his classmates couldn't read his notebooks. *Little Twelvetoes* inspired him to do the same with numbers – and he adopted base 12 for personal use.

Base 12 has 12 digits: 0 to 9 and two extra ones to represent ten and eleven. The standard notation for each of these two 'transdecimal' digits is χ and Ɛ. So, counting to 12 now goes: 0, 1, 2, 3, 4, 5, 6, 7, 8, 9, χ, Ɛ, 10. (See table opposite.)

The new single digits are given new names to avoid confusion, so χ is called *dek* and Ɛ is called *el.* Also, we give 10 the name *do*, pronounced *doh*, and short for dozen, to avoid confusion with 10 in base ten. Counting upwards from *do* in base 12, or 'dozenal', we have *do one* for 11, *do two* for 12, *do three* for 13 all the way up to *twodo* for 20.

Michael devised a private calendar using base 12. Each date in this calendar was the number of days, counted in base 12, from the day he was born. He still uses it, and he told me later that I visited him on the 80Ɛ9th day of his life.

Michael adopted base 12 for reasons of personal security, but he is not alone in having fallen for its charms. Many serious thinkers have argued that 12 is a better base for a number system because the number is more versatile than 10. In fact, base 12 is more than a number system – it is a politico-mathematical cause. One of its earliest champions was Joshua Jordaine, who in 1687 self-published *Duodecimal Arithmetick*. He claimed that 'nothing was more natural and genuine' than counting in twelves. In the nineteenth century high-profile duodeciphiles included Isaac Pitman, who had gained considerable fame for inventing a widespread system of shorthand,

1 *one*	2 *two*	3 *three*	4 *four*	5 *five*	6 *six*	7 *seven*	8 *eight*	9 *nine*	χ *dek*	Ɛ *el*	10 *do*
11 *do* *one*	12 *do* *two*	13 *do* *three*	14 *do* *four*	15 *do* *five*	16 *do* *six*	17 *do* *seven*	18 *do* *eight*	19 *do* *nine*	1χ *do* *dek*	1Ɛ *do* *el*	20 *twodo*
21 *twodo* *one*	22 *twodo* *two*	23 *twodo* *three*	24 *twodo* *four*	25 *twodo* *five*	26 *twodo* *six*	27 *twodo* *seven*	28 *twodo* *eight*	29 *twodo* *nine*	2χ *twodo* *dek*	2Ɛ *twodo* *el*	30 *threedo*
31 *threedo* *one*	32 *threedo* *two*	33 *threedo* *three*	34 *threedo* *four*	35 *threedo* *five*	36 *threedo* *six*	37 *threedo* *seven*	38 *threedo* *eight*	39 *threedo* *nine*	3χ *threedo* *dek*	3Ɛ *threedo* *el*	40 *fourdo*
41 *fourdo* *one*	42 *fourdo* *two*	43 *fourdo* *three*	44 *fourdo* *four*	45 *fourdo* *five*	46 *fourdo* *six*	47 *fourdo* *seven*	48 *fourdo* *eight*	49 *fourdo* *nine*	4χ *fourdo* *dek*	4Ɛ *fourdo* *el*	50 *fivedo*
51 *fivedo* *one*	52 *fivedo* *two*	53 *fivedo* *three*	54 *fivedo* *four*	55 *fivedo* *five*	56 *fivedo* *six*	57 *fivedo* *seven*	58 *fivedo* *eight*	59 *fivedo* *nine*	5χ *fivedo* *dek*	5Ɛ *fivedo* *el*	60 *sixdo*
61 *sixdo* *one*	62 *sixdo* *two*	63 *sixdo* *three*	64 *sixdo* *four*	65 *sixdo* *five*	66 *sixdo* *six*	67 *sixdo* *seven*	68 *sixdo* *eight*	69 *sixdo* *nine*	6χ *sixdo* *dek*	6Ɛ *sixdo* *el*	70 *sevendo*
71 *sevendo* *one*	72 *sevendo* *two*	73 *sevendo* *three*	74 *sevendo* *four*	75 *sevendo* *five*	76 *sevendo* *six*	77 *sevendo* *seven*	78 *sevendo* *eight*	79 *sevendo* *nine*	7χ *sevendo* *dek*	7Ɛ *sevendo* *el*	80 *eightdo*
81 *eightdo* *one*	82 *eightdo* *two*	83 *eightdo* *three*	84 *eightdo* *four*	85 *eightdo* *five*	86 *eightdo* *six*	87 *eightdo* *seven*	88 *eightdo* *eight*	89 *eightdo* *nine*	8χ *eightdo* *dek*	8Ɛ *eightdo* *el*	90 *ninedo*
91 *ninedo* *one*	92 *ninedo* *two*	93 *ninedo* *three*	94 *ninedo* *four*	95 *ninedo* *five*	96 *ninedo* *six*	97 *ninedo* *seven*	98 *ninedo* *eight*	99 *ninedo* *nine*	9χ *ninedo* *dek*	9Ɛ *ninedo* *el*	χ0 *dekdo*
χ1 *dekdo* *one*	χ2 *dekdo* *two*	χ3 *dekdo* *three*	χ4 *dekdo* *four*	χ5 *dekdo* *five*	χ6 *dekdo* *six*	χ7 *dekdo* *seven*	χ8 *dekdo* *eight*	χ9 *dekdo* *nine*	χχ *dekdo* *dek*	χƐ *dekdo* *el*	Ɛ0 *eldo*
Ɛ1 *eldo* *one*	Ɛ2 *eldo* *two*	Ɛ3 *eldo* *three*	Ɛ4 *eldo* *four*	Ɛ5 *eldo* *five*	Ɛ6 *eldo* *six*	Ɛ7 *eldo* *seven*	Ɛ8 *eldo* *eight*	Ɛ9 *eldo* *nine*	Ɛχ *eldo* *dek*	ƐƐ *eldo* *el*	100 *gro*

Dozenal numbers from 1 to 100.

and Herbert Spencer, the Victorian social theorist. Spencer urged base-system reform on behalf of 'working people, people of narrow incomes and the minor shopkeepers who minister to their wants'. The American inventor and engineer John W. Nystrom was also a fan. He described base 12 as 'duodenal' – perhaps the most unfortunate double entendre in the history of science.

The reason that 12 might be considered superior to ten is because of its divisibility. Twelve can be divided by 2, 3, 4 and 6, whereas ten can be divided only by 2 and 5. Advocates of base 12 argue that we are much more likely to want to divide by 3 or 4 than divide by 5 in our daily lives. Consider a shopkeeper. If you have 12 apples, then you can divide them up into two bags of 6, three bags of 4, four bags of 3, or six bags of 2. This is much more user-friendly than 10, which can only be cleanly divided into two bags of 5, or five of 2. The word 'grocer', in fact, is a relic of a retailer's preference for 12 – it comes from 'gross', meaning a dozen dozen, or 144. The multi-divisibility of 12 also explains the utility of imperial measure: a foot, which is 12 inches, can be cleanly divided by 2, 3 and 4 – which is quite a plus for carpenters and tailors.

Divisibility is also relevant to multiplication tables. The easiest tables to learn in any base are the ones of numbers that divide that base. This is why, in base ten, the 2 and 5 times tables – which are just the even numbers and the numbers ending in 5 or 0 – are so painless to recite. Likewise, in base 12 the simplest times tables are also those of its divisors: 2, 3, 4 and 6.

2 × 1 = 2	3 × 1 = 3	4 × 1 = 4	6 × 1 = 6
2 × 2 = 4	3 × 2 = 6	4 × 2 = 8	6 × 2 = 10
2 × 3 = 6	3 × 3 = 9	4 × 3 = 10	6 × 3 = 16
2 × 4 = 8	3 × 4 = 10	4 × 4 = 14	6 × 4 = 20
2 × 5 = χ	3 × 5 = 13	4 × 5 = 18	6 × 5 = 26
2 × 6 = 10	3 × 6 = 16	4 × 6 = 20	6 × 6 = 30
2 × 7 = 12	3 × 7 = 19	4 × 7 = 24	6 × 7 = 36
2 × 8 = 14	3 × 8 = 20	4 × 8 = 28	6 × 8 = 40
2 × 9 = 16	3 × 9 = 23	4 × 9 = 30	6 × 9 = 46
2 × χ = 18	3 × χ = 26	4 × χ = 34	6 × χ = 50
2 × Ɛ = 1χ	3 × Ɛ = 29	4 × Ɛ = 38	6 × Ɛ = 56
2 × 10 = 20	3 × 10 = 30	4 × 10 = 40	6 × 10 = 60

If you look at the final digits of each column, you see a striking pattern. The two times table is, again, all the even numbers. The three times table is all the numbers ending in 3, 6, 9 and 0. The four times table is the numbers ending in 4, 8 and 0, and the six times table all the numbers ending 6 or 0. In other words, in base 12 we get the 2, 3, 4 and 6 times tables for free. Since many children have difficulty in learning their times tables, if we converted to base 12 we would be carrying out a great humanitarian act. Or so the argument goes.

The campaign for base 12 should not be conflated with the crusade against metric by fans of imperial measure. Those people who prefer feet and inches over metres and centimetres have no issue as to whether one foot should be 12 inches, or 10 inches, as it would be in dozenal. Historically, however, an underlying theme of the campaign for base 12 has been a jingoistic anti-Frenchness. Perhaps the finest example of such a view was a pamphlet from 1913 by engineer Rear-Admiral G. Elbrow, in which he called the French metric system 'retrograde'. He published a list of the dates, in base 12, of the kings and queens of England. He also noticed that Britain had been invaded shortly after each decimal millennium – by the Romans in 43 CE and the Normans in 1066. 'What if, at the beginning of the [third millennium],' he prophesized, 'these two [countries] may again appear in the same direction, and this time in conjunction?' Invasion by France and Italy might be averted, he argued, simply by rewriting the year 1913 as 1135, as it would be in dozenal, thus delaying the third millennium by several centuries.

The most famous dozenalist call-to-arms, though, was an article in *The Atlantic Monthly* in October 1934 by the writer F. Emerson Andrews, which led to the formation of the Duodecimal Society of America, or DSA. (It later changed its name to the Dozenal Society of America since 'duodecimal' was deemed to be overly reminiscent of the system they were aiming to replace.) Andrews claimed that base ten had been adopted with 'inexcusable shortsightedness' and wondered whether it 'would be so tremendous a sacrifice' to abandon it. The DSA initially insisted prospective members pass four tests in dozenal arithmetic, although this requirement was quickly dropped. The *Duodecimal Bulletin*, which continues to this day, is an excellent publication and the only place outside medical literature with articles on hexadactyly, the condition of being born with six fingers. (Which is more common than you might think. About one in every 500 people is born with at least an extra finger or toe.) In 1959 a sister organization, the Dozenal Society of Great Britain, was founded, and a year later the First International Duodecimal Conference was held in France. It was also the last. Still, both societies continue to battle for a dozenal future, seeing themselves as downtrodden militants rallying against the 'tyranny of ten'.

Michael de Vlieger's youthful infatuation with base 12 was not a passing phase; he is the current president of the DSA. In fact, he is so committed to the system that he uses it in his job as a designer of digital architectural models.

While base 12 certainly makes times tables easier to learn, its greatest advantage is how it cleans up fractions. Base ten is often messy when you want to divide. For example, a third of 10 is 3.33..., where the threes go on for ever. A quarter of 10 is 2.5, which needs a decimal place. In base 12, however, a third of 10 is 4 and a quarter of 10 is 3. Nice. Expressed as a percentage, a third is 40 per cent, and a quarter 30 per cent. (Although, of course, the correct phrase would now be per *gross*.) In fact if you look at how 100 is divided by the numbers 1 to 12, base 12 provides more concise numbers (note that the semi-colons in the right column stand for the 'dozenal' point).

Fraction of 100	**Decimal**	**Dozenal**
One	100	100
Half	50	60
Third	33.333...	40
Quarter	25	30
Fifth	20	24;97...
Sixth	16.666...	20
Seventh	14.285...	18;6Χ35...
Eighth	12.5	16
Ninth	11.111...	14
Tenth	10	12;497...
Eleventh	9.09...	11;11...
Twelfth	8.333...	10

It is this increased precision that makes base 12 better suited to Michael's needs. Even though his clients supply him with dimensions

in decimal, he prefers to translate them into dozenal. 'It gives me more choices when dividing into simple ratios,' he said.

'Avoiding [messy] fractions helps ensure things fit. Sometimes, because of time constraints or late-breaking changes, I will need to quickly apply a lot of change at a location that doesn't jive with the grid I initially set up. Thus it's important to have predictable simple ratios. I've got more and cleaner choices with dozenal, and it's faster.' Michael even believes that using base 12 gives his business an edge, comparing it to cyclists and swimmers who shave their legs.

The DSA used to want to replace decimal with dozenal, and its fundamentalist wing still does, but Michael's ambitions are more modest. He wants simply to show people that there is an alternative to the decimal system, and that perhaps it suits their needs better. He knows that the chances of the world abandoning *dix* for *douze* are non-existent. The change would be both confusing and expensive. And decimal works well enough for most people – especially in the computer age, where mental arithmetic skills are less required generally. 'I would say that dozenal is the optimum base for general computation, for everyday use,' he added, 'but I am not here to convert anybody.'

An immediate goal of the DSA is to get the numerals for dek and el into Unicode, the repertoire of text characters used by most computers. In fact, a major debate in dozenal society is which symbols to use. The DSA standard χ and Ɛ were designed in the 1940s by William Addison Dwiggins, one of the US's foremost font designers and the creator of the typefaces Caledonia and Electra. Isaac Pitman preferred ↊ and ↋. Jean Essig, a French enthusiast, preferred ∠ and ↊. Some practical members would prefer * and # since they are already on the 12 buttons of a touch-tone phone. The number words are also an issue. The *Manual of the Dozen System* (written in 1960, or 1174 in dozenal) recommends the terms dek, el and do (with gro for 100, mo for 1000, and do-mo, gro-mo, bi-mo and tri-mo for the next highest powers of do). Another suggestion is to keep ten, eleven and twelve and continue with twel-one, twel-two, and so on. Such is the sensitivity over terminology that the DSA is careful not to push any system. Great care is needed not to marginalize devotees of any particular symbol or term.

Michael's love of avant-garde bases did not stop at 12. He has toyed with eight, which he sometimes uses when doing DIY at home. 'I use bases as tools,' he said. And he has gone up to base 60. For this he had to design 50 extra symbols to add to the ten digits we have already. His purpose was not practical. He described working in base 60 as like going up a high mountain. 'I can't live up there. It is too big a grouping. In the valley it is decimal, and there I can breathe. But I can visit the mountain to see what the view is like.' He has written out tables of factors in a base 60, or sexagesimal, system, and stared in wonder at the patterns they revealed. 'There definitely is a beauty there,' he told me.

While base 60 seems like the product of an extraordinarily fertile imagination, sexagesimal has historical pedigree. It is actually the most ancient base system that we know of.

The simplest form of numerical notation is the tally. It has been used in different forms across the world. The Incas kept count by tying knots on ropes, while cave dwellers painted marks on rocks and, since the invention of wooden furniture, bedposts have – figuratively, at least – been marked with notches. The oldest discovered 'mathematical artefact' is believed to be a tally stick: a 35,000-year-old baboon fibula found in a Swaziland cave. The 'Lebombo bone' has 29 lines scraped on it, which possibly denote a lunar cycle.

As we saw in the previous chapter, humans can instantly tell the difference between one item and two, between two items and three, but beyond four it gets difficult. This is true of notches as well. For any convenient system of tally-keeping, the tallies need to be grouped. In Britain, tally convention is to mark four vertical lines and then make the fifth a diagonal crossing through them – the so-called 'five-bar gate'. In South America, the preferred style is for the first four lines to mark a square and the fifth is a diagonal in the square. The Japanese, Chinese and Koreans use a more elaborate method, constructing the character 正, which means 'correct' or

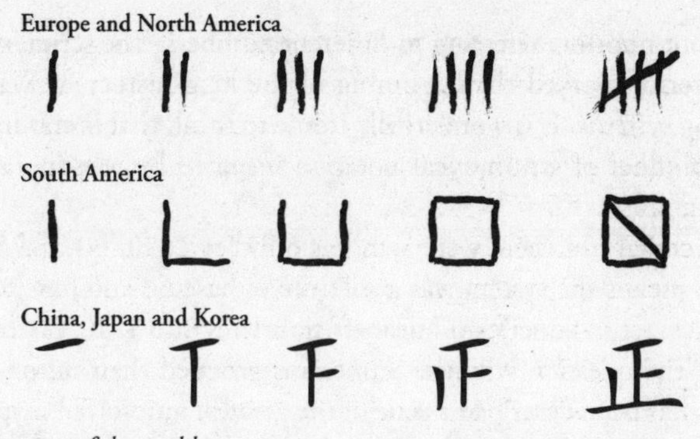

Tally systems of the world.

'proper'. (The next time you have sushi, ask the waiter to show you how he is tallying your dishes.)

Around 8000 BC a practice of using small clay pieces with markings to refer to objects emerged throughout the ancient world. These tokens primarily recorded numbers of things, such as sheep to be bought and sold. Different clay pieces referred to different objects or numbers of objects. From that moment sheep could be counted without actually being there, which made trade and stock-keeping much easier. It was the birth of what we understand now as numbers.

In the fourth millennium BC in Sumer, an area now in present-day Iraq, this token system evolved into a script in which a pointed reed was pressed into soft clay. Numbers were first represented by circles or fingernail shapes. By around 2700 BC the stylus had a flat edge and the imprints looked rather like bird footprints, with

	1	10	60	3600
Archaic Sumerian numbers **Fourth century BC**				
Cuneiform numbers **Third century BC**				

different imprints referring to different numbers. The script, called cuneiform, marked the beginning of the long history of Western writing systems. It is wonderfully ironic to think that literature was a by-product of a numerical notation invented by Mesopotamian accountants.

In cuneiform there were symbols only for 1, 10, 60 and 3600, which means the system was a mixture of base 60 and base ten, as the basic set of cuneiform numbers translates into 1, 10, 60 and 60 × 60. The question why the Sumerians grouped their numbers in sixties has been described as one of the greatest unresolved mysteries in the history of arithmetic. Some have suggested it was the result of the fusion of two previous systems, with bases five and 12, though no conclusive evidence of this has been found.

The Babylonians, who made great advances in maths and astronomy, embraced the Sumerian sexagesimal base, and later the Egyptians, followed by the Greeks, based their time-counting methods on the Babylonian way – which is why, to this day, there are 60 seconds in a minute and 60 minutes in an hour. We are so used to telling the time in base 60 that we never question it, even though it really is quite unexplained. Revolutionary France, however, wanted to iron out what they saw as an inconsistency in the decimal system. When the National Convention introduced the metric system for weights and measures in 1793, it also tried to decimalize time. A decree was signed establishing that every day would be divided into ten hours, each containing 100 minutes, each of which contained 100 seconds. This worked out neatly, making 100,000 seconds in the day – compared to 86,400 (60 × 60 × 24) seconds. The revolutionary second was, therefore, a fraction shorter than the normal second. Decimal time became mandatory in 1794 and watches were produced with the numbers going up to ten. Yet the new system was completely bewildering to the populace and abandoned after little more than six months. An hour with 100 minutes is also not as convenient as an hour with 60 minutes, since 100 does not have as many divisors as 60. You can divide 100 by 2, 4, 5, 10, 20, 25 and 50, but you can divide 60 by 2, 3, 4, 5, 6, 10, 12, 15, 20 and 30. The failure of decimal time was a small victory for dozenal thinking. Not only does 12 divide into 60 but it also divides into 24, the number of hours in a day.

Revolutionary watch with decimal and traditional clock face.

A more recent campaign to decimalize time also flopped. In 1998 the Swiss conglomerate Swatch launched Swatch Internet Time, which divided the day into 1000 parts called beats (equivalent to 1min 26.4secs). The manufacturer sold watches that displayed its 'revolutionary vision of time' for a year or so before sheepishly removing them from its catalogue.

The French and Swiss, however, are not the only Western nations to have had barmy counting procedures in the not too distant past. The tally stick, which became outdated the moment the first Sumerian printed his first cuneiform tablet, was used as a form of British currency until 1826. The Bank of England used to issue souped-up tally sticks that were worth a monetary value based on the distance of a mark from the base. A document written in 1186 by the Lord Treasurer Richard Fitzneal set out the values as:

£1000	thickness of the palm of the hand
£100	breadth of a thumb
£20	breadth of a little finger
£1	width of a swollen barleycorn

The procedure the Treasury used was, in fact, a system of 'double tallies'. A piece of wood was split down the middle, giving two parts – the stock and the foil. A value was marked – tallied – on the stock and was also marked on the foil, which acted like a receipt. If I lent some money to the Bank of England, I would be given a stock with a notch indicating the amount – which explains the origin of the words *stockholder* and *stockbroker* – while the bank kept the foil, which had a matching notch.

This practice was abandoned barely two centuries ago. In 1834, the Treasury decided to incinerate the obsolete pieces of wood in a furnace under the Palace of Westminster, the seat of British government. The fire, however, spread out of control. Charles Dickens wrote: 'The stove, overgorged with these preposterous sticks, set fire to the panelling; the panelling set fire to the House of Commons; the two houses [of government] were reduced to ashes.' Obscure financial instruments have often impacted on the work of government, but only the tally stick has brought down a parliament. When the palace was rebuilt it had a brand new clock tower, Big Ben, which quickly became the most recognizable landmark in London.

An argument often used in favour of the imperial system over metric is that the words sound better. A case in point is the measures for wine:

2 gills = 1 chopin
2 chopins = 1 pint
2 pints = 1 quart
2 quarts = 1 pottle
2 pottles = 1 gallon
2 gallons = 1 peck
2 pecks = 1 demibushel
2 demibushels = 1 bushel (or firkin)
2 firkins = 1 kilderkin
2 kilderkins = 1 barrel

2 barrels = 1 hogshead
2 hogsheads = 1 pipe
2 pipes = 1 tun

This system is base two, or binary, which is usually expressed using the digits 0 and 1. Numbers in binary are the numbers you would use in base ten when only 0 and 1 appear. In other words, the sequence that begins 0, 1, 10, 11, 100, 101, 110, 111, 1000. So, 10 is two, 100 is four, 1000 is eight and so on, with each extra 0 on the end representing multiplication by two. (Which is just like base ten – adding a 0 on the end of a number is multiplication by ten.) In the wine measures, the smallest unit is a gill. Two gills makes a chopin, 4 gills a pint, 8 gills a quart, 16 gills a pottle, etc. The measures replicate perfectly the binary numerals. If a gill is represented by 1, then a chopin is 10, a pint is 100, a quart is 1000 and this carried on all the way to a tun, which is 10,000,000,000,000.

Binary can claim as its cheerleader the greatest mathematician ever to have fallen in love with a non-standard base. Gottfried Leibniz was one of the most important thinkers of the late seventeenth century, a scientist, philosopher and statesman. One of his duties was as librarian to the court of the Duke of Brunswick in Hanover. Leibniz was so excited with base two that he once wrote a letter to the Duke urging him to cast a silver medallion inscribed with the

Design for Leibniz's binary medallion, in Johann Bernard Wiedeburg's Dissertatio mathematica de praestantia arithmeticae binaria prae decimali *(1718). As well as the words* Imago Creationis, *the Latin reads 'From nothing comes one and everything, but the one is necessary'.*

words *Imago Creationis* – 'in the image of the world' – as a tribute to the binary system. For Leibniz, binary had practical and spiritual relevance. First, he thought that its capacity for describing every number in terms of doubles facilitated a variety of operations. '[It] permits the Assayer to weigh all sorts of masses with few weights and could serve in coinage to give more value with fewer pieces,' he wrote in 1703. Leibniz did admit that binary had some practical drawbacks. The numbers are much longer when written out: 1000 in decimal, for example, is 1,111,101,000 in binary. But he added: 'In recompense for its length, [binary] is more fundamental to science and gives new discoveries.' By looking at the symmetries and patterns in binary notation, he claimed, new mathematical insights are revealed, and number theory is richer and more versatile because of it.

Second, Leibniz marvelled at how the binary system chimed with his religious views. He believed that the cosmos was composed of being, or substance, and non-being, or nothingness. The duality was perfectly symbolized by the numbers 1 and 0. In the same way that God creates all beings from the void, all numbers can be written in terms of 1s and 0s. Leibniz's conviction that binary exemplified a fundamental metaphysical truth was – to his great delight – strengthened when later in life he was shown the *I Ching*, the ancient Chinese mystical text. The *I Ching* is a book of divination. It contains 64 different symbols, each of which comes with an accompanying commentary. The reader randomly selects a symbol (traditionally by casting yarrow sticks) and interprets the related text – a little like one might read an astrological chart. Each symbol in the *I Ching* is a hexagram, which means it is composed of six horizontal lines. The lines are either broken or unbroken, corresponding to a *yin* or a *yang*. The 64 hexagrams in the *I Ching* are the full set of combinations of *yins* and *yangs* when taken in groups of six at a time.

A particularly elegant way of ordering the hexagrams is shown opposite. If each *yang* is written 0 and each *yin* is 1, then the sequence matches precisely the binary digits from 0 to 63.

This way of ordering is known as the Fu Hsi sequence. (Strictly speaking, it is the inverse of Fu Hsi, but they are mathematically equivalent.) When Leibniz was made aware of the binary nature of Fu Hsi, it gave him 'a high opinion of [the *I Ching*'s] profundity'. Since he thought that the binary system mirrored Creation, his

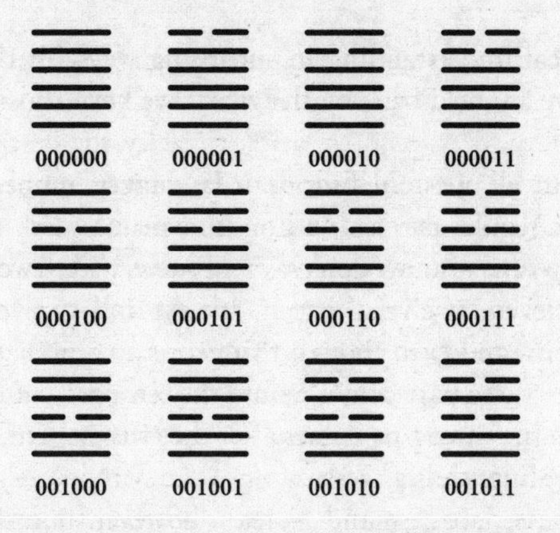

Part of the Fu Hsi sequence of the I Ching *and its binary equivalent.*

discovery that it also underlay Taoist wisdom meant that Eastern mysticism could now be accommodated within his own Western beliefs. 'The substance of the ancient theology of the Chinese is intact and, purged of additional errors, can be harnessed to the great truths of the Christian religion,' he wrote.

Leibniz's panegyrics on base two were a rather eccentric preoccupation of the pre-eminent polymath of his day. Yet in ascribing a fundamental importance to the system, he was more prescient than even he could ever have imagined. The digital age runs on binary, as computer technology relies at a most basic level on a language comprised of 0s and 1s. 'Alas!' wrote mathematician Tobias Dantzig. 'What was once hailed as a monument to monotheism ended in the bowels of a robot.'

'Freedom is the freedom to say two plus two equals four,' wrote Winston Smith, the protagonist of George Orwell's *Nineteen Eighty-Four*. Orwell was making a comment not only about freedom of speech in the Soviet Union, but also about mathematics. Two plus two is always four. No one can tell you it isn't. Mathematical truths cannot be influenced by culture or ideology.

On the other hand, our approach to mathematics is very much influenced by culture. The selection of base ten, for example, was not premised on mathematical reasons but on physiological ones, the number of our fingers and toes. Language also shapes mathematical understanding in surprising ways. In the West, for example, we are held back by the words we have chosen to express numbers.

In almost all Western European languages, number words do not follow a regular pattern. In English we say twenty-one, twenty-two, twenty-three. But we don't say tenty-one, tenty-two, tenty-three – we say eleven, twelve, thirteen. Eleven and twelve are unique constructions, and even though thirteen is a combination of three and ten, the three part comes before the ten part – unlike twenty-three, when the three part comes after the twenty part. Between ten and twenty, English is a mess.

In Chinese, Japanese and Korean, however, number words do follow a regular pattern. Eleven is written ten one. Twelve is ten two, and so on with ten three, ten four up to ten nine for nineteen. Twenty is two ten, and twenty-one is two ten one. You pronounce numbers in all cases just as you see them written down. So what? Well, it does make a difference at a young age. Experiments have repeatedly shown that Asian children find it easier to learn to count than Europeans. In one study with Chinese and American four- and five-year-olds, the two nationalities performed similarly when learning to count to 12, but the Chinese were about a year ahead with higher numbers. A regular system also makes arithmetic clearer to understand. A simple sum such as twenty-five plus thirty-two when expressed as two ten five plus three ten two is one step closer to the answer already: five ten seven.

Not all European languages are irregular. Welsh, for example, is just like Chinese. Eleven in Welsh is *un deg un* (one ten one); twelve is *un deg dau* (one ten two), and so on. Ann Dowker and Delyth Lloyd at the University of Oxford tested the maths abilities of Welsh- and English-speaking kids from the same Welsh village. While Asian children may be better than American children because of many cultural factors, such as hours spent practising or attitudes towards maths, cultural factors can be eliminated if the children are all living in the same place. Dowker and Lloyd concluded that while general arithmetic performance was more or less equal between Welsh- and English-speakers, the Welsh-speakers did demonstrate better mathematical skills in specific areas – such as reading, comparing and manipulating two-digit numbers.

German is even more irregular than English. In German, twenty-one is *einundzwanzig*, or one-and-twenty, twenty-two is *zweiundzwanzig*, or two-and-twenty, and this continues with the unit value preceding the tens value all the way up to 99. This means that when a German says a number over 100, the digits are not pronounced in a consecutive order: three hundred and forty-five is *dreihundertfünfundvierzig*, or three-hundred-five-and-forty, which lists the numbers in the higgledy-piggledy form 3-5-4. Such is the level of concern in Germany that this makes numbers more confusing than they have to be, that a campaign group *Zwanzigeins* (Twenty-one) has been set up to push for a change to a more regular system.

And it's not just the positioning of number words, or their irregular forms between eleven and nineteen, that puts the speakers of the main Western European languages at a disadvantage with some Asian language-speakers. We are also handicapped by how long it takes us to say numbers. In *The Number Sense*, Stanislas Dehaene writes down the list 4, 8, 5, 3, 9, 7, 6 and asks us to spend 20 seconds memorizing it. English-speakers have a 50 per cent chance of remembering the seven numbers correctly. By contrast, Chinese-speakers can memorize nine digits in this way. Dehaene says that this is because the number of digits we can hold in our heads at any one time is determined by how many we can say in a two-second loop. The Chinese words for one to nine are all concise single syllables: *yi, er, san, si, wu, liu, qi, ba, jiu*. They can be uttered in less than a quarter of a second, so in a two-second span a Chinese-speaker can rattle through nine of them. English number words, by contrast, take just under a third of a second to say (thanks to the frankly cumbersome 'seven', with two syllables, and the extended syllable 'three'), so our limit in two seconds is seven. The record, however, goes to the Cantonese, whose digits are spoken with even more brevity. They can remember ten of them in a two-second period.

While Western languages seem to be working against any mathematical ease of understanding, in Japan language is recruited as an ally. Words and phrases are modified in order to make their multiplication tables, called *kuku*, easier to learn. The tradition of these tables originated in ancient China, spreading to Japan around the eighth century. *Ku* in Japanese is nine, and the name comes from the fact that the tables used to begin at the end, with $9 \times 9 = 81$.

Around 400 years ago they were changed so that the *kuku* now begins 'one one is one'.

The words of the *kuku* are simply:

One one is one
One two is two
One three is three …

This carries on to 'One nine is nine', and then the twos begin with:

Two one is two
Two two is four

And so on to nine nine is eighty-one.

So far, this seems very similar to the plain British style of reciting the times tables. In the *kuku*, however, whenever there are two ways to pronounce a word, the way that flows better is used. For example, the word for one can be *in* or *ichi*, and rather than starting the *kuku* with either *in in* or *ichi ichi*, the more sonorous combination *in ichi* is used. The word for eight is *ha*. Eight eights should be *ha ha*. Yet the line in the *kuku* for 8×8 is *happa* since it rolls quicker off the tongue. The result is that the *kuku* is rather like a piece of poetry, or a nursery rhyme. When I visited an elementary school in Tokyo and watched a class of seven- and eight-year-olds practise their *kuku*, I was struck by how much it sounded like a rap – the phrases were syncopated and jolly. Certainly it bore no relation to how I remember reciting my times tables at school, which was with the metronomic delivery of a steam train going up a hill. Makiko Kondo, the teacher, said that she teaches her pupils *kuku* with an uptempo rhythm because it makes it fun to learn. 'First we get them to recite it, and only some time later do they come to understand the real meaning.' The poetry of the *kuku* seems to embed the times tables in Japanese brains. Adults told me that they know, for example, that seven times seven is 49 not because they remember the maths but because the music of 'seven seven forty-nine' sounds right.

While the irregularities of Western number words may be unfortunate for budding arithmeticians, they are of extreme interest to mathematical historians. The French for eighty is *quatre-vingts*, or four-twenties, indicating that ancestors of the French once used a base-20 system. It has also been suggested that the reason why the words for 'nine' and 'new' are identical or similar in many Indo-European languages, including French (*neuf, neuf*), Spanish (*nueve, nuevo*), German (*neun, neu*) and Norwegian (*ni, ny*) is a legacy of a long-forgotten base-eight system, where the ninth unit would be the first of a new set of eight. (Excluding thumbs, we have eight fingers, which could be how such a base developed. Or possibly from counting the gaps between the fingers.) Number words are also a reminder of how close we are to the numberless tribes of the Amazon and Australia. In English, *thrice* can mean both three times and many times; in French, *trois* is three and *très* is very: shadows, perhaps, of our own 'one, two, many' past.

Whereas certain aspects of number – such as the base, the style of numeral or the form of the words used – have differed widely between cultures, the early civilizations were surprisingly unified in the mechanics of how they counted and calculated. The general method they used is called 'place value' and is the principle by which different positions are used to represent different orders of number. Let's consider what this means in the context of shepherds in medieval Lincolnshire. As I wrote earlier, they had 20 numbers from *yan* to *figgit*. Once a shepherd counted 20 sheep, he put a pebble aside and started counting from *yan* to *figgit* all over again. If he had 400 sheep, he would have 20 pebbles, since 20 × 20 = 400. Now imagine the shepherd had a thousand sheep. If he counted them all he would have 50 pebbles, since 50 × 20 = 1000. Yet the problem with the shepherd having 50 pebbles is that he has no way to count them, since he cannot count higher than 20!

A way to solve this is to draw parallel furrows on the ground, as in the figure overleaf. When the shepherd counts 20 sheep he puts a pebble in the first furrow. When he counts another 20 sheep, he puts another pebble in the first furrow. Slowly the first furrow fills up with pebbles. When the time comes to put a twentieth pebble in the furrow, however, he instead puts a single pebble in the second furrow, and clears the first furrow of all pebbles. In other words, one

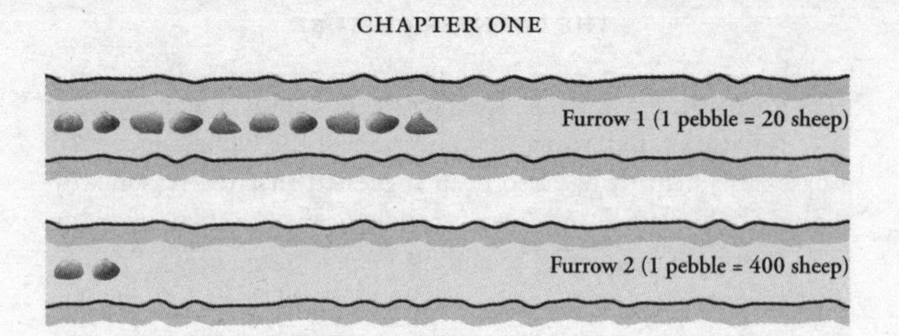

Total sheep = $(10 \times 20) + (2 \times 400) = 1000$

pebble in the second furrow means 20 pebbles in the first – just as one pebble in the first furrow means 20 sheep. A pebble in the second row stands for 400 sheep. A shepherd who has a thousand sheep and uses this procedure will have two pebbles in the second furrow, and ten in the first. By using a place-value system like this one – by which each furrow confers a different value to the pebble in it – he has used only 12 pebbles to count 1000 sheep rather than the 50 he would have needed without it.

Place-value counting systems have been used all over the world. Instead of pebbles in furrows, the Incas used beans or grains of maize in trays. North American Indians threaded pearls or shells on different-coloured string. The Greeks and Romans used counters of bone, ivory or metal on tables that had different columns marked out. In India they used marks on sand.

The Romans also made a mechanical version, with beads sliding in slots, called an abacus. These portable versions spread through the civilized world, though different countries preferred different versions. The Russian *schoty* has ten beads per rod (except on the row that has four beads, used by cashiers to denote quarter roubles). The Chinese *suan-pan* has seven, while the Japanese *soroban*, like the Roman abacus, has just five.

About a million children annually in Japan learn the abacus, attending one of 20,000 after-school abacus clubs. One evening in Tokyo I visited one in a west Tokyo suburb. The club was a short walk from a local train line, on the corner of a residential block. Thirty brightly coloured bicycles were parked outside. A large window displayed trophies, abacuses and a line of wooden slats with the calligraphied names of its star pupils.

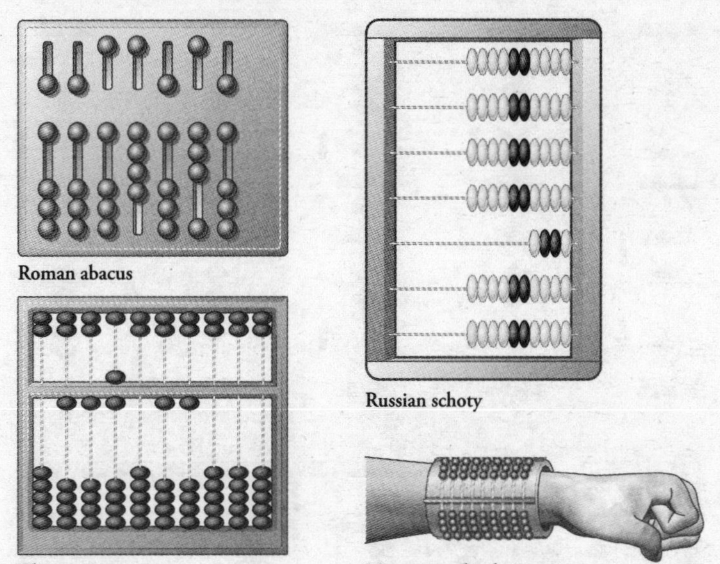

Roman abacus

Chinese suan-pan

Russian schoty

Mayan nepohualtzintzin

The Japanese equivalent of 'reading, 'riting and 'rithmetic' is *yomi, kaki, soroban*, or reading, writing, abacus. The phrase dates from the period in Japanese history between the seventeenth and nineteenth centuries when the country was almost totally isolated from the rest of the world. As a new merchant class emerged, which required skills other than proficiency with a samurai sword, so did a culture of private community-run schools that taught language and arithmetic – with the focus on abacus training.

Yuji Miyamoto's abacus club is a modern descendant of these older *soroban* establishments. When I walked in, Miyamoto, who was wearing a dark blue suit and white shirt, was standing in front of a small classroom of five girls and nine boys. He was reading out numbers in Japanese with the breathless syncopation of a horseracing commentator. As the children added them all up, the clatter of beads sounded like a swarm of cicadas.

In a *soroban*, there are exactly ten positions of beads per column, representing the numbers from 0 to 9, as shown overleaf.

When a number is displayed on the *soroban*, each digit of the number is represented on a separate column using one of the ten positions.

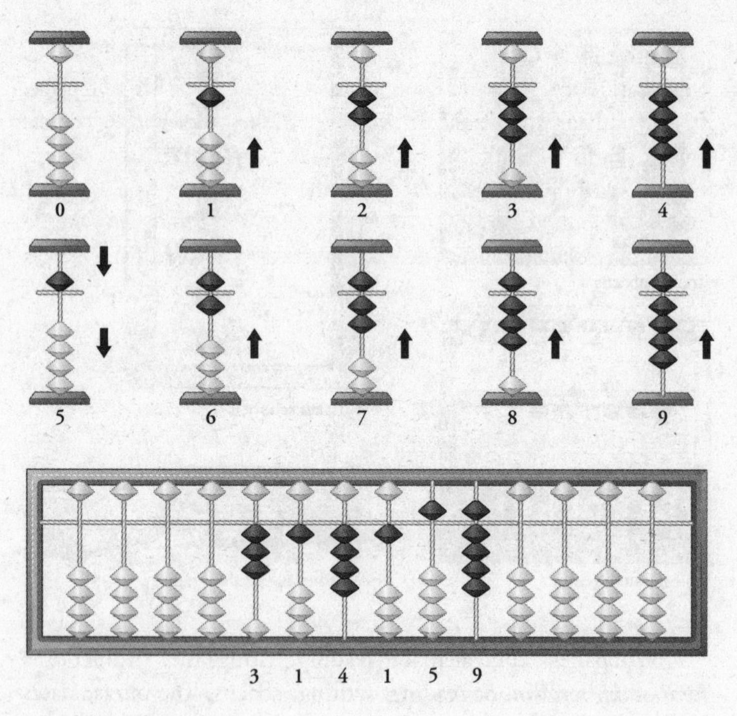

Numbers on the soroban.

The abacus was invented as a way of counting, but it really came into its own as a method for calculation. Arithmetic became much easier to do when helped by the flicking of beads. For example, to calculate 3 plus 1 you start with 3 beads, move 1 bead and the answer is right before your eyes – 4 beads. To calculate, say, 31 plus 45 you start with two columns marking 3 and 1, move 4 bead positions up the left column and 5 up the right column. The columns now read 7 and 6, which is the answer, 76. With a little bit of practice and application, it becomes easy to add numbers of any length so long as there are enough columns to accommodate them. If on any one column the two numbers add up to more than ten, then you will need to move the beads on the column to the left up one position. For example, 9 plus 2 on one column moves to a 1 on the column to the left and a 1 on the original column, expressing the answer, 11. Subtraction, multiplication and division are a little more complicated, but once mastered can be done extremely quickly.

Until the availability of cheap calculators in the 1980s abacuses were commonly seen on shop counters from Moscow to Tokyo. In fact, during the transition between the manual and electronic eras, a product combining both calculator and abacus was sold in Japan. Addition is usually faster on the abacus since you get your answer as soon as you input the numbers. With multiplication the electronic calculator gives you a slight speed advantage. (The abacus was also a way for the sceptical abacist to check the calculator's result, just in case he didn't believe it.)

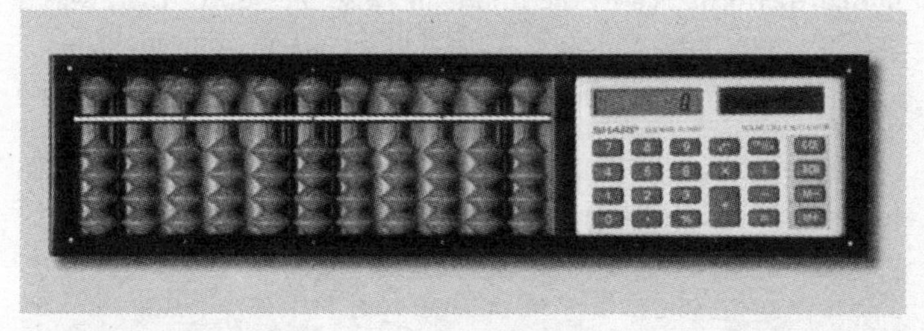

Sharp's abacus-calculator.

Abacus use has dropped in Japan since the 1970s when, at its peak, 3.2 million pupils a year sat the national *soroban* proficiency exam. Yet the abacus still remains a defining aspect of growing up, a mainstream extra-curricular activity like swimming, violin or judo. Abacus training, in fact, is run like a martial art. Levels of ability are measured in dans, and there is a competitive structure of local, provincial and national competitions. One Sunday I went to see a regional event. Almost 300 children, aged between 5 and 12, sat at desks in a conference hall with an array of special *soroban* accessories, like sleek abacus bags. An announcer stood at the front of the hall and dictated, with the intonation of an impatient muezzin, numbers to be added, subtracted or multiplied. It was a knock-out competition that lasted several hours. A chorus of military brass-band music was pumped through the sound system when the trophies – each with a winged figure holding an abacus aloft – were presented to the victors.

At Miyamoto's school he introduced me to one of his best pupils. Naoki Furuyama, aged 19, is a former national *soroban* champion. He was dressed casually, with a light checked shirt over a black T-shirt, and seemed a relaxed and well-adjusted teenager – certainly not the cliché of a socially awkward übergeek. Furuyama can multiply two six-digit numbers together in about four seconds, which is about as long as it takes to say the problem. I asked him what the point was of being able to calculate so fast, since there is no need for such skills in daily life. He replied that it helped his powers of concentration and self-discipline. Miyamoto was standing with us, and he interrupted. What was the point of running 26 miles, he asked me? There was never any need to run 26 miles, but people did it as a way of pushing human performance to the limit. Likewise, he added, there was a nobility in training one's arithmetical brain as far as one could.

Some parents send their children to abacus club because it is a way to improve school maths results. But that does not completely explain the abacus's popularity. Other after-school clubs provide more targeted maths tuition – Kumon, for example, a method of ploughing through worksheets that started in Osaka in the early 1950s, is now followed by more than four million children around the world. Abacus club is fun. I saw that in the faces of the pupils at Miyamoto's school. They clearly enjoyed their dexterity at flicking the beads with speed and precision. The Japanese heritage of the *soroban* generates national pride. Yet the real joy of the abacus, I thought, is more primal: it has been used for thousands of years and, in some cases, is still the fastest way to do sums.

After a few years of using an abacus, when you are so familiar with the positioning of the beads, it becomes possible to perform calculations simply by visualizing an abacus in your head. This is called *anzan*, and Miyamoto's top pupils have all learned it. The feat was amazing to watch – even though there was nothing to see. Miyamoto read out numbers to a totally silent, still classroom and within seconds the students raised their hands with the answers. Naoki Furuyama told me that he visualizes an abacus with eight columns. In other words, his imaginary abacus can display every number from 0 to 99,999,999.

Miyamoto's abacus club is one of the best in the country in terms of the dans of its pupils and their achievements in national tournaments. Its speciality, however, is *anzan*. A few years ago Miyamoto decided to devise a type of arithmetical challenge that could only be answered using *anzan*. When you read out a sum to a pupil, for example, it can be answered in many different ways: using a calculator, pencil and paper, an abacus or *anzan*. Miyamoto wanted to show that there were some circumstances when *anzan* was the only possible method.

His solution was the computer game *Flash Anzan*, which he demonstrated for me. He told the class to get ready, pressed play and the pupils stared at a TV screen at the front of the room. The machine beeped three times to indicate it was about to start, and then the following 15 numbers appeared, one at a time. Each number appeared for only 0.2 seconds, so the whole thing was over in three seconds:

164
597
320
872
913
450
568
370
619
482
749
123
310
809
561

The numbers flashed by so quickly I barely had time to register them. Yet as soon as the last number flashed, Naoki Furuyama smiled and said the sum of the numbers was 7907.

It is impossible to solve a *Flash Anzan* challenge with a calculator or an abacus since there is no time to remember the digits being flashed at you, let alone type them into a machine or arrange beads. *Anzan* does not require you to remember the digits. All you do is shift the beads in your brain whenever you see a new number. You start with 0, then on seeing 164 instantly visualize the abacus on 164. On seeing 597 the internal abacus rearranges to the sum, which is 761. After 15 additions you cannot remember any of the flashed numbers nor the intermediate sums, but the imaginary abacus in your head will show the answer: 7907.

The wow factor of *Flash Anzan* has made it a national fad, and Nintendo has even released a *Flash Anzan* game for its DS consoles. Miyamoto showed me some clips from a *Flash Anzan* TV game show in which teenage *anzan* stars battled it out in front of screaming fans. Miyamoto says his game has helped recruit many new pupils to abacus clubs all over Japan. 'People didn't realize what you could do with *soroban* skills,' he said. 'With all this coverage, now they do.'

Neural imaging scans show that the parts of the brain activated by the abacus, or *anzan*, are different from the parts activated by normal arithmetical calculations and language. Traditional 'pen and paper' arithmetic depends on neural networks associated with linguistic processing. The *soroban* relies on networks associated with visuospatial information. Miyamoto simplifies this as '*soroban* uses the right brain, normal maths uses the left brain'. Not enough scientific research has been done to understand what benefits this segregation brings, or how it relates to general intelligence, concentration or other skills. Yet it does explain an astonishing phenomenon: that *soroban* experts are able to multitask in the most incredible way.

Miyamoto met his wife, a former national *soroban* champion, when they frequented the same abacus club as youngsters. Their daughter, Rikako, is a *soroban* prodigy. Pity her if she wasn't. At age eight, she completed her top dan – a level that only one in 100,000 people ever achieve in their lifetimes. Rikako, who is now aged nine, was in class. She was wearing a pastel-blue top, and her fringe came down to her glasses. She looked very alert and pursed her lips as a sign of concentration.

Shiritori is a Japanese word game that starts with a person saying *shiritori* and each subsequent person must say a word that starts with the last syllable of the previous word. So, a possible second word would be *ringo* (apple), because it begins with *ri*. Miyamoto asked Rikako and the girl next to her to play *shiritori* with each other at the same time as playing a game of *Flash Anzan* in which 30 three-digit numbers were to be displayed in 20 seconds. The machine sounded its introductory pips and the girls' dialogue went:

Ringo
Gorira (gorilla)
Rappa (trumpet)
Panda (panda bear)
Dachou (ostrich)
Ushi (cow)
Shika (deer)
Karasu (crow)
Suzume (sparrow)
Medaka (killifish)
Kame (turtle)
Medama yaki (fried egg)

At the end of the 20 seconds, Rikako said: 17,602. She had been able to add up the 30 numbers and play *shiritori* simultaneously.

CHAPTER TWO

Behold!

I do not consider my birth date an especially riveting conversational opener. That may be, however, because I have not spent enough time in the company of men like Jerome Carter. I had just sat down for lunch with him and his wife Pamela at their home in Scottsdale, Arizona, when it came out: 22 November.

'Waaaoooooow!' said Pamela, a 57-year-old former air stewardess, who was wearing a pretty pink top and a denim skirt.

Jerome looked at me. In a serious tone he confirmed her enthusiasm: 'You have got a very good number there.'

Jerome, who is aged 53, does not look like your average mystic. He was dressed in an orange Hawaiian shirt and white shorts, his strong frame reflecting previous careers as a karate champion and international bodyguard. So what was so good about 22/11, I asked?

'Well, 22 is a master number. So is 11. There are only four master numbers: 11, 22, 33 and 44.'

Jerome has a distinguished face, with strong smile lines and a shiny bald dome. He also has a wonderfully musical voice, part sports commentator, part rapping MC: 'You were born on the twenty-second,' he said. 'It is no accident that our first president was born on the twenty-second. Two and two equals what? Four. We elect our presidents when? Every four years. We pay our taxes in the fourth month. Everything in the United States is four. Everything. Our first navy had 13 ships, 1 and 3 equals 4. We used to have 13 colonies, 1 and 3 equals 4. There were 13 signers of the Declaration of Independence. Four. Where were they standing at? 1300 Locust Street. Four!

'Number four controls money. You were born under it. It's a very powerful number. The number four is the square, so it involves law, structure, government, organization, journalism, construction.'

He was beginning to hit his stride: 'That's how I told O.J. he was going to walk. I looked at his lawyers. All of his lawyers were born under the number four. Johnny Cochran, born on the twenty-second, 2 and 2 equals 4. F. Lee Bailey, born on the thirteenth. 1 and 3 equals 4. Barry Scheck, born on the fourth. Robert Shapiro, born on the thirty-first, 3 and 1 equals 4. He had four lawyers born under the number four. The verdict came down when? Four p.m. OK? Hitler would have walked!

'As Mike Tyson said when I did his numbers, when it's the time with these numbers, even your mistakes turn out good.'

Jerome is a professional numerologist. He believes numbers express *qualities*, not just *quantities*. His gift, he said, is that he can use this understanding to gather insights about people's personalities and even predict the future. Actors, musicians, athletes and corporations pay good money for his advice. 'Most numerologists are poor. Most psychics are poor,' he said. 'Which makes no sense.' Jerome, on the other hand, lives in a beautiful home in a luxury condo with three $25,000 motorbikes in the garage.

Birth dates are an obvious source of numbers from which to derive character traits. So are names, since words can be broken down into letters each assigned a number value. 'Puff Daddy was about to go to jail,' he said. 'Puff Daddy had bad relationships. I changed his name to P. Diddy. Then when he wanted to settle down, I changed his name to – Diddy. These were my suggestions and he took them. Jay-Z wanted to marry Beyoncé. I told him he needs to go back to his original name. He went back to Shawn Carter.'

I asked Jerome if he had any recommendations for me.

'What's your full name?' he said.

'Alexander Bellos, but everyone calls me Alex.'

'What a bummer.' He paused for dramatic effect.

'Is Alexander better?' I asked.

He boomed: 'Let us just say that one of the greatest men that walked this Earth was not called Alex the Great.

'I'm just telling you. I have talked to people named Alex before. Just on a simple basis: the first letter of the name is very important. "A" is 1. You've already got that with Alex. But with Alexander you end with an "r". "R" equals 9. So the first and the last letters of your name are 1 and 9. Alpha and omega. The beginning and ending. Now let's take the first and the last letter of Alex. Just the sound of "x".' He pronounced it 'ekkss' with a grimace that looked like he was about to vomit. 'Do you want to use that? I wouldn't. I would never go by Alex.

'God said a good name is rather to be chosen than riches of gold! He didn't say a nickname is rather to be chosen!'

'Alex is not a nickname,' I protested. 'It's an abbreviation.'

'Why are you fighting it, Alexander?'

Jerome then asked for my pad and scribbled out the following table:

1	2	3	4	5	6	7	8	9
A	B	C	D	E	F	G	H	I
J	K	L	M	N	O	P	Q	R
S	T	U	V	W	X	Y	Z	

This, he explained, showed which numbers corresponded to which letters. He took his fingers to the first column: 'Letters that equal one are A, J, S. Allah, Jehovah, Jesus, Saviour, Salvation. Two is the number of diplomats, ambassadors. Two gives good advice, two you love, you're a team player, that's B, K and T, that's why if you go to a Burger King you can have it your way. Number three controls radio, TV, entertainment and numerology. C, L, U. Of course, you go into radio and television, you don't have a clue.' He gave me an ironic wink. 'But if you learn numerology, it will open you up to the clue of life. Number four: D, M, V. How many wheels on a car? Where do you get the licence? The Department of Motor Vehicles. Five is halfway between 1 and 10: E, N and W. Five is the number of change. If you scramble the letters you get 'new'. Six is the number of Venus, love, family, community. When you see a beautiful woman, what do you see? A FOX. Seven is the number of spirituality. Jesus was born on the twenty-fifth, 2 and 5 equals 7. Eight is the number of business, finance, commerce, money. Where do you keep the money? In the headquarters. Nine is the only one that has two letters. I and R. You ever talked to a Jamaican? Everything is irie, man.'

On conclusion he put down his pen and looked me full in the face: 'This,' he said, 'is Jerome Carter's method of the Pythagorean system.'

Pythagoras is the most famous name in mathematics, entirely due to his theorem about triangles. (More about that later.) He is credited with other contributions, though, such as the discovery of 'square numbers'. Imagine, as was common practice, counting with pebbles. (The Latin for pebble is *calculus*, which explains the origin of the word 'calculate'.) When you make a square array in which pebbles are placed equally apart in rows and columns, a two-row/column square requires four pebbles, and a three-row/column square requires nine. In other words, multiplying the number *n* by itself is equivalent to working out the number of pebbles in a square array with *n* rows and columns. The idea is so instinctive that the term 'square' to describe self-multiplication has stuck.

Pythagoras observed some excellent patterns in his squares. He saw that the number of pebbles used in the 2 square, 4, was the sum of 1 and 3, while the number used in the 3 square, 9, was the sum of 1 and 3 and 5. The 4 square has 16 pebbles – or, 1 + 3 + 5 + 7. In other words, the square of the number *n* is the sum of the first *n* odd numbers. This can be seen by looking at how you construct a pebble square:

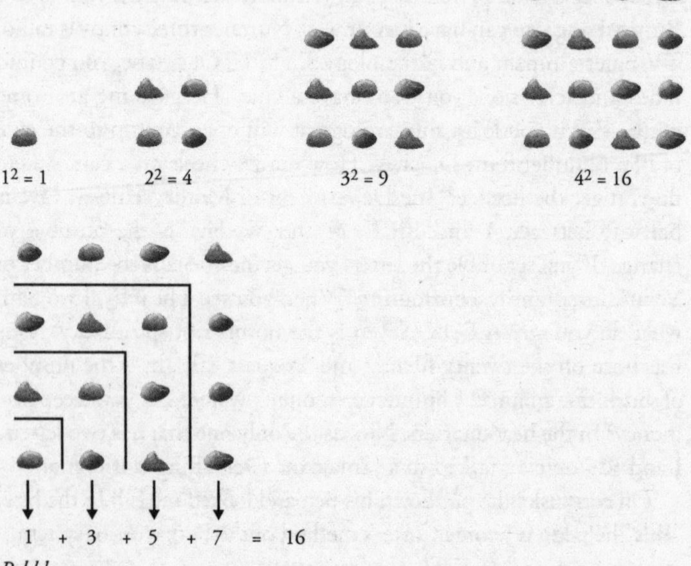

Pebble squares.

Another pattern Pythagoras discovered relates to music. One day, according to legend, as he walked past a smithy and heard clinking hammer sounds coming from inside, he noticed that the pitch of the clinking changed depending on the weights of the anvils. This provoked him to investigate the relationship between the pitch of a vibrating string and its length. This in turn led him to the realization that if the length of a string is halved, the pitch increases by an octave. Other harmonies occur when the string is divided in the ratios 3:2 and 4:3, and so on.

Pythagoras was entranced by the numerical patterns he found in nature, believing that the secrets of the universe could be understood only through mathematics. Yet rather than seeing maths merely as a tool to describe nature, he saw numbers as somehow the essence of nature – and he tutored his flock to revere them. For Pythagoras was not just a scholar. He was the charismatic leader of a mystical sect devoted to philosophical and mathematical contemplation, the Pythagorean Brotherhood, which was a combination of health farm, boot camp and ashram. Disciples had to obey strict rules, such as never urinating towards the sun, never marrying a woman who wears gold jewellery, and never passing an ass lying in the street. So select was the group that those wishing to join the Brotherhood had to go through a five-year probationary period, during which they were allowed to see Pythagoras only from behind a curtain.

In the Pythagorean spiritual cosmos, ten was divine not for any reason to do with fingers or toes, but because it was the sum of the first four numbers (1 + 2 + 3 + 4 = 10), each of which symbolized one of the four elements: fire, air, water and earth. The number 2 was female, 3 was male, and 5 – their union – was sacred. The crest of the Brotherhood was the pentagram, or five-pointed star. While the idea of worshipping numbers may now seem bizarre, it perhaps reflects the scale of wonderment at the discovery of the first fragments of abstract mathematical knowledge. The excitement of learning that there is order in nature, when previously you were not aware that there was any at all, must have felt like a religious awakening.

Pythagoras's spiritual teachings were more than just numerological. They included a belief in reincarnation, and he was probably a vegetarian. In fact, his dietary requirements have been hotly debated for more than 2000 years. The Brotherhood famously forbade ingestion of the small, round, black fava bean, and one account of Pythagoras's death has him fleeing attackers when he came to a field of fava beans. As the story goes, he preferred to be captured and killed rather than tread on them. The reason the beans were sacred, according to one ancient source, was that they sprouted from the same primordial muck as humans did. Pythagoras had proved this by showing that if you chew up a bean, crush it with your teeth, and then put it for a short while in the sun, it will begin to smell like semen. A more recent hypothesis was that the Brotherhood was just a colony for those with hereditary fava-bean allergies.

Pythagoras lived in the sixth century BC. He did not write any books. All we know about him was written many years after he died. Though the Brotherhood was lampooned in ancient Athenian comic theatre, by the beginning of the Christian Era Pythagoras himself was seen in a rather favourable light, viewed as being a unique genius; his mathematical insights making him the intellectual forefather of the great Greek philosophers. Miracles were attributed to him, and some authors, rather oddly, claimed that he had a thigh made of gold. Others wrote that he once walked across a river, and the river called out to him, loud enough for all to hear, 'Greetings, Pythagoras'. This posthumous myth-making has parallels with the story of another Mediterranean spiritual leader and, in fact, Pythagoras and Jesus were temporarily religious rivals. As Christianity was taking root in Rome in the second century CE, the empress Julia Domna encouraged her citizens to worship Apollonius of Tyna, who claimed he was Pythagoras reincarnated.

Pythagoras has a dual and contradictory legacy: his mathematics and his anti-mathematics. Maybe, in fact, as some academics now suggest, the only ideas that can be correctly attributed to him are the mystical ones. Pythagorean esotericism has been a constant presence in Western thought since antiquity, but was especially in vogue during the Renaissance, thanks to the rediscovery of a poem of 'self-help' maxims written around the fourth century BC called *The Golden Verses of Pythagoras*. The Pythagorean Brotherhood was the model for many occult secret societies and influenced the creation of freemasonry, a fraternal organization with elaborate rituals that is believed to have almost half a million members in the UK alone. Pythagoras also inspired the 'founding mother' of Western numerology, Mrs L. Dow Balliett, an Atlantic City housewife who wrote the book *The Philosophy of Numbers* in 1908. 'Pythagoras said the Heavens and Earth vibrate to the single numbers or digits of numbers,' she wrote, and she proposed a system of fortune-telling in which each letter of the alphabet corresponded with a number from 1 to 9. By adding up the numbers of the letters in a name, she asserted, personality traits could be divined. I tested this idea on myself. 'Alex' is 1 + 3 + 5 + 6 = 15. I completed the process by adding the two digits of the answer, getting 1 + 5 = 6. This gives me a name vibration of six, which means that I 'should always be dressed with care and precision; be fond of dainty effects and colors, lifting your especial colors of orange, scarlet and heliotrope into their lighter shades, yet always keeping their true tones'. My gems are topaz, diamond, onyx and jasper, while my mineral is borax, and my flowers are tuberose, laurel and chrysanthemum. My odour is japonica.

Numerology, of course, is now an established dish on the buffet of modern mysticism, with no shortage of gurus willing to advise on lottery numbers or speculate on the portent of a prospective date. It sounds like harmless fun – and I enjoyed my conversation with Jerome Carter immensely – yet giving numbers spiritual significance can also have sinister consequences. In 1987, for example, the military government in Burma issued new banknotes with a face value divisible by nine – for the sole reason that nine was the ruling general's favourite number. The new notes helped precipitate an economic crisis, which led to an uprising on 8 August 1988 – the eighth of the eighth of '88. (Eight was the anti-dictatorship movement's favourite number.) The protest was violently put down, however, on 18 September: in the ninth month, on a day divisible by nine.

Pythagoras's Theorem states that *for any right-angled triangle, the square of the hypotenuse is equal to the sum of the squares of the other two sides*. Its words are imprinted in my brain like an old nursery rhyme or Christmas carol; the phrase is nostalgic and comforting independently of its meaning.

The hypotenuse is the side opposite the right angle, and a right angle is a quarter turn. The theorem is the smash hit of basic geometry, the first truly thought-provoking mathematical concept we are taught at school. What I find exciting about it is how it reveals a deep connection between numbers and space. Not all triangles are right-angled, but when they are, the squares of two of the sides must equal the square of the third. Likewise, the theorem holds in the other direction too. Take any three numbers. If the square of two of them equals the square of the third then you can construct a right-angled triangle out of lengths of those sizes.

In some commentaries about Pythagoras it is said that before he founded the Brotherhood he travelled on a fact-finding mission to Egypt. If he had spent any time on an Egyptian building site he would have seen that the labourers used a trick to create a right angle that was an application of the theorem that would later gain his name. A rope was marked with knots spread out at a distance of 3, 4 and 5 units. Since $3^2 + 4^2 = 5^2$, when the rope was stretched around three posts, with a knot at each post, it formed a triangle with one right angle.

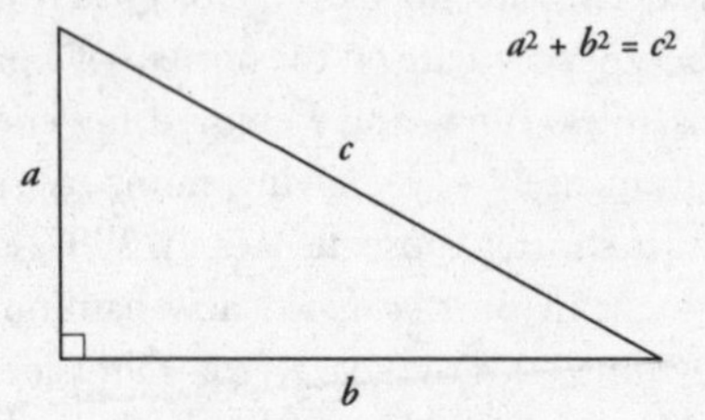

Pythagoras's Theorem.

Rope-stretching was the most convenient way to achieve right angles, which were needed so that bricks, or giant stone blocks such as those used to construct the Pyramids, could be stacked next to and on top of each other. (The word hypotenuse comes from the Greek for 'stretched against'.) The Egyptians could have used many other numbers in addition to 3, 4 and 5 to get real right angles. In fact, there is an infinite number of numbers *a, b* and *c* such that $a^2 + b^2 = c^2$. They could have marked out their rope into sections of 5, 12 and 13, for example since 25 + 144 = 169, or 8, 15 and 17, since 64 + 225 = 289, or even 2772, 9605 and 9997 (7,683,984 +

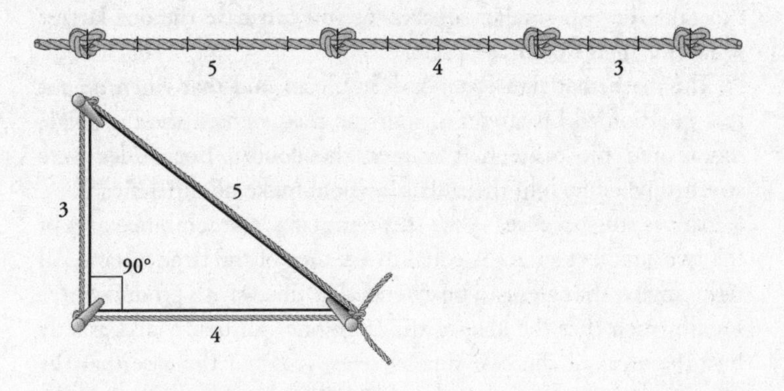

The Egyptian equivalent of a set square was a rope divided in the ratio 3:4:5, which provides a right angle when tied around three posts.

92,256,025 = 99,940,009) though that would hardly have been practical. The numbers 3, 4, 5 are best suited to the task. As well as being the triple with the lowest value, it is also the only one whose digits are consecutive integers. Due to its rope-stretching heritage, the right-angled triangle with sides that are in the ratio 3:4:5 is known as an Egyptian triangle. It is a pocket-sized right-angle-generating machine that is a jewel of our mathematical patrimony, an intellectual artefact of great power, elegance and concision.

The squares mentioned in Pythagoras's Theorem can be understood as numbers and also as pictures – literally the squares drawn on the sides of the triangle. Imagine that in the following image the squares are made of gold. You are not engaged to a member of the Pythagorean Brotherhood, so acquiring gold is desirable. Either you

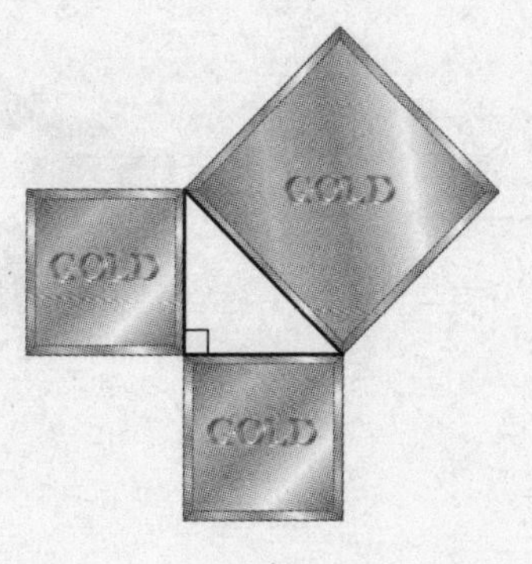

can take the two smaller squares, or you can have the one largest square. Which would you prefer?

The mathematician Raymond Smullyan said that when he put this question to his students, half the class wanted the big single square and the other half wanted the double. Both sides were stunned when he told them that it would make no difference.

This is true because, as the theorem states, the combined area of the two smaller squares is equal to the area of the large square. All right-angled triangles can be extended in this way to produce three squares such that the area of the large one can be divided exactly into the areas of the two smaller ones. It is not the case that the square on the hypotenuse is sometimes the sum of the squares of the other two sides, and sometimes not. The fit is perfect at all times.

It is not known if Pythagoras really discovered his theorem, even though his name has been attached to it since classical times. Whether or not he did, it vindicates his world-view, demonstrating a remarkable harmony in the mathematical universe. In fact, the theorem reveals a relationship between more than just the squares on the sides of a right-angled triangle. The area of a semicircle on the hypotenuse, for example, is equal to the sum of the areas of the semicircles on the other two sides. A pentagon on the hypotenuse is equal to the sum of pentagons on the other two sides, and this holds for hexagons, octagons and, indeed, any regular or irregular shape. If, say, three Mona Lisas were drawn on a right-angled triangle, then the area of big Mona is equal to the combined area of the two smaller ones.

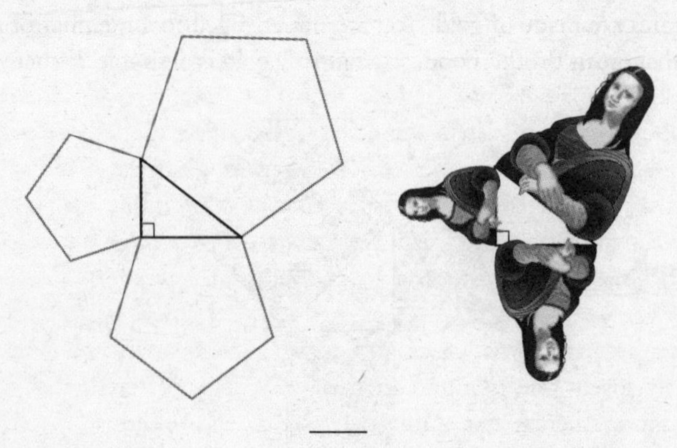

For me, the real delight in Pythagoras's Theorem is in the realization of why it must be true. The simplest proof is as follows. It dates back to the Chinese, possibly to before even Pythagoras was born, and is one of the reasons why many doubt he came up with the theorem in the first place.

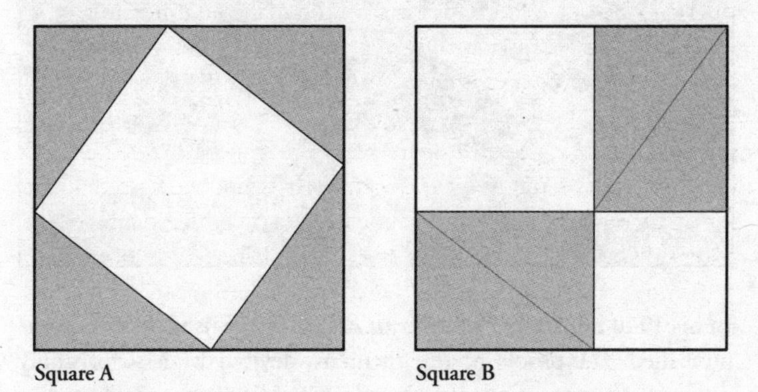

Square A **Square B**

Stare at the two squares for a while before reading on. Square A is the same size as square B, and all the right-angled triangles inside the square are also the same size. Because the squares are equal, the white area inside them is also equal. Now, note that the big white square inside square A is the square of the hypotenuse of the right-angled triangle. And the smaller white squares inside square B are the squares of the other two sides of the triangle. In other words, the square of the hypotenuse is equal to the square of the other two sides. Voilà.

Since we can construct a square like A and B for any shape or size of right-angled triangle, the theorem must be true in all cases.

The thrill of maths is the moment of instant revelation, from proofs such as this, when suddenly everything makes sense. It is immensely satisfying, an almost physical pleasure. The Indian mathematician Bhaskara was so taken by a similar Pythagoras proof that underneath a picture of it in his twelfth-century maths book *Lilavati*, he wrote no explanation, just the word 'Behold!'

There are many other proofs of Pythagoras's Theorem, and a particularly lovely one is found in the figure overleaf, credited to the Arabic mathematician Annairizi, and dated around 900 CE. The theorem is contained within the repeating pattern. Can you spot it? (If you can't, some help is included as an appendix, p. 418.)

In his 1940 book *The Pythagorean Proposition*, Elisha Scott Loomis published 371 proofs of the theorem, devised by a surprisingly diverse collection of people. One dating to 1888 was attributed to E.A. Coolidge, a blind girl, another to Ann Condit, a 16-year-old high-school student, dating to 1938, while others were attributed to Leonardo da Vinci and US President James A. Garfield. Garfield had stumbled on his proof during some mathematical amusements with colleagues when he was a Republican congressman. 'We think it something on which the members of both houses can unite without distinction of party,' he said when the proof was first published in 1876.

The diversity of proofs is a testament to the vitality of maths. There is never a 'right' way to attack a maths problem, and it's intriguing to chart the different routes that different minds have taken in finding solutions. Above opposite are three different proofs from three different eras: one by Liu Hui, a Chinese mathematician from the third century CE, one by Leonardo da Vinci (1452–1519) and the third by Henry Dudeney, Britain's most famous puzzlist, dated 1917. Both Liu Hui and Dudeney's are 'dissection proofs' in which the two small squares are divided into shapes that can be reassembled perfectly in the big square. Leonardo's needs a little more thought. (If you need help, see the appendix on p. 418 again.)

A particularly dynamic proof was devised by the mathematician Hermann Baravalle, shown below opposite. There is something

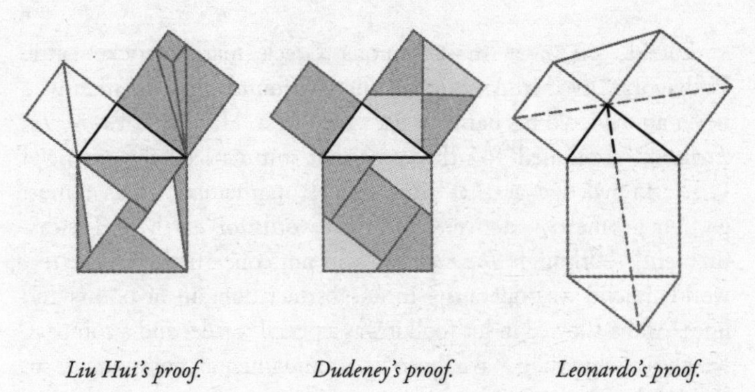

Liu Hui's proof. *Dudeney's proof.* *Leonardo's proof.*

more organic about this one – it shows how the big square, like an amoeba, divides itself into the two smaller ones. At each stage, the area shaded is the same. The only step that isn't obvious is step 4. When a parallelogram is 'sheared', or moved in such a way that preserves the base and altitude, its area stays the same.

Baravalle's proof is similar to the most established one in mathematical literature, that set out by Euclid around 300 BC.

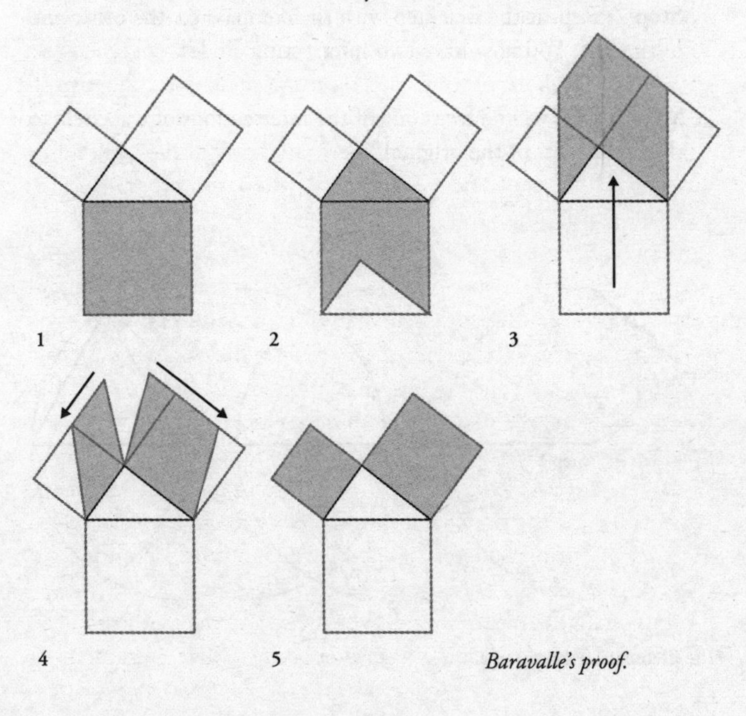

Baravalle's proof.

Euclid, the next most famous Greek mathematician after Pythagoras, lived in Alexandria, the city founded by the man who never abbreviated his name to Alex the Great. His *chef-d'œuvre*, *The Elements*, contained 465 theorems that summarized the extent of Greek knowledge at that time. Greek mathematics was almost entirely geometry – derived from their words for 'earth' and 'measurement' – although *The Elements* was not concerned with the real world. Euclid was operating in an abstract domain of points and lines. All he allowed in his toolkit was a pencil, a ruler and a compass, which is why they have been the fundamental components of children's pencil cases for centuries.

Euclid's first task (Book 1, Proposition 1) was to show that, given any line, he could make an equilateral triangle, i.e. a triangle with three equal sides, with that line as one side:

Step 1: Put the compass point on one end of the given line and draw a circle that passes through the other end of the line.

Step 2: Repeat the first step with the compass on the other end of the line. You now have two intersecting circles.

Step 3: Draw a line from one of the intersections of the circle to the end points of the original line.

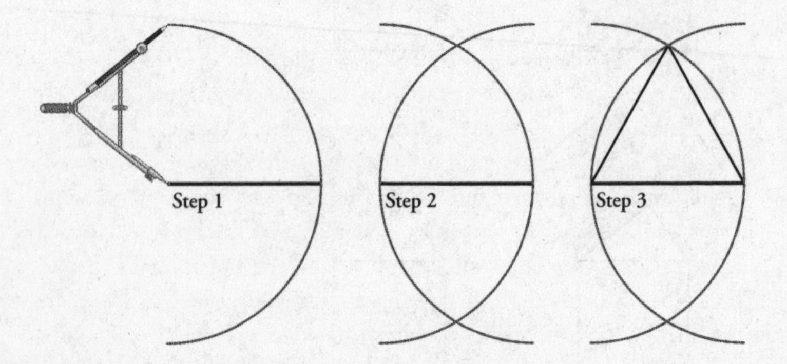

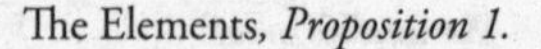
The Elements, *Proposition 1.*

He then meticulously progressed from proposition to proposition, revealing a host of properties of lines, triangles and circles. For example, Proposition 9 shows how to 'bisect' an angle, that is construct an angle that is exactly half of a given angle. Proposition 32 states that the internal angles of a triangle always add up to two right angles, or 180 degrees. *The Elements* is a *magnum opus* of pedantry and rigour. Nothing is ever assumed. Every line follows logically from the line before. Yet from only a few basic axioms, Euclid amassed an impressive body of compelling results.

The grand finale of the first book is Proposition 47. The commentary from a 1570 edition of the first English translation reads: 'This most excellent and notable Theoreme was first invented of the greate philosopher Pithagoras, who for the exceeding ioy conceived of the invention thereof, offered in sacrifice an Oxe, as recorde Hierone, Proclus, Lycius, & Vitruvius. And it hath bene commonly called of barbarous writers of the latter time Dulcarnon.' Dulcarnon means two-horned, or 'at wit's end' – possibly because the diagram of the proof has two horn-like squares, or possibly because understanding it is indeed horribly difficult.

There is nothing pretty about Euclid's proof of Pythagoras's Theorem. It is long, meticulous and convoluted, and requires a diagram full of lines and superimposed triangles. Arthur Schopenhauer, the nineteenth-century German philosopher, said it was so unnecessarily complicated that it was a 'brilliant piece of perversity'. To be fair to Euclid, he was not trying to be playful (as was Dudeney), or aesthetic (as was Annairizi) or intuitive (as was Baravalle). Euclid's driving concern was the rigour of his deductive system.

While Pythagoras saw wonder in numbers, Euclid in *The Elements* reveals a deeper beauty, a watertight system of mathematical truths. On page after page he demonstrates that mathematical knowledge is of a different order than any other. The propositions of *The Elements* are true in perpetuity. They do not become less certain, or indeed less relevant with time (which is why Euclid is still taught at school and why Greek playwrights, poets and historians are not). The Euclidean method is awe-inspiring. The seventeenth-century English polymath Thomas Hobbes is said to have glanced at a copy of *The Elements* that lay open in a library when he was a 40-year-old man. He read one of the propositions and exclaimed: 'By God, this is impossible!' So, he read the previous proposition, and then the one before that, and so on, until he was convinced it all made sense. In the process, he fell in love with geometry for the certainty it prescribed, and the deductive approach influenced his most famous works of political philosophy. Since *The Elements*, logical reasoning has been the gold standard of all human enquiry.

Euclid started off by carving up two-dimensional space into the family of shapes known as polygons, which are those shapes made from only straight lines. With his compass and straightedge he was able to construct not just an equilateral triangle, but also a square, a pentagon and a hexagon. Polygons for which every side has the same length and the angles between the sides are all equal are called *regular polygons*. Interestingly, Euclid's method, however, is not effective for all of them. The heptagon (seven sides), for example, cannot be constructed with a compass and straightedge. The octagon *is* constructible, but then the nonagon again is not. Meanwhile, the staggeringly complex regular polygon that has 65,537 sides is constructible, and in fact has been constructed. (It was chosen because the number is equal to $2^{16} + 1$). In 1894 it took Johann Gustav Hermes, a German mathematician, ten years to do it.

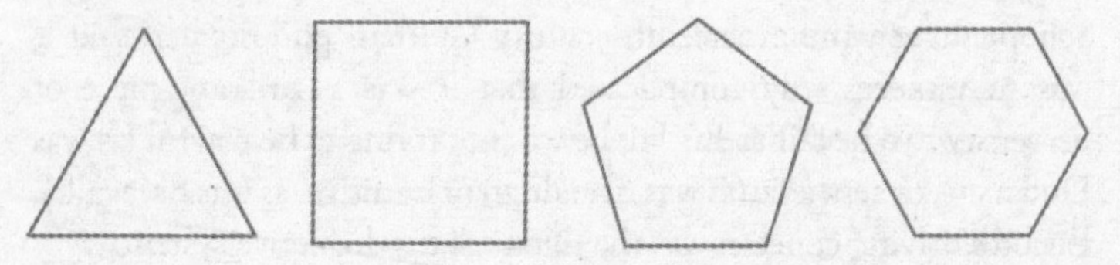

Regular polygons.

One of Euclid's pursuits was to investigate the three-dimensional shapes that can be made from joining identical regular polygons together. Only five shapes fit the bill: the tetrahedron, the cube, the octahedron, the icosahedron and the dodecahedron, the quintet known as the Platonic solids since Plato wrote about them in the *Timaeus*. He equated them with the four elements of the universe plus the heavenly space that surrounds them all. The tetrahedron was fire; the cube, earth; the octahedron, air; the icosahedron, water; and the dodecahedron, the encompassing dome. The Platonic

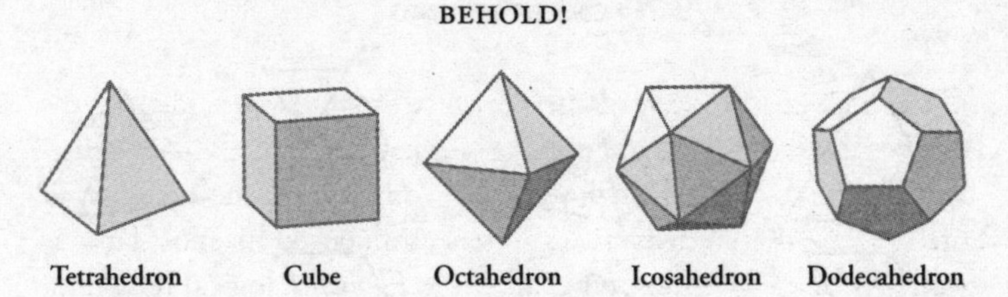

The Platonic solids.

solids are particularly interesting because they are perfectly symmetrical. Twist them, roll them, invert them or flip them and they always stay the same.

In the thirteenth and final book of *The Elements*, Euclid proved why there are only five Platonic solids by working out all the possible solid objects that can be made from regular polygons, starting with the equilateral triangle, and then moving on to squares, pentagons, hexagons and so on. The diagram overleaf shows how he reached his conclusion. To make a solid object from polygons you must always have a point where three sides meet, a corner, or what's called a vertex. When you join three equilateral triangles at a vertex, for example, you get a tetrahedron (A). When you join four, you get a pyramid (B). A pyramid is not a Platonic solid because not all the sides are the same, but by sticking an inverted pyramid on the bottom you get an octahedron, which is. Join five equilateral triangles together and you have the first part of an icosahedron (C). But join six and you get … a flat piece of paper (D). You cannot make a solid angle with six equilateral triangles, so there are no other ways to create a different Platonic solid made up of them. Continuing this procedure with squares, it is evident that there is only one way to join three squares at a corner (E). This will end up as a cube. Join four squares and you get … a flat piece of paper (F). No more Platonic solids here. Similarly, three pentagons together give a solid angle, which becomes the dodecahedron (G). It is impossible to join four pentagons. Three hexagons meeting at the same point lie flat alongside each other (H), so it is impossible to make a solid object out of them. There are no more Platonic solids since it is impossible to join three regular polygons of more than six sides at a vertex.

Using Euclid's method, many mathematicians after him have asked new questions and made new discoveries. For example, in 1471 the German mathematician Regiomontanus wrote a letter to a friend

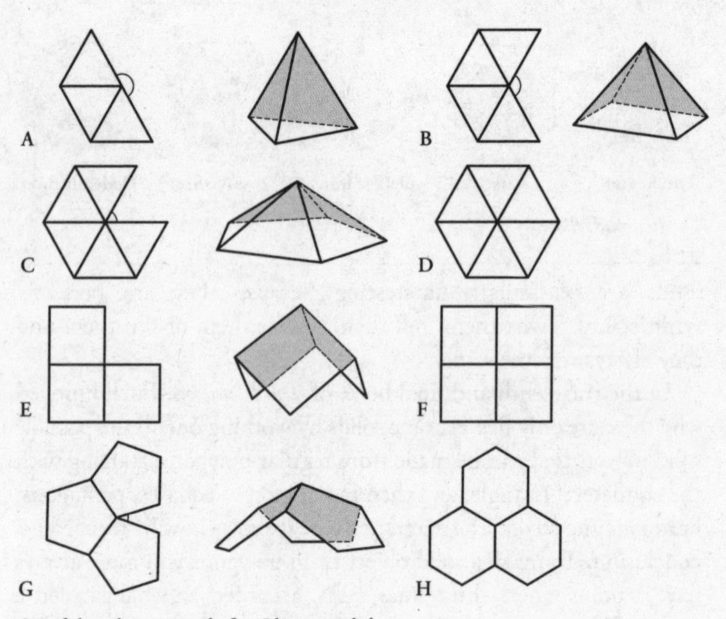

Proof that there are only five Platonic solids.

in which he posed the following problem: 'At what point on the ground does a perpendicularly suspended rod appear largest?' This has been rephrased as the 'statue problem'. Imagine there is a statue on a plinth in front of you. When you are too close you have to crick your neck and you have a very narrow angle to look up at it. When you are far away, you have to strain your eyes and, again, see it through a very narrow angle. Where, then, is the best place to view it?

Consider a side-on view of the situation, as in the diagram opposite. We want to find the point on the dotted line, which represents eye level, such that the angle from the point to the statue is greatest. The solution follows from the third book of *The Elements*, on circles. The angle is largest when a circle that goes through the top and bottom of the statue rests on the dotted line.

The problem is equivalent to that faced by rugby players wanting to know the best distance from the tryline to kick a conversion. If you are too near the opposing tryline the angle is too tight, but if you are too far the angle is also reduced. Where is the optimum position? Here we need to take an aerial view of the pitch and draw a similar diagram. The point on the dotted line of allowed kick

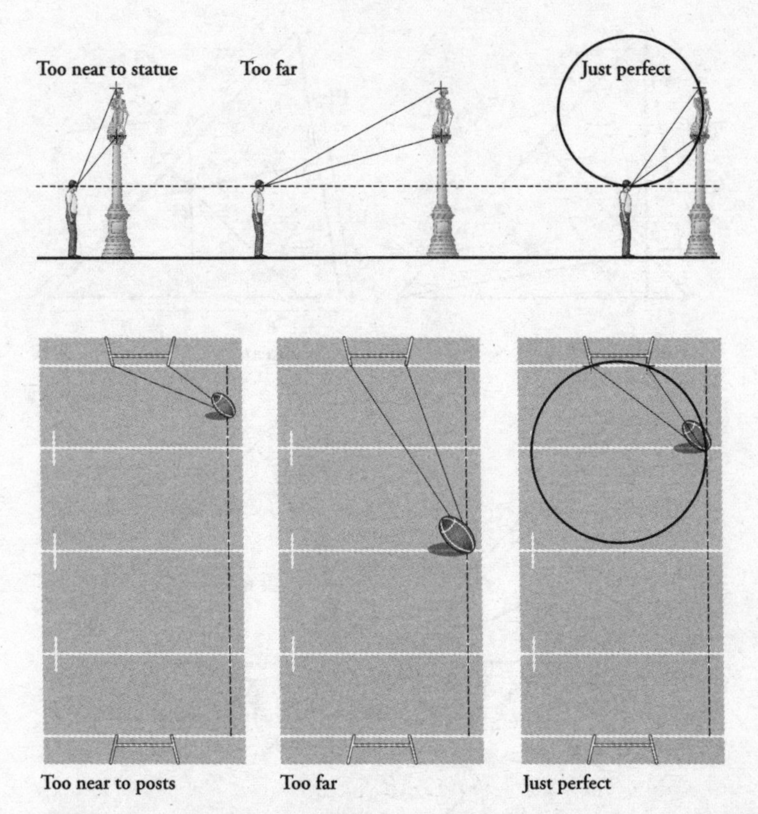

Regiomontanus's statue and rugby problem.

spots that subtends the greatest angle to the posts is precisely the point touched by a circle that passes through both posts and rests on the line.

Perhaps the most stunning result in Euclidean geometry, though, is one that reveals an astonishing property about triangles. Let's first consider where the centre of a triangle is. This is a surprisingly unclear issue. In fact, there are four ways we can define the centre of a triangle and they all represent different points (unless the triangle is equilateral, when they all coincide). The first is called the *orthocentre*, and is the intersection of the lines from each vertex that meet the opposing sides perpendicularly, which are called the *altitudes*. It is already pretty amazing to think that, for any triangle, the altitudes always meet at the same point.

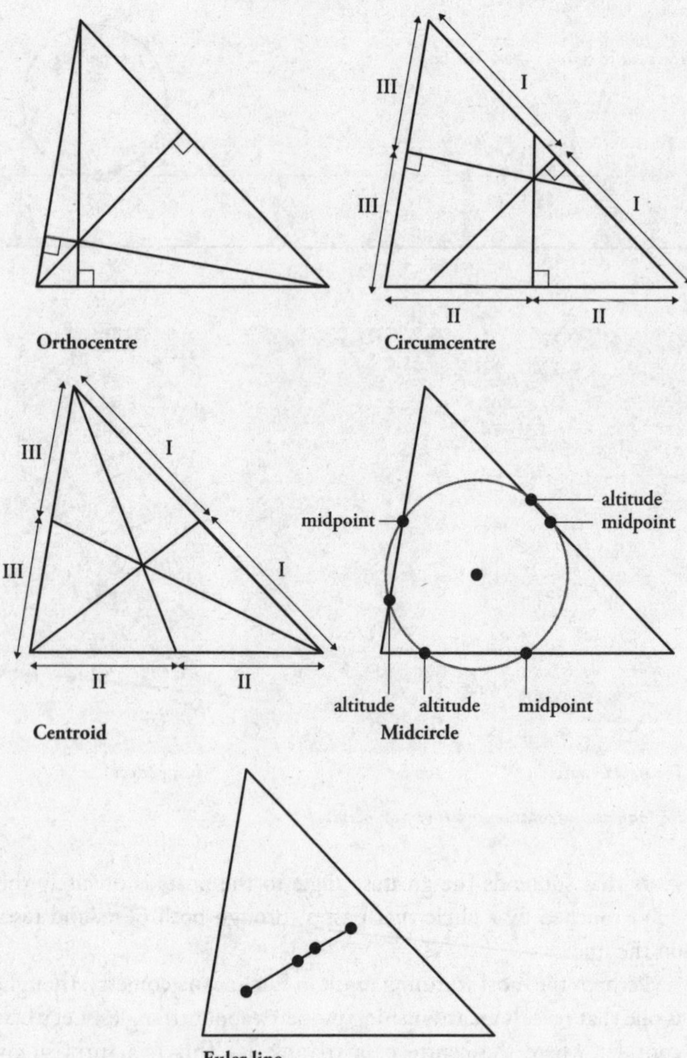

Construction of the Euler line.

The second is the *circumcentre*, which is the intersection of the perpendiculars drawn from the halfway points of the sides. Again, it is very neat that these lines will also always meet, whatever triangle you choose.

The third is the *centroid*, which is the intersection of the lines that go from the vertex to the midpoints of opposing lines. They always meet too. Finally, the *midcircle* is a circle that passes through the midpoints of each side and also the intersections of the sides and the altitudes. Every triangle has a midcircle, and its centre is the fourth type of middle point a triangle can have.

In 1767 Leonhard Euler proved that for all triangles the orthocentre, the circumcentre, the centroid and the centre of the midcircle are always on the same line. This is mind-blowing. Whatever the shape of a triangle these four points have a dazzlingly uniform relation to each other. The harmony is truly wondrous. Pythagoras would have been most awed.

Though this may be hard to fathom these days, *The Elements* was a literary sensation. Until the twentieth century, it is said to have had more editions printed than any other book except the Bible. This was all the more remarkable since *The Elements* is no easy read. One edition, however, is worth mentioning for its unorthodox approach in making the text accessible. Oliver Byrne, whose day job was Surveyor of Her Majesty's Settlements in the Falkland Islands, rewrote Euclid in colour. Instead of the long proofs, he drew illustrations in which angles, lines and areas were marked in geometrical blocks of red, yellow, blue or black. His *Elements … in which coloured diagrams and symbols are used instead of letters for the greater ease of learning* was published in 1847 and has been described as 'one of the oddest and most beautiful books of the whole nineteenth century'. In 1851 it was one of the few British books on display at the Great Exhibition, though the public failed to see the excitement. Indeed, Byrne's publishers went bust in 1853, with more than 75 per cent of stock of *The Elements* unsold. Its high production costs had contributed to the bankruptcy.

Byrne's illustrated proofs did make Euclid more intuitive, predating colour-coordinated textbooks of recent years. Aesthetically it was also ahead of its time. The gaudy primary colours, asymmetrical layout, angularity, abstract shapes and plentiful empty space anticipated the paintings of many twentieth-century artists. Byrne's book looks like a tribute to Piet Mondrian, published 25 years before Mondrian was even born.

As masterful as the Euclidean method was, it could not solve all problems; some quite simple ones, in fact, are unsolvable using just a compass and ruler. The Greeks suffered for this. In 430 BC Athens was struck by a plague of typhoid. Its citizens consulted the oracle at Delos, who advised them to double the size of Apollo's altar, which was cube-shaped. Relieved that such an apparently easy task would save them, they constructed a new altar where the sides of the cube were double the length of those of the original altar. Yet by doubling the side of a cube, the volume of the cube increases by two cubed, or eight. Apollo was not happy and made the pestilence worse. The challenge that the god set – that is, given a cube, construct a second cube that has twice the volume – is called the Delian Problem, and it is one of the three classic problems of antiquity that cannot be solved by Euclidean tools. The other two are *the squaring of the circle*, which is the construction of a square that has the same area as a given circle, and *the trisection of an angle*, which is the construction of an angle that is a third of a given angle. Realizing why Euclidean geometry cannot solve these problems – and why other methods can – has been a long-time preoccupation of mathematics.

The Greeks were not the only people intrigued by the wonders of geometrical shapes. The most sacred object in Islam is a Platonic solid: the Ka'ba, or Cube, is the black palladium at the centre of Mecca's Sacred Mosque, around which pilgrims walk anticlockwise during the Hajj. (In fact, its dimensions make it just off a perfect cube.) The Ka'ba also marks the point that worshippers must face during daily prayer, wherever they are in the world. Mathematics plays more of a role in Islam than in any other major religion. More than a millennium before the advent of GPS technology, the requirement to face Mecca relied on complicated astronomical calculations – which is one reason why Islamic science was unequalled for almost a thousand years.

Islamic art is epitomized by the ingenious geometrical mosaic arrangements on the walls, ceilings and floors of its sacred buildings, a consequence of the ban on images of people and animals in holy sites. Geometry was thought to express truth beyond what was merely human, much in keeping with the Pythagorean position that the universe reveals itself through mathematics. The symmetries and endless loops that Islamic craftsmen created in their patterns were an allegory of the Infinite, an expression of the sacred, mathematical order of the world.

The beauty of a repeating mosaic pattern lies not so much in the aesthetic appeal of the replicated image as in the effortlessness with which the tiles perfectly fill the space. The better the geometry, the better the art. Working out what shapes will tile a wall so that there are no gaps and no overlaps is quite a mathematical challenge, familiar to anyone who has ever tiled a bathroom floor. It turns out that only three of the regular polygons are able to 'tessellate', which is the technical word for covering a plane so that no region is uncovered. These are the equilateral triangle, the square and the hexagon. In fact, a triangle is not required to be equilateral in order to tessellate. The sides can be of any size. For any triangle, all you need to do is join it to an identical triangle placed upside-down, as in the diagram below. The combined shape is a parallelogram. The parallelogram can be joined with identical ones to form a row, and these rows can fit together side by side. This type of tessellation – in which the same pattern repeats endlessly – is called periodic.

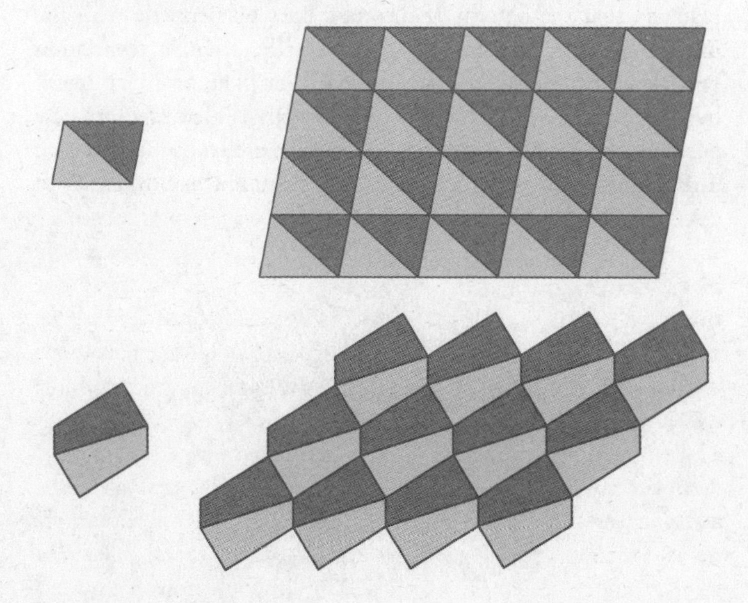

Triangle and quadrilateral tessellations.

A square tile will fill a flat surface. That's obvious. So does any rectangle. That is also a trivial observation – staring at a brick wall is equally the scrutiny of a tessellation of rectangles. What is surprising, however, is that any shape with four sides can also produce periodic tessellations. Draw any four-sided shape. Join this shape with one upside-down, as we did with the triangle above, and you create a six-sided shape, an irregular hexagon, with the property that each of the opposite edges are equal. Since the opposite edges are equal, the shape can be laid in a row such that the edges fit perfectly alongside each other. As the diagram on the previous page shows, this fit works in the direction of each of the sides, and the repeated hexagons cover a plane perfectly.

I said that a periodic tessellation is one that repeats endlessly. There is a more practical definition of periodicity. Imagine a plane extending infinitely in all directions and covered with the triangle tessellation on the previous page. Now imagine making an identical copy of the tessellation on tracing paper and placing it on the plane. Periodicity can be defined as the capacity to lift up the copy, move it along to another position and then to put it back down on the plane so that the pattern of the copy lines up perfectly with the original pattern. We can do this with the triangle tessellation because we can move the copy to the left (or right, or up, or down) by any number of triangles. When the copy is aligned to its new position, the copy is a perfect fit for the tessellation underneath. This definition of periodicity is helpful because it is now easier to

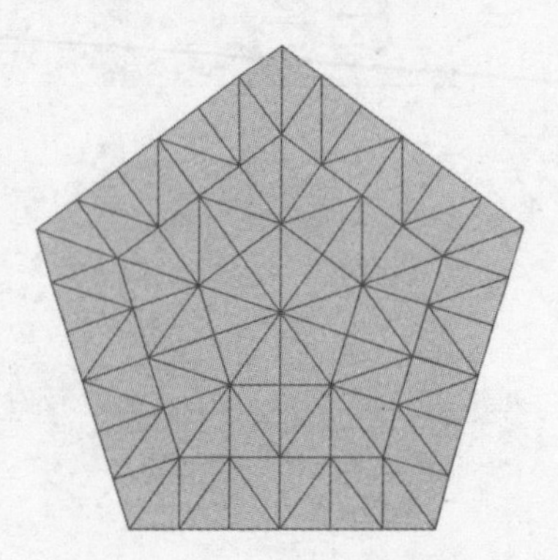

Nonperiodic tessellation.

explain the concept of *nonperiodicity*. A nonperiodic tessellation is one that when a copy is made, there is only one position where the copy fits perfectly over the plane – the original position. For example, the tessellation below opposite is nonperiodic. (Imagine that the tessellation goes on for ever, in widening concentric pentagons.) If you made a copy of it, the copy coincides only with the underlying tessellation in one position.

Many types of tile that can be arranged periodically can also be arranged nonperiodically. The question that tantalized mathematicians in the second half of the twentieth century, however, was whether or not there existed any sets of tiles that could be tiled *only* nonperiodically. These would be tiles that could cover a plane surface but were incapable of producing repeated patterns. The idea is counter-intuitive – if tiles are so well suited and harmonious that they can tile a plane without leaving any gaps, then it would seem natural that they are able to do so in a regular, repeating way. For a long time it was believed that nonperiodic tiles did not exist.

Then along came Roger Penrose with his kites and darts. In the 1970s, Penrose – a cosmologist – thrilled the maths world when he developed several types of nonperiodic tiles. The simplest were created by judiciously cutting a rhombus in two, to form two different shapes, which he called a kite and a dart. Since any four-sided shape can produce a periodic tessellation, Penrose then had to formulate a rule for how these tiles could be joined that would restrict the patterns they could make to being nonperiodic. He did this by

Penrose's dart and kite can tile only nonperiodically.

drawing two arcs on each kite and dart and stipulating that tiles must be connected so that like arc always joins with like.

The discovery of nonperiodic tiling was an exciting breakthrough for maths, but not as exciting as it would later be for physics and chemistry. In the 1980s researchers were amazed to discover a type of crystal that they did not believe existed. The tiny structure displayed a nonperiodic pattern, behaving in three dimensions just like Penrose's tiles did in two. The existence of these structures – called quasicrystals – changed the way scientists understood the nature of matter, since it contradicted classical theory that all crystals must have symmetrical lattices derived from the Platonic solids. Penrose may have invented his tiles for fun, but they were unduly prophetic about the natural world.

Half a millennium ago, Islamic geometers might also have understood about non-periodic tessellations. In 2007 Peter J. Lu from Harvard University and Paul J. Steinhardt from Princeton claimed that their studies of mosaics in Uzbekistan, Afghanistan, Iran, Iraq and Turkey showed that the craftsmen had made 'nearly perfect quasicrystalline Penrose patterns, five centuries before discovery in the West'. It is possible, therefore, that Islamic mathematics may have been even more advanced than historians of science have traditionally thought.

Hinduism also used geometry to illustrate the divine. Mandalas are symbolic representations of deities and the cosmos. The most complex of these is the Sri Yantra, a figure made up of five triangles pointing down and four pointing up, all overlapping a central point, or *bindu*. It is said to represent the essential outline of the universal processes of emanation and reabsorption, and is used as a focus for meditation and worship. Constructing it is very imprecise – its structure is enigmatically described in a long poem, but the sacred texts do not give enough detail. Mathematicians are baffled to this day by exactly how it is properly constructed.

Another Eastern culture has long embraced the joys of geometric shapes. Origami, the art of paper-folding, evolved from the custom of Japanese farmers thanking the gods at harvest time by making an offering of some of their crops on a piece of paper. Rather than placing the produce on a flat sheet, they would make one diagonal fold in the paper to give the offering a human touch. Origami

The Sri Yantra.

flourished in Japan over the last few hundred years as an informal pastime, the kind of thing parents did with their children for fun. It fitted in perfectly with the Japanese love of artistic understatement, attention to detail and economy of form.

Business-card origami sounds like the ultimate Japanese invention, uniting two national passions. In fact, the practice is abhorrent to them. The Japanese see business cards as an extension of the individual, so playing with them is considered grossly offensive, even with origamic intent. When I tried to fold one in a Tokyo restaurant I was almost ejected for my antisocial behaviour. In the rest of the world, however, business-card origami is a modern paper-folding subgenre. It dates back more than a hundred years, to the (now obsolete) practice of visiting-card origami.

A simple example is to fold a business card so that the bottom right corner meets the top left corner, and then fold the overlaps, as shown overleaf. Repeat this with another business card, except this time fold the bottom left to the top right. You now have two pieces that can be slotted together to form a tetrahedron. It is a winning way, so I am told, to hand over your business card during mathematics conferences.

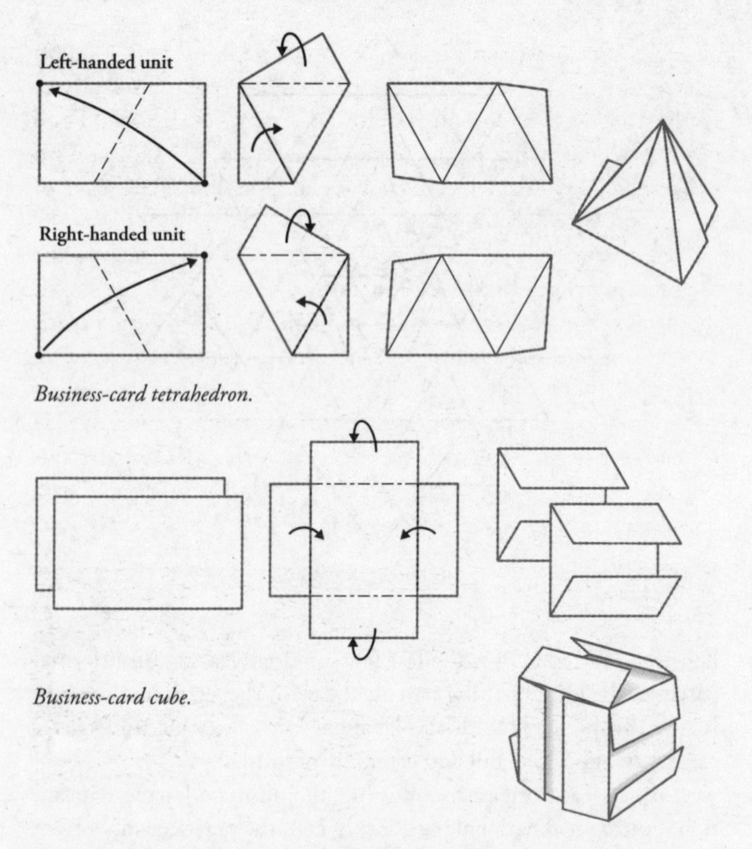

Business-card tetrahedron.

Business-card cube.

The octahedron can be made from four cards, and an icosahedron with ten of them. It is also easy to make a fourth Platonic solid – the cube. Put two cards on top of each other like a plus sign and fold the flaps as shown above. This creates the shape of a square. Six cards folded in this way slot together to form a cube, although the flaps are on the outside. You need another six cards to slide on to each face in order to make the cube clean.

The Zen master of business-card origami is Jeannine Mosely, a software developer from Massachussetts. A few years ago she found herself with 100,000 cards in her garage – she inherited three batches from her colleagues at work, the first time when the company changed its name, the second when the company moved addresses, and then again when it was discovered that the new cards all had a typo. You can make a lot of tetrahedrons with 100,000 business cards. Yet Mosely had much grander ambitions than the Platonic solids. Why limit herself to the ancient Greeks? Had 2000 years of geometry not produced a more exciting 3-D shape? With her resources Mosely felt ready to tackle the ultimate challenge of her art, the Menger sponge.

Before we get to the Menger sponge, I need to introduce the Sierpinski carpet. The bizarre shape was invented by the Polish mathematician Waclaw Sierpinski in 1916. You start with a black square. Imagine it is made of nine identical subsquares, and remove the central one (figure A). Now for each of the remaining subsquares, repeat the operation – that is, imagine they are made of nine subsquares and remove the central one (B). Repeat this process again (C). The Sierpinski carpet is what you get if you continue this process ad infinitum.

In 1926 the Austrian mathematician Karl Menger came up with the idea of a three-dimensional version of the Sierpinski carpet, which is now known as the Menger sponge. You start with a cube. Imagine it is made up of 27 identical subcubes and remove the subcube at the very centre, as well as the 6 subcubes that are in the

Thinking inside the box: Jeannine Mosely and her Menger Sponge.

centre of each of the sides of the original cube. You are left with a cube that looks like it has three square holes drilled through it (D). Treat each of the 20 subcubes that are left like the original cube, and extract 7 of the 27 subcubes (E). Repeat the process again (F) and the block starts to look like it has been devoured by a cluster of geometrically obsessed woodworm.

The Menger sponge is a brilliantly paradoxical object. As you continue the iterations of taking out smaller and smaller cubes the volume of the sponge gets smaller and smaller, eventually becoming invisible – as though the woodworm have eaten the whole lot. Yet each iteration of cube removal also makes the surface area of the sponge increase. By taking more and more iterations you can make the surface area larger than any area you want, meaning that as the number of iterations approaches infinity, the surface area of the sponge also approaches infinity. In the limit, the Menger sponge is an object with an infinitely large surface area that is also invisible.

Mosely constructed a level-three Menger sponge – in other words, a sponge after three iterations of cube removal (F). The project took her ten years. She enlisted the help of about 200 people, and used 66,048 cards. The finished sponge was 4ft 8in high, wide and deep.

'For a long time I wondered if I was doing something completely and utterly ridiculous,' she told me. 'But when I had got it done I stood next to it and realized the scale gave it a grandeur. A particularly wonderful thing is that you can stick your head and shoulders into the model and see this amazing figure from a viewpoint that you have never seen before.' It was endlessly fascinating to look at because the more she zoomed in, the more she saw the patterns repeating themselves. 'You simply look at it and it doesn't need to be explained. You can understand it just by looking at it. It is an idea made solid; math made visual.' The business-card Menger sponge is a beautifully crafted object that creates an emotional and intellectual response. It belongs just as much to geometry as it does to art.

Although origami was originally a Japanese invention, paper-folding techniques also developed independently in other countries. A European pioneer was the German educator Friedrich Fröbel, who used paper-folding in the mid nineteenth century as a way of teaching young children geometry. Origami had the advantage of allowing kindergarten pupils to feel the objects created, rather than just see them in drawings. Fröbel inspired the Indian mathematician T. Sundara Row to publish *Geometric Exercises in Paper Folding* in 1901, in which he argued that origami was a mathematical method that in some cases was more powerful than Euclid's. He wrote that 'several important geometric processes ... can be effected much more easily than with the compass and ruler'. But even he did not anticipate just how powerful origami can actually be.

In 1936 Margherita P. Beloch, an Italian mathematician at the University of Ferrara, published a paper that proved that starting with a length L on a piece of paper, she could fold a length that was the cube root of L. She might not have realized it at the time, but this meant that origami could solve the problem given to the Greeks at Delos, where the oracle demanded that the Athenians double the volume of a cube. The Delian Problem can be rephrased as the challenge to create a cube with sides that are $\sqrt[3]{2}$ – the cube root of two – times the side of a given cube. In origami terms, the challenge is reduced to folding the length $\sqrt[3]{2}$ from the length 1. Since we can double 1 to get 2 by folding the length 1 on itself, and we can find the cube root of this new length following Beloch's steps, the problem was solved. It also followed from Beloch's proof that any angle could be trisected – which cracked the second great unsolvable problem of antiquity. Beloch's paper, however, remained in obscurity for decades, until, in the 1970s, the maths world began to take origami seriously.

The first published origami proof of the Delian Problem was by a Japanese mathematician in 1980, and angle trisection followed by an American in 1986. The boom of interest stemmed in part from frustration with more than two millennia of Euclidean orthodoxy. The restrictions imposed by Euclid's limitation to working with only a ruler and compass had narrowed the scope of mathematical enquiry. As it turns out, origami is more versatile than a ruler and compass, for example, in constructing the regular polygons. Euclid was able to draw an equilateral triangle, square, pentagon and hexagon, but recall that the heptagon (which has seven sides) and nonagon (nine) eluded him. Origami can fold heptagons and nonagons relatively easily, although it meets its match with the 11-agon. (Strictly speaking, this is one-fold-at-a-time origami. If multiple folds are allowed any polygon can in theory be constructed, even though a physical demonstration may be so hard as to be impossible.)

Far from being child's play, origami is now at the cutting edge of maths. Literally. When Erik Demaine was 17 he and his collaborators proved that it is possible to create any shape with straight sides by folding a piece of paper and making just one cut. Once you decide on the shape you want to make, you work out the fold pattern, fold the paper, make the single cut, unfold the paper and the detached shape will fall out. While it might appear that such a result would be of interest only to schoolchildren making increasingly complex Christmas decorations, Demaine's work has found uses in industry, especially in car airbag design. Origami has connections to protein folding, and now has applications in the most unexpected spheres: in creating arterial stents, robotics, and in the solar panels of satellites.

A guru of modern origami is Robert Lang, who as well as advancing the theory behind paper-folding has turned the pastime into a sculptural art form. A former NASA physicist, Lang has pioneered the use of computers in designing fold patterns to create new and increasingly complex figures. His original figures include bugs, scorpions, dinosaurs and a man playing a grand piano. The fold patterns are almost as beautiful as the finished design.

The United States now has as good a claim as Japan does to being at the forefront of origami, partly because origami is so embedded within Japanese society as an informal pursuit that there is more of a barrier to taking it seriously as a science. The cause is not helped by a Monty Pythonesque factionalization between different organizations, each claiming exclusive access to origami's soul. I was surprised to hear Kazuo Kobayashi, chairman of the International Origami Association, dismiss the work of Robert Lang as elitist: 'He is doing it for himself,' he tut-tutted. 'My origami is about the rehabilitation of the sick and educating children.'

Nevertheless, there are many Japanese origami enthusiasts doing interesting new work, and I travelled to Tsukuba, a modern university town just north of Tokyo, to meet one of them. Kazuo Haga is a retired entomologist, whose professional expertise is in the embryonic development of insect eggs. His tiny office was stacked with books and display cases of butterflies. Haga, who is aged 74, was wearing large glasses with a thin black rim that framed his face geometrically. I noticed immediately that he is a very shy man, soft and modest – and was rather nervous of being interviewed.

But Haga's timidity is only social. In origami he is a rebel. Choosing to stay out of the origami mainstream, he has never felt constrained by any conventions. For example, according to the rules of traditional Japanese origami, there are only two ways to make the first fold. Both are folding it in half – either folding along a diagonal, bringing two opposing corners together, or along the midline, bringing adjacent corners together. These are known as the 'primary creases' – the diagonals and the midlines of the square.

Haga decided to be different. What if he folded a corner on to the midpoint of a side? Ker-azee! He did this for the first time in

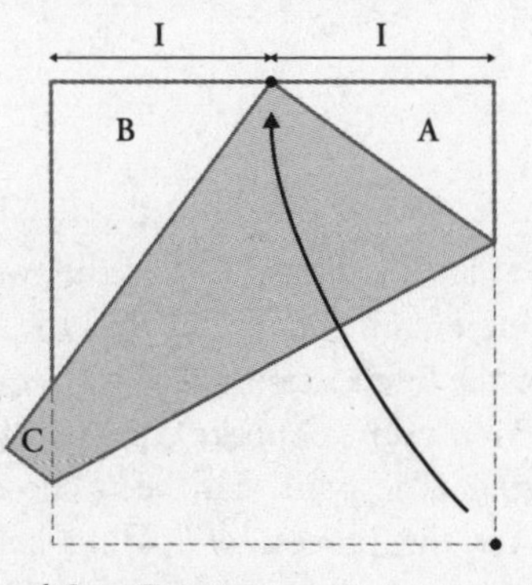

Haga's theorem: A, B and C are Egyptian.

1978. This simple fold had the effect of opening the curtains on to a sublime new world. Haga had created three right-angled triangles. Yet these were not any old right-angled triangles. All three were Egyptian, the most historic and iconic triangle of them all.

Feeling the thrill of discovery, but having no one to share it with, he sent a letter about the fold to Professor Koji Fushimi, a theoretical physicist known to have an interest in origami. 'I never got a reply,' said Haga, 'but then all of a sudden he wrote an article in a magazine called *Mathematics Seminar* referring to Haga's Theorem. That was the substitute for his reply.' Since then Haga has given his name to two other origami 'theorems', although he says he has another 50. He tells me this not as evidence of arrogance but as a measure of how the area is so rich and untapped.

In Haga's Theorem a corner is folded on to the midpoint of a side. Haga wondered if anything interesting might be revealed if he folded a corner on to a random point on the side. Deciding to demonstrate this to me, he took a blue piece of square origami paper and marked an arbitrary point on one of the sides with a red pen. He folded one of the opposite corners on to this mark, leaving a crease, and then unfolded it. He then folded the other opposite corner on to the mark to make a second crease, leaving the square now with two separate intersecting lines.

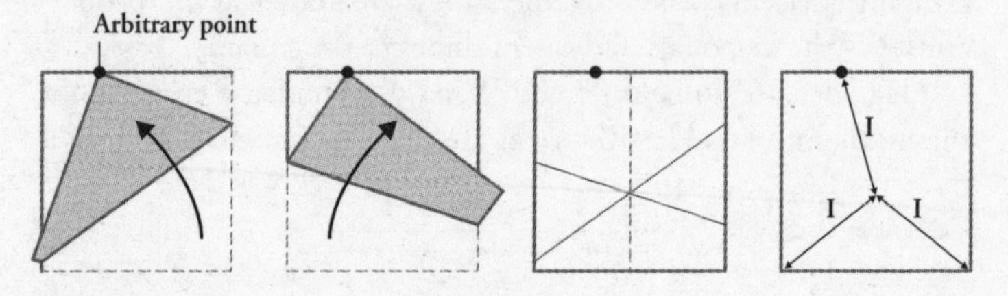

Haga's other theorem.

Haga showed me that the intersection of the two folds was always on the middle line of the paper, and that the distance from the arbitrary point to the intersection was always equal to the distance from the intersection to the opposite corners. I found Haga's folds totally mesmerizing. The point had been chosen randomly, and was off-centre. Yet the process of folding behaved like a self-correcting mechanism.

Haga wanted to show me a final pattern. The name he chose for this discovery sounded like a haiku: *an arbitrarily made 'mother line' bears eleven wonder babies.*

Step 1: Make an arbitrary fold in a square piece of paper.

Step 2: Fold each edge along that fold separately, always unfolding to leave a crease, as demonstrated below in A to E.

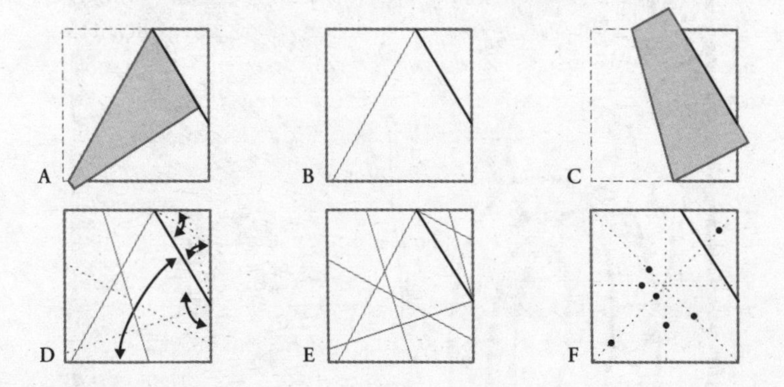

Mother line showing seven of her eleven wonder babies.

Again, this is very simple to carry out, and reveals a beautiful geometric order. Every intersection is on the primary creases, as F shows. (The diagram shows the seven intersections that are inside the original square; the other four are on extensions of the folds.) The first fold was random, yet all the folds meet with perfect concision and regularity on diagonals and midlines.

It struck me that if any man can be said to embody the soul of Pythagoras in the modern world, it is surely Kazuo Haga. Both men share a passion for mathematical discovery based on a wonder for the simple harmonies of geometry. The experience seemed to have touched Haga spiritually, the same way it did Pythagoras two millennia ago. 'Most of the Japanese are trying to create new shapes in origami,' said Haga. 'My aim was to escape from the idea that you have to create something physical, and instead discover mathematical phenomena. That is why I find it so interesting. You find that in the very, very simple world you can still discover fascinating things.'

CHAPTER THREE

Something about Nothing

Every year the Indian seaside town of Puri fills with a million pilgrims. They come for one of the most spectacular festivals in the Hindu calendar – the Rath Yatra, in which three chariots the size of carnival floats are pulled through the town. When I visited, the streets were crowded with cymbal-crashing, mantra-chanting devotees, barefoot holy men with long beards and middle-class Indian tourists with fashionable T-shirts and neon saris. It was mid-summer, the beginning of the monsoon season, and in between downpours festival-workers sprayed the faces of passers-by with water to cool them down. Smaller Rath Yatra processions take place simultaneously all over India, although Puri's is the focal event and its chariots the biggest.

The festival gets under way only when the local holy man, the Shankaracharya of Puri, stands in front of the crowds and blesses them. The Shankaracharya is one of Hinduism's most important sages, the head of a monastic order that dates back more than a thousand years. He was also the reason I had travelled to Puri. As well as being a spiritual leader, the Shankaracharya is a published mathematician. I was also a pilgrim in search of enlightenment.

Right away in India, I noticed something unfamiliar about their use of numbers. In my hotel reception I picked up a copy of *The Times of India*. With the paper's corners flapping from gusts of competing metal fans, the front-page headline caught my eye:

5 crore more Indians
than govt thought

Crore is the Indian-English word for ten million, so the article was saying that India had just discovered 50 million citizens it never knew it had – a number roughly comparable to the population of England.

It was startling that a country could overlook such a large number, even if it represented less than 5 per cent of the overall population. Yet I was more puzzled by the word *crore*. Indian English has different words for high numbers than British or American English. For example, the word 'million' is not used. A million is instead expressed as *ten lakh*, where *lakh* is 100,000. Since 'million' is unheard of in India, the Oscar-winning film *Slumdog Millionaire* was released there as *Slumdog Crorepati*. A very rich person is someone who has a crore of dollars or rupees, not a million of them. The table of Indian equivalents for British/American number words is as follows:

British/American	**Notation**	**Indian English**	**Notation**
Ten	10	Ten	10
Hundred	100	Hundred	100
Thousand	1,000	Thousand	1,000
Ten thousand	10,000	Ten thousand	10,000
Hundred thousand	100,000	Lakh	1,00,000
One million	1,000,000	Ten lakh	10,00,000
Ten million	10,000,000	Crore	1,00,00,000
Hundred million	100,000,000	Ten crore	10,00,00,000

Note that above a thousand, Indians introduce a comma after every two digits, while in the rest of the world the convention is every three.

The use of *lakh* and *crore* is a legacy of the mathematics of ancient India. The words come from the Hindi *lakh* and *karod*, which in turn came from the Sanskrit words for those numbers, *laksh* and *koti*. In ancient India coining words for large numbers was a scientific and religious preoccupation. For example, in the *Lalitavistara Sutra*, a Sanskrit text that dates from the beginning of the fourth century at the latest, the Buddha is challenged to express numbers higher than a hundred *koti*. He replies:

> One hundred *koti* are called an *ayuta*, a hundred *ayuta* make a *niyuta*, a hundred *niyuta* make a *kankara*, a hundred *kankara* make a *vivara*, a hundred *vivara* are a *kshobhya*, a hundred *kshobhya* make a *vivaha*, a hundred *vivaha* make a *utsanga*, a hundred *utsanga* make a *bahula*, a hundred *bahula* make a *nâgabala*, a

> hundred *nâgabala* make a *titilambha*, a hundred *titilambha* make a *vyavasthânaprajñapati*, a hundred *vyavasthânaprajñapati* make a *hetuhila*, a hundred *hetuhila* make a *karahu*, a hundred *karahu* make a *hetvindriya*, a hundred *hetvindriya* make a *samâptalambha*, a hundred *samâptalambha* make a *gananâgati*, a hundred *gananâgati* make a *niravadya*, a hundred *niravadya* make a *mudrâbala*, a hundred *mudrâbala* make a *sarvabala*, a hundred *sarvabala* make a *visamjñagati*, a hundred *visamjñagati* make a *sarvajña*, a hundred *sarvajña* make a *vibhutangamâ*, a hundred *vibhutangamâ* make a *tallakshana*.

Just as in contemporary India, the Buddha went up the list in multiples of a hundred. Since a *koti* is ten million, the value of a *tallakshana* is ten million multiplied by a hundred 23 times, which works out as 10 followed by 52 zeros, or 10^{53}. This is a phenomenally large number, so large, in fact, that if you measure the entire universe from end to end in metres, and then square that number, you are roughly around 10^{53}.

But Buddha didn't stop there. He was just warming up. He explained that he had described only the *tallakshana* counting system, and above it there was another one, the *dhvajâgravati* system, made up of the same number of terms. And above that, another one, the *dhavjâgranishâmani*, again with 24 number words. In fact, there were another six systems to go – which the Buddha, of course, listed perfectly. The last number in the final system is equivalent to 10^{421}: one followed by 421 zeros.

It is worth taking a breath to consider the view. There are an estimated 10^{80} atoms in the universe. If we take the smallest measurable unit of time – known as Planck time, which is a second divided into 10^{43} parts – then there have been about 10^{60} units of Planck time since the Big Bang. If we multiply the number of atoms in the universe by the number of Planck times since the Big Bang – which gives us the number of unique positions of every particle since time began – we are still only on 10^{140}, which is way, way smaller than 10^{421}. The Buddha's big number has no practical application – at least not for counting things that exist.

Not only was the Buddha able to fathom the impossibly large, he was also proficient in the realm of the impossibly tiny, explaining how many atoms there were in the *yojana*, an ancient unit of length around 10km. A *yojana*, he said, was equivalent to:

Four *krosha*, each of which was the length of
One thousand arcs, each of which was the length of
Four cubits, each of which was the length of
Two spans, each of which was the length of
Twelve phalanges of fingers, each of which was the length of
Seven grains of barley, each of which was the length of
Seven mustard seeds, each of which was the length of
Seven poppy seeds, each of which was the length of
Seven particles of dust stirred up by a cow, each of which was the length of
Seven specks of dust disturbed by a ram, each of which was the length of
Seven specks of dust stirred up by a hare, each of which was the length of
Seven specks of dust carried away by the wind, each of which was the length of
Seven tiny specks of dust, each of which was the length of
Seven minute specks of dust, each of which was the length of
Seven particles of the first atoms.

This was, in fact, a pretty good estimate. Just say that a finger is 4cm long. The Buddha's 'first atoms' are, therefore, 4cm divided by seven ten times, which is $0.04\text{m} \times 7^{-10}$ or 0.0000000001416m, which is more or less the size of a carbon atom.

The Buddha was by no means the only ancient Indian interested in the incredibly large and the unfeasibly small. Sanskrit literature is full of astronomically high numbers. Followers of Jainism, a sister religion to Hinduism, defined a *raju* as the distance covered by a god in six months if he covers 100,000 yojana in each blink of his eye. A *palya* was the amount of time it takes to empty a giant *yojana*-sized cube filled with the wool of newborn lambs if one strand is removed every century. The obsession with high (and low) numbers was metaphysical in nature, a way of groping towards the infinite and of grappling with life's big existential questions.

Before Arabic numerals became an international lingua franca, humans had many other ways of writing down numbers. The first number symbols that emerged in the West were notches, cuneiform bird tracks and hieroglyphics. When languages developed their own alphabets, cultures began to use letters to represent numbers. The Jews used the Hebrew aleph (א) to mean one, bet (ב) to be two and so on. The tenth letter, yod (י), was ten, after which each letter went up in tens, and on reaching 100 went up in hundreds. The twenty-second and final letter of the Hebrew alphabet, tav (ת), was 400. Using letters for numbers was confusing and also encouraged a numerological approach to counting. Gematria, for example, was the practice of adding up the numbers of the letters in Hebrew words to find a value and using this number for speculations and divinations.

The Greeks used a similar system, with alpha (α) being one, beta (β) being two, and so on to the twenty-seventh letter of their alphabet, sampi (ϡ), which was 900. Greek mathematical culture, the most advanced in the classical world, did not share the Indian hunger for jumbo numbers. The highest-value number word they had was *myriad*, meaning 10,000, which they wrote as a capital M.

Roman numerals were also alphabetic, although their system had more ancient roots than those of either the Greeks or the Jews. The symbol for one was I, probably a relic of a notch on a tally stick. Five was V, maybe because it looked like a hand. The other numbers were X, L, C, D and M for 10, 50, 100, 500 and 1000. All the other numbers were generated using these seven capital letters. The Roman system's provenance from the tally stick made it a very intuitive way of writing out numbers. It was also efficient – using just seven symbols compared to 22 in Hebrew and 27 in Greek – and Roman numerals were the predominant number system in Europe for well over a thousand years.

Roman numerals, however, were very poorly suited to arithmetic. Let's try to calculate 57 × 43. The best way to do this is with a method known as Egyptian or peasant multiplication, because it dates back at least to ancient Egypt. It is an ingenious method, though slow.

You first decompose one of the numbers being multiplied into powers of two (which are 1, 2, 4, 8, 16, 32 and so on, doubling each time) and make a table of the doubles of the other number. So, for the example 57 × 43, let us decompose 57 and draw up a table of doubles of 43. I'm using Arabic numerals to show how it's done, and will translate into Roman numerals afterwards.

Decomposition: 57 = 32 + 16 + 8 + 1

Table of doubles:

1 × 43 =	43
2 × 43 =	86
4 × 43 =	172
8 × 43 =	344
16 × 43 =	688
32 × 43 =	1376

The multiplication of 57 × 43 is equivalent to the addition of the numbers in the table of doubles that correspond to amounts in the decomposition. This sounds like a mouthful but is fairly straightforward. The decomposition contains a 32, a 16, an 8 and a 1. In the table, 32 corresponds to 1376, 16 corresponds to 688, 8 corresponds to 344 and 1 corresponds to 43. So, we can rewrite the initial multiplication as 1376 + 688 + 344 + 43, which equals 2451.

Now for the Roman numerals: 57 is LVII and 43 is XLIII. The decomposition and the table becomes:

LVII = XXXII + XVI + VIII + I

and

XLIII
LXXXVI
CLXXII
CCCXLIV
DCLXXXVIII
MCCCLXXVI

so,

LVII × XLIII = MCCCLXXVI + DCLXXXVIII + CCCXLIV + XLIII = MMCDLI

Oof! By breaking down the calculation into digestible morsels involving only doubling and adding, Roman numerals are just about up to the task. Still, we did much more work than we needed to. I mentioned earlier that the Roman system was intuitive and efficient. I'm taking that back. The Roman system quickly becomes counter-intuitive since the length of the number is not dependent on value. MMCDLI is larger than DCLXXXVIII, but uses fewer numerals, which goes against common sense. And any efficiency gained by using only seven symbols is forfeited by the inefficiency of how they are used. Often long strings are required to signify small numbers: LXXXVI uses six symbols, compared to the Arabic equivalent, 86, which uses two.

Compare the calculation above with the method of 'long' multiplication we all learned at school:

$$\begin{array}{r} 57 \\ \underline{\times\ 43} \\ 0171 \\ \underline{2280} \\ \underline{2451} \end{array}$$

There is a very simple reason why our method is easier and quicker. Neither the Romans nor the Greeks or Jews had a symbol for zero. When it comes to sums, nothing makes all the difference.

The Vedas are Hinduism's sacred texts. They have been passed down orally for generations until being transcribed into Sanskrit about 2000 years ago. In one of the Vedas a passage about the construction of altars lists the following number words:

Dasa	10
Sata	100
Sahasra	1000
Ayuta	10,000
Niyuta	100,000
Prayuta	1,000,000
Arbuda	10,000,000
Nyarbuda	100,000,000

Samudra	1,000,000,000
Madhya	10,000,000,000
Anta	100,000,000,000
Parârdha	1,000,000,000,000

With names for every multiple of ten, large numbers can be described very efficiently, which provided astronomers and astrologers (and, presumably, altar builders) with a suitable vocabulary for referring to the enormous quantities required in their calculations. This is one reason why Indian astronomy was ahead of its time. Consider the number 422,396. The Indians started at the smallest digit, at the right, and enumerated the others successively from right to left: *Six and nine dasa and three sata and two sahasra and two ayuta and four niyuta*. It is not too much of a step to realize that you can leave out the names for the powers of ten, since the position of the number in the list defines its value. In other words, the number above could be written: *six, nine, three, two, two, four*.

This type of enumeration is known as a 'place-value' system, which we discussed earlier. An abacus bead has a value dependent on which column it is in. Likewise, each number in the above list has a value dependent on its position in the list. Place-value systems, however, require the concept of a 'place-holder'. For example, if a number has two units, no *dasa* and three *sata* it cannot be written *two, three* since that refers to two units and three *dasa*. A place-holder is needed to maintain the correct positions, to make it clear that there are no *dasa*, and the Indians used the word *shunya* – meaning 'void' – to refer to this place-holder. The number that is just two units and three *sata* would be written as *two, shunya, three*.

The Indians were not the first to introduce a place-holder. That honour probably went to the Babylonians, who wrote their number symbols in columns with a base 60 system. One column was for units, the next column was for 60s, the next for 3600s and so on. If a number had no value for that column, it was initially left blank. This, however, led to confusion, so they eventually introduced a symbol that denoted the absence of a value. This symbol, however, was used only as a marker.

After adopting *shunya* as a place-holder, the Indians took the idea and ran with it, upgrading *shunya* into a fully fledged number of its own: zero. Nowadays, we have no difficulty in understanding that zero is a number. But the idea was far from obvious. The Western civilizations, for example, failed to come up with it even after thousands of years of mathematical enquiry. Indeed, the scale of the conceptual leap achieved by India is illustrated by the fact that the classical world was staring zero in the face and still saw right through it. The abacus contained the concept of zero because it relied on place value. For example, when a Roman wanted to express 101, he would push a bead in the first column to signify 100, move no beads in the second column, indicating no tens, and push a bead in the third column to signify a single unit. The second, untouched column was expressing nothing. In calculations, the abacist knew he had to respect untouched columns just as he had to respect ones in which the beads were moved. But he never gave the value expressed by the untouched column a numerical name or symbol.

Zero took its first tentative steps as a bonafide number under the tutelage of Indian mathematicians such as Brahmagupta, who in the seventh century showed how *shunya* behaved towards its number siblings:

A debt minus *shunya* is a debt
A fortune minus *shunya* is a fortune
Shunya minus *shunya* is *shunya*
A debt subtracted from *shunya* is a fortune
A fortune subtracted from *shunya* is a debt
The product of *shunya* multiplied by a debt or fortune is shunya
The product of *shunya* multiplied by *shunya* is *shunya*

If 'fortune' is understood as a positive number, a, and 'debt' as a negative number, $-a$, Brahmagupta has written out the statements:

$-a - 0 = -a$
$a - 0 = a$
$0 - 0 = 0$
$0 - (-a) = a$
$0 - a = -a$
$0 \times a = 0, 0 \times -a = 0$
$0 \times 0 = 0$

Numbers had emerged as tools for counting, as abstractions that described amounts. But zero was not a counting number in the same way; understanding its value required a further level of abstraction. Yet the less that maths was tied to actual things, the more powerful it became.

Treating zero as a number meant that the place-value system that had made the abacus the best way to calculate could be properly exploited using written symbols. Zero would enhance mathematics in other ways too, by leading to the 'invention' of negative numbers and decimal fractions – concepts we learn effortlessly at school and are intrinsic to our needs in daily life, but which were in no way self-evident. The Greeks made fantastic mathematical discoveries without a zero, negative numbers or decimal fractions. This was because they had an essentially spatial understanding of mathematics. To them it was nonsensical that nothing could be 'something'. Pythagoras was no more able to imagine a negative number than a negative triangle.

India 1st century	—	=	≡	[illegible]	[illegible]	[illegible]	[illegible]	[illegible]	[illegible]	
India 9th century	[illegible]	[illegible]	[illegible]	[illegible]	[illegible]	[illegible]	[illegible]	[illegible]	[illegible]	o
North Africa 14th century	[illegible]	2	[illegible]	[illegible]	[illegible]	6	[illegible]	8	9	
Spain 10th century	I	[illegible]	[illegible]	[illegible]	[illegible]	[illegible]	[illegible]	8	9	
England 14th century	1	2	3	[illegible]	[illegible]	[illegible]	[illegible]	8	[illegible]	0
France 16th century	1	2	3	4	5	6	7	8	9	0

Evolution of modern numerals.

Of all the innovative ways that numbers were treated in ancient India, perhaps none was more curious than the vocabulary used to describe the numbers from zero to nine. Rather than each digit having a unique name, there was a colourful lexicon of synonyms. Zero, for example, was *shunya*, but it was also 'ether', 'dot', 'hole' or the 'serpent of eternity'. One could be 'earth', 'moon', 'the pole star' or 'curdled milk'. Two was interchangeable with 'arms', three with 'fire' and four with 'vulva'. The names that were chosen depended on context and conformed to Sanskrit's strict rules of versification and prosody. For example, the following verse is a piece of number-crunching from an ancient astrological text:

> The apsides of the moon in a yuga
> Fire. Vacuum. Horsemen. *Vasu**. Serpent. Ocean,
> and of its waning node
> *Vasu*. Fire. Primordial Couple. Horsemen. Fire. Twins.
>
> The translation is:
> [The number of revolutions] of the apsides of the moon in a [cosmic cycle is]
> Three. Zero. Two. Eight. Eight. Four, [or 488,203]
> and of its waning node
> Eight. Three. Two. Two. Three. Two. [or 232,238]

While at first it might seem confusing to have flowery alternatives for each digit, it actually makes perfect sense. During a period in history when manuscripts were flimsy and easily spoiled, astronomers and astrologers needed a backup method to remember significant numbers accurately. Strings of digits were more easily memorized when described in verse with varied names, rather than when using the same number names repeatedly.

Another reason why numbers were passed down orally was that the numerals that were emerging in different regions of India for the numbers from one to nine (zero, I will come to later) were not the same. Two people who did not understand each other's number symbols could at least communicate numbers using words. By 500 CE,

---

* A group of eight divinities in the Indian epic the *Mahabharata*.

however, there was greater uniformity in the numerals used, and India had the three elements that were required for a modern decimal number system: ten numerals, place value, and an all-singing, all-dancing zero.

Owing to its ease of use, the Indian method spread to the Middle East, where it was embraced by the Islamic world, which accounts for why the numerals have come to be known, erroneously, as Arabic. From there they were brought to Europe by an enterprising Italian, Leonardo Fibonacci, his last name meaning 'son of Bonacci'. Fibonacci was first exposed to the Indian numerals while growing up in what is now the Algerian city of Béjaïa, where his father was a Pisan customs official. Realizing that they were much better than Roman ones, Fibonacci wrote a book about the decimal place-value system called the *Liber Abaci*, published in 1202. It opens with the happy news:

> The nine Indian figures are:
>
> 9 8 7 6 5 4 3 2 1
>
> With these nine figures, and with the sign 0, which the Arabs call *zephyr*, any number whatsoever is written, as will be demonstrated.

More than any other book, the *Liber Abaci* introduced the Indian system to the West. In it Fibonacci demonstrated ways to do arithmetic that were quicker, easier and more elegant than the methods the Europeans had been using. Long multiplication and long division might seem dreary to us now, but at the beginning of the thirteenth century they were the latest technological novelty.

Not everyone, however, was convinced to switch immediately. Professional abacus operators felt threatened by the easier counting method, for one thing. (They would have been the first to realize that the decimal system was essentially the abacus with written symbols.) On top of that, Fibonacci's book appeared during the period of the Crusades against Islam, and the clergy was suspicious of anything with Arab connotations. Some, in fact, considered the new arithmetic the Devil's work precisely because

Arithmetica, the spirit of arithmetic, adjudicates between Boethius, who is using Arabic numerals, and Pythagoras, who has a counting board. Her adoring gaze and the numbers on her dress give away which method she prefers. From a woodblock engraving in Gregorius Reisch's Margarita Philosophica *(1503).*

it was so ingenious. A fear of Arabic numerals is revealed through the etymology of some modern words. From *zephyr* came 'zero' but also the Portuguese word *chifre*, which means '[Devil] horns', and the English word *cipher*, meaning 'code'. It has been argued that this was because using numbers with a *zephyr*, or zero, was done in hiding, against the wishes of the Church.

In 1299 Florence banned Arabic numerals because, it was said, the slinky symbols were easier to falsify than solid Roman Vs and Is. A 0 could easily become a 6 or 9, and a 1 morph seamlessly into a 7. As a consequence, it was only around the end of the fifteenth century that Roman numerals were finally superseded, though negative numbers took much longer to catch on in Europe, gaining acceptance only in the seventeenth century, because they were said to be used in calculations of illegal money-lending, or usury, which was associated with blasphemy. In places where no calculation is needed, however, such as legal documents, chapters in books and dates at the end of BBC programmes, Roman numerals still live on.

With the adoption of Arabic numbers, arithmetic joined geometry to become part of mathematics in earnest, having previously been more of a tool used by shopkeepers, and the new system helped open the door to the scientific revolution.

A more recent Indian contribution to the world of numbers is a set of arithmetical tricks collectively known as Vedic Mathematics. It was discovered at the beginning of the twentieth century by a young swami, Bharati Krishna Tirthaji, who claimed to have found them in the Vedas, which was rather like, say, a vicar announcing he had found a method for solving quadratic equations in the Bible. Vedic Mathematics is based on the following list of 16 aphorisms, or sutras, which Tirthaji said were not actually written in any passage of the Vedas, instead being detectable only 'on the basis of intuitive revelation'.

1. By one more than the one before
2. All from 9 and the last from 10
3. Vertically and Cross-wise
4. Transpose and Apply
5. If the Samuccaya is the Same it is Zero

6. If One is in Ratio the Other is Zero
7. By Addition and by Subtraction
8. By the Completion or Non-Completion
9. Differential Calculus
10. By the Deficiency
11. Specific and General
12. The Remainders by the Last Digit
13. The Ultimate and Twice the Penultimate
14. By One Less than the One Before
15. The Product of the Sum
16. All the Multipliers

Was he serious? Yes, and very much so too. Tirthaji was one of the most respected holy men of his generation. A former child prodigy, graduating in Sanskrit, philosophy, English, maths, history and science at the age of 20, he was also a talented orator who, it became clear early into adulthood, was destined to take a prominent role in Indian religious life. In 1925 Tirthaji was indeed made a Shankaracharya, one of the senior positions in traditional Hindu society, in charge of a nationally important monastery in Puri, Orissa, on the Bay of Bengal. This is the town that I was visiting, the focus of the Rath Yatra chariot festival, where I was hoping to meet the incumbent Shankaracharya, who is the current ambassador for Vedic Mathematics.

In his role as Shankaracharya in the 1930s and 1940s, Tirthaji regularly toured India, giving sermons to crowds of tens of thousands, usually dispensing spiritual guidance but also promoting his new way of calculation. The 16 sutras, he taught, were to be used as if they were mathematical formulae. While they might have sounded ambiguous, like chapter titles in an engineering book or numerological mantras, they in fact referred to specific rules. One of the most straightforward is the second, *All from 9 and the last from 10*. This is to be implemented whenever you are subtracting a number from a power of ten, such as 1000. If I want to calculate 1000 – 456, for example, then I subtract 4 from 9, 5 from 9 and 6 from 10. In other words, the first two numbers from 9 and the last from 10. The answer is 544. (The other sutras are applications for other situations, more of which I will introduce later.)

Tirthaji promoted Vedic Mathematics as a gift to the nation, arguing that maths that usually took schoolchildren 15 years to learn could, with the sutras, be learned in just eight months. He even went as far as claiming that the system could be expanded to cover not just arithmetic but algebra, geometry, calculus and astronomy. Due to Tirtharji's moral authority and charisma as a public speaker, audiences loved him. The general public, he wrote, were 'highly impressed, nay, thrilled, wonder-struck and flabbergasted!' at Vedic Mathematics. To those who asked whether the method was maths or magic, he had a set reply: 'It is both. It is magic until you understand it; and it is mathematics thereafter.'

In 1958, when he was 82 years old, Tirthaji visited the United States, which caused much controversy back home because Hindu spiritual leaders are forbidden from travelling abroad, and it was the first time that a Shankaracharya had ever left India. His trip provoked great curiosity in the US. The West Coast would later become a focus for flower power, gurus and meditation, but back then no one had seen anyone like him. When Tirthaji arrived in California the *Los Angeles Times* called him 'one of the most important – and least-known – personages in the world'.

Tirthaji had a full schedule of talks and TV appearances. Though he spoke mostly about world peace, he devoted one lecture entirely to Vedic Mathematics. The venue was the California Institute of Technology, one of the most prestigious scientific institutions in the world. Tirthaji, who weighed not much more than seven stone and was wearing traditional robes, sat in a chair at the front of a wood-panelled classroom. In a quiet voice, but with a commanding presence, he told his audience: 'I have been, from my childhood, equally fond of metaphysics on the one side and mathematics on the other. And I've found no difficulty at all.'

He went on to explain exactly how he had found the sutras, asserting that there was a wealth of hidden knowledge in the Vedic texts that came from the many double meanings of words and phrases. These mystical 'puns', he added, had been totally lost on Western Indologists. 'The supposition is that mathematics was not part of the Vedic literature,' he said, 'but when I was able to find it, well, it was easy sailing.'

Tirthaji's opening trick was to demonstrate how to multiply 9 × 8 without using a multiplication table. This uses the sutra *All from 9 and the last from 10*, although why it does so only becomes clear later.

First, he drew a 9 on the blackboard, followed by the difference of 9 from 10, which is –1. Underneath he drew 8 and next to it the difference of 8 from 10, which is –2.

$$\begin{array}{ll} 9 & -1 \\ 8 & -2 \end{array}$$

The first number of the answer can be derived in four different ways. Either add the numbers in the first column and subtract ten (9 + 8 – 10 = 7). Or add the numbers in the second column and add ten (–1 – 2 + 10 = 7) or add either of the diagonals (9 – 2 = 7, and 8 – 1 = 7). The answer is always seven.

$$\begin{array}{ll} 9 & -1 \\ \hline 8 & -2 \\ \hline 7 & \end{array}$$

The second part of the answer is calculated by multiplying the two numbers in the second column (–1 × –2 = 2). The complete answer is 72.

$$\begin{array}{ll} 9 & -1 \\ \hline 8 & -2 \\ \hline 7 & 2 \end{array}$$

I find this trick immensely satisfying. Writing out a single-digit number next to its difference from ten is somehow like pulling it apart to reveal its inner personality, lining up the ego and alter ego. We get a deeper understanding of how the number behaves. A sum such as 9 × 8 is about as mundane as they come, yet scratch the surface and we see unexpected elegance and order. And the method works not just for 9 × 8 but for any two numbers. Tirthaji chalked up another example: 8 × 7.

$$\begin{array}{rr} 8 & -2 \\ \underline{7} & \underline{-3} \\ \underline{5} & \underline{6} \end{array}$$

Again, the first digit can be derived in any one of the four ways: $8 + 7 - 10 = 5$, or $-2 - 3 + 10 = 5$, or $8 - 3 = 5$ or $7 - 2 = 5$. The second digit is the product of the digits in the second column, $-2 \times -3 = 6$.

Tirthaji's tactic reduces the multiplication of two single-digit numbers to an addition and the multiplication of the differences of the original numbers from ten. In other words, it reduces the multiplication of two single digit numbers larger than five to an addition and the multiplication of two numbers less than five. Which means that it is possible to multiply by six, seven, eight and nine without going higher than our five-times table. This is useful to people who find learning their times tables difficult.

In fact, the technique explained by Tirthaji is the same as a method of finger calculation used at least since the Renaissance in Europe, and still used by farmhands in parts of France and Russia as late as the 1950s. On each hand the fingers are assigned the numbers from 6 to 10. To multiply two numbers together, say 8 and 7, touch the 8 finger to the 7 finger. The number of digits above the linking fingers on one side is subtracted from the linking finger on the other side (either 7 – 2 or 8 – 3) to give 5. The number of digits above the linking fingers on each side, 2 and 3, are multiplied to make 6. The answer, as above, is 56.

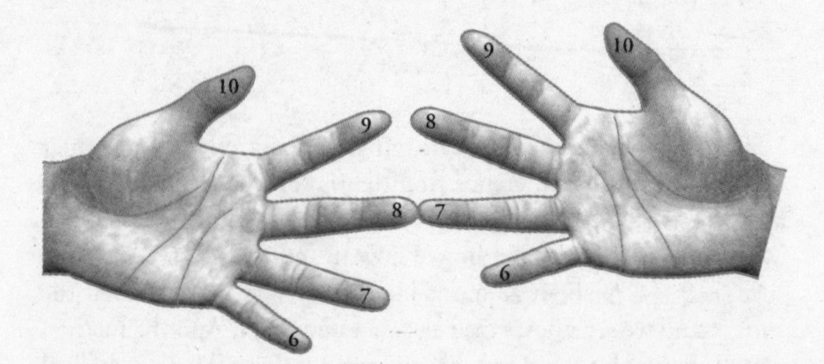

How to calculate 8 × 7 with 'peasant' finger multiplication.

Tirthaji continued his talk by demonstrating that the method also works when multiplying two-digit numbers, this time using the example 77 × 97. He wrote on the board:

77
97

Then, instead of writing the difference of 77 from 10, he wrote the difference of each number from 100. (This is where the second sutra comes in. When subtracting a number from 100, or any larger power of 10, all the digits of the number are subtracted from 9 apart from the last one, which is subtracted from 10, as I showed on p. 127):

77 –23
97 –3

As before, in order to get the first part of the answer there are four options. He chose to show the two diagonal additions: 77 – 3 = 97 – 23 = 74.

77 –23
97 –3
74

The second part is derived by multiplying both digits in the right-hand column: –23 × –3 = 69.

77 –23
97 –3
74 69

The answer is 7469.

Tirthaji then proceeded to an example with three figures: 888 × 997. This time the difference is calculated from 1000.

888 –112
997 –003
885, 336

Diagonal addition gives 885 for the first part, and multiplication of the right column gives 336 for the second, for an answer of 885,336.

'Equations are rendered much easier by these formulae,' Tirthaji commented. The students reacted with spontaneous hearty laughter. Perhaps the chuckles came from the absurdity of an 82-year-old guru in a robe teaching basic arithmetic to some of the smartest maths students in the US. Or perhaps it was in appreciation of the playfulness of Tirthaji's arithmetical tricks. Arabic numerals are a mine of hidden patterns, even at such a simple level as multiplying two single digits together. Tirthaji then continued his talk with techniques for squaring, dividing and algebra. The response seems to have been enthusiastic, judging by a transcription made of the lecture notes: 'Immediately following the end of the demonstration, one student was heard to ask his friend beside him, "What do you think?" His friend's reply, "Fantastic!"'

On his return to India, Tirthaji was summoned to the holy city of Varanasi, where a special council of Hindu elders discussed his breach of protocol in leaving the country. It was decided that his trip was to be the first and last time that a Shankaracharya was allowed to travel abroad, and Tirthaji undertook a purification ritual just in case he had consumed unHindu food while on his travels. Two years later, he passed away.

In my hotel in Puri I met up with two leading proponents of Vedic Mathematics to learn more about it. Kenneth Williams is a 62-year-old former maths teacher from southern Scotland who has written several books about the method. 'It is so beautifully presented and so unified as a system,' he told me. 'When I first found out about it I thought this is the way mathematics ought to be.' Williams was a subdued, gentle man with a priestly forehead, trim salt-and-pepper beard and heavy-lidded blue eyes. With him was the much more talkative Gaurav Tekriwal, a 29-year-old stockbroker from Kolkata, who was wearing a crisp white shirt and Armani shades. Tekriwal is president of the Vedic Maths Forum India, an organization that runs a website, arranges talks and sells DVDs.

Tekriwal had helped me secure an audience with the Shankaracharya, and he and Williams wanted to accompany me. We hailed a motor rickshaw and set off to the Govardhan Math, an auspicious name but one that, unfortunately, has nothing to do with maths. It means monastery, or temple. We passed the seafront and small streets lined with stalls selling food and patterned silk. The Math is a plain brick and concrete building the size of a small country church, surrounded by palm trees and a sand garden planted with basil, aloe vera and mango. In the courtyard is a banyan tree, its trunk decorated with ochre cloth, where Shankara, the eighth-century Hindu sage who founded the order, is believed to have sat and meditated. The only modern touch was a shiny black frontage on the first floor – a bullet-proof façade that was installed to protect the Shankaracharya's room after the Math received Muslim terrorist threats.

The current Shankaracharya of Puri, Nischalananda Sarasvati, inherited the position from the man who inherited it from Tirthaji. He is proud of Tirthaji's mathematical legacy, and has published five books on the Vedic approach to numbers and calculation. On reaching the Math we were ushered into the room the Shankaracharya uses for his audiences. The only pieces of furniture were an antique sofa with deep red upholstery and, immediately in front of it, a low chair with a large seat and wooden back covered in a red shawl: the Shankaracharya's throne. We sat facing it, cross-legged on the floor, and waited for the holy man to arrive.

Sarasvati entered the room, wearing a faded pink robe. His senior disciple stood up to recite some religious verses, and then Sarasvati clasped his hands in prayer, touching the image of Shankara on the back wall. He had blue eyes, a white beard and a light-skinned, bald pate. Sitting down in a half lotus on his throne, he assumed an expression between serene and glum. As the session was about to begin, a man in blue robes dived in front of me, throwing himself prostrate before the throne with outstretched hands. Sighing like an exasperated grandfather, the Shankaracharya nonchalantly shooed him away.

Religious procedure requires the Shankaracharya to speak Hindi, so I used his senior disciple as an interpreter. My first question was 'How is maths linked to spirituality?' After several minutes the reply came back. 'In my opinion, the creation, the standing and the destruction of this whole universe happens in a very mathematical form. We do not differentiate between mathematics and spirituality. We see mathematics as the fountainhead of Indian philosophies.'

Sarasvati then told a story about two kings who met in a forest. The first king told the other that he could count all the leaves on a tree just by looking at it. The second king doubted him and started tearing off the leaves to count them one by one. When he had finished he arrived at the precise number given by the first king. Sarasvati said the story was evidence that the ancient Indians had the ability to count large numbers of objects by looking at them as a whole instead of enumerating them individually. This and many other skills from that era, he added, had been lost. 'All these lost sciences can be regained by the help of serious contemplation, serious meditation and serious effort,' he said. The process of studying the ancient scriptures with the intention of rescuing ancient knowledge, he added, is exactly what Tirthaji had done with mathematics.

During the interview the room filled up with about 20 people, who sat silently as the Shankaracharya spoke. As the session drew to a close, a middle-aged software consultant from Bangalore asked a question about the significance of the number 10^{62}. The number was in the Vedas, he said, so it had to mean something. The Shankaracharya agreed. It was in the Vedas and, yes, it had to mean something. This prompted a discussion about how the Indian government is neglecting the country's heritage, and the Shankaracharya lamented that he spent most of his energy in trying to protect traditional culture so he could not devote more time to maths. This year he had managed only 15 days.

Over breakfast the next day I asked the computer consultant about his interest in the number 10^{62}, and he answered with a lecture on the scientific achievements of ancient India. Thousands of years ago, he said, Indians understood more about the world than what is known today. He mentioned that they flew aeroplanes. When I asked if there was any proof of this, he replied that stone engravings of millennia-old aircraft have been found. Did these planes use the jet engine? No, he said, they were powered using the Earth's magnetic field. They were made from a composite material and flew at a low speed, between 100 and 150kmph. He then became increasingly annoyed by my questions, interpreting my desire for proper scientific explanations as an affront to Indian heritage. Eventually, he refused to speak to me.

While Vedic science is fantastical, occultist and barely credible, Vedic Mathematics stands up to scrutiny, even though the sutras are mostly so vague as to be meaningless and to accept the story of their origin in the Vedas requires the suspension of disbelief. Some of the techniques are so specific as to be nothing more than curiosities – such as a tip for calculating the fraction $\frac{1}{19}$ in decimal. But some are very neat indeed.

Consider the example of 57 × 43 from earlier. The standard method of multiplying these numbers is to write down two intermediary lines, and then add them:

$$\begin{array}{r} 57 \\ \underline{\times\ 43} \\ 0171 \\ \underline{2280} \\ \underline{2451} \end{array}$$

Using the third sutra, *Vertically and Cross-wise*, we can find the answer quite handily as follows.

Step 1: Write the numbers on top of each other:

$$\begin{array}{cc} 5 & 7 \\ & \\ 4 & 3 \end{array}$$

Step 2: Multiply the digits in the right-hand column: 7 × 3 = 21. The final digit of this number is the final digit of the answer. Write it below the right-hand column, and carry the 2.

$$\begin{array}{cc} 5 & 7 \\ & | \\ \underline{4} & \underline{3} \\ & {}_{2}1 \end{array}$$

Step 3: Find the sum of the cross-wise products: $(5 \times 3) + (7 \times 4) = 15 + 28 = 43$. Add the 2 that is carried from the previous step to get 45. The final digit of this number, the 5, is written underneath the left-hand column, with 4 carried.

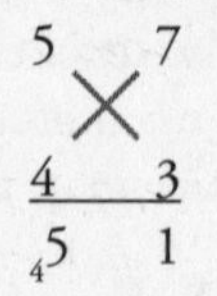

Step 4: Multiply the digits in the left-hand column, $5 \times 4 = 20$. Add the 4 that has been carried to make 24, to give the final answer:

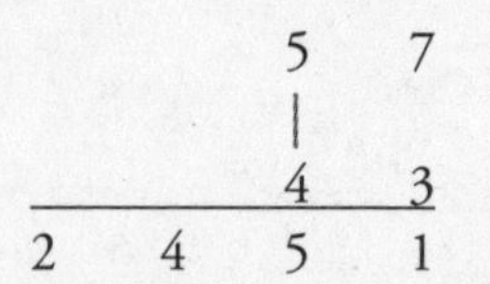

The numbers have been multiplied vertically and cross-wise, as the sutra said on the tin. This method can be generalized to multiplications of numbers of any size. All that changes is that more numbers need to be vertically and cross-multiplied.

For example, 376×852:

3 7 6

8 5 2

Step 1: We start with the right column: $6 \times 2 = 12$

3 7 6

8 5 2

$_{1}2$

Step 2: Then the sum of the cross-products between the units and the tens column, $(7 \times 2) + (6 \times 5) = 44$, plus the 1 carried above. This is 45.

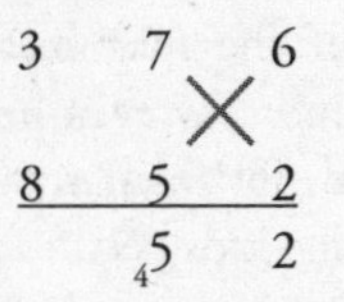

Step 3: Now we move to the cross-products between the units and the hundreds column, and add them to the vertical product of the tens column, (3 × 2) + (8 × 6) + (7 × 5) = 89, plus the 4 carried above. This is 93.

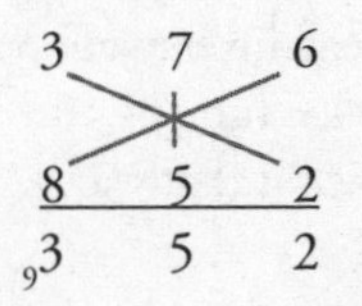

Step 4: Moving leftwards, we now cross-multiply the first two columns: (3 × 5) + (7 × 8) = 71, plus the 9 carried above. This is 80.

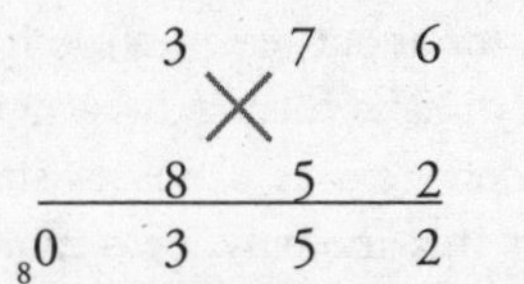

Step 5: Finally, we find the vertical product of the left column, 3 × 8 = 24, plus 8 carried above. This is 32. The final answer: 320,352.

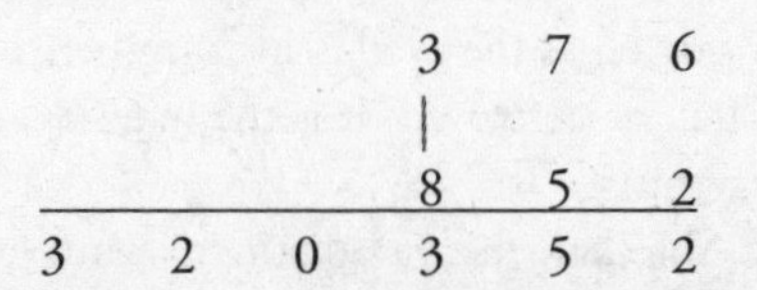

Vertically and Cross-wise, or 'cross-multiplication', is faster, uses less space and is less laborious than long multiplication. Kenneth Williams told me that whenever he explains the Vedic method to school pupils they find it easy to understand. 'They can't believe they weren't taught it before,' he said. Schools favour long multiplication because it spells out every stage of the calculation. *Vertically*

and Cross-wise keeps some of the machinery hidden. Williams thinks this is no bad thing, and may even help less bright pupils. 'We have to steer a path and not insist that kids have to know everything all of the time. Some kids need to know how [multiplication] works. Some don't want to know how it works. They just want to be able to do it.' If a child ends up not being able to multiply because the teacher insists on teaching a general rule that he or she cannot grasp, he said, then the child is not being educated. For the smarter kids, added Williams, Vedic Maths brings arithmetic alive. 'Mathematics is a creative subject. Once you have a variety of methods, children realize you can invent your own and they become inventive too. Maths is a really fun, playful subject and [Vedic Maths] brings out a way to teach it that way.'

My first audience with the Shankaracharya had not covered all intended topics of discussion, so I was granted a second one. At the beginning of the session, the senior disciple had an announcement to make: 'We would like to say something about zero,' he said. The Shankaracharya then spoke for about ten minutes in Hindi in an animated manner, with the disciple then translating: 'The present mathematical system considers zero as a non-existent entity,' he declared. 'We want to rectify this anomaly. Zero cannot be considered a non-existent entity. The same entity cannot be existing in one place and non-existing somewhere else.' The thrust of the Shankaracharya's argument was, I think, the following: people consider the 0 in 10 to exist, but 0 on its own not to exist. This is a contradiction – either something exists or it does not. So zero exists. 'In Vedic literature zero is considered as the everlasting number,' he said. 'Zero cannot be annihilated or destroyed. It is the indestructible base. It is the basis of everything.'

By now I was used to the Shankaracharya's distinctive mix of mathematics and metaphysics. I had given up asking him to clarify certain points since by the time my comments had been translated into Hindi, discussed and then translated back, the answers inevitably added to my confusion. I decided to stop concentrating on the details of his speech, and let the translated words just float over me. I looked at the Shankaracharya closely. He was wearing an orange robe today, tied with a big knot behind his neck, and his forehead had been daubed with beige paint. I wondered what it would be like to live the way he did. I had been told that he sleeps in an unfurnished room, eats the same bland curry every day, and that he has no need or desire for possessions. Indeed, at the beginning of the session a pilgrim had approached him to give him a bowl of fruit, and as soon as he received it, he had given the fruit away to the rest of us. I got a mango, which was by my feet.

Trying to experience the Shankaracharya's wisdom in a different way, I thought of the phrase 'zero is an existent entity' and repeated it like a mantra in my head. I let go. Suddenly I am lost in my thoughts. And it all makes sense. 'Zero is an existent entity' is not just the Shankaracharya's mathematical point of view, but a pithy phrase of self-description. Sitting in front of me is Mr Zero himself, the embodiment of *shunya* in flesh and bone.

It was a moment of clarity, maybe even of enlightenment. Nothing was not nothing in Hindu thought. Nothing was everything. And the monastic, self-abnegating Shankaracharya was a perfect ambassador for this nothingness. I thought about the deep connection between Eastern spirituality and mathematics. Indian philosophy had embraced the concept of nothingness just as Indian maths had embraced the concept of zero. The conceptual leap that led to the invention of zero happened in a culture that accepted the void as the essence of the universe.

The symbol that emerged in ancient India for zero perfectly encapsulated the Shankaracharya's overriding message that mathematics cannot be separated from spirituality. The circle, 0, was chosen because it portrays the cyclical movements of the face of heaven. Zero means nothing, and it means eternity.

Pride in the invention of zero has helped make mathematical excellence an aspect of Indian national identity. Schoolchildren must learn their times tables up to 20, which is twice as high as I was taught growing up in the UK. In previous decades Indians were required to learn their tables up to 30. One of India's top non-Vedic mathematicians, S.G. Dani, attested to this: 'As a child I did have this impression of mathematics being extremely important,' he told me. It was always common for elder people to set children mathematical challenges, and it was greatly appreciated if they got the answers right. 'Irrespective of whether it is useful or not, maths is something that is valued in India by one's peer group and friends.'

Dani is senior professor of mathematics at the Tata Institute of Fundamental Research in Bombay. He has an academic's comb-over frizz, rimmed tortoiseshell glasses and a moustache that frames the length of his upper lip. And he is no fan of Vedic Mathematics; he neither believes that Tirthaji's arithmetical methods can be found in the Vedas nor does he believe it is particularly helpful to say that they do. 'There are many better ways to bringing interest into mathematics than resorting to inputting them into ancient texts,' he said. 'I don't believe that they are making mathematics interesting. The selling point is that these algorithms make you fast, not that it makes it interesting, not that it makes you internalize what is going on. The interest is in the end, not the process.' He is doubtful they do make calculation quicker, since real life does not throw up such perfectly formed problems as finding the decimal breakdown of $\frac{1}{19}$. At the end of the day, he added, the conventional method is more convenient.

So, I was surprised that Dani spoke empathetically of Tirthaji's mission with Vedic Maths. Dani related to Tirthaji on an emotional level. 'The dominant feeling that I had for him is that he had this inferiority complex that he was trying to conquer. As a child I also had this kind of attitude. In India in those days [shortly after Independence] there was a strong feeling that we needed to get back [from the British] what we lost by hook or by crook. It was mostly in terms of artefacts, stuff that the British might have taken away. Because we lost such a lot, I thought we should have the equivalent back of what we lost.

'Vedic Mathematics is a misguided attempt to claim arithmetic back for India.'

Some of the tricks of Vedic Mathematics are so simple that I wondered if I might come across them anywhere else in arithmetical literature. I thought that Fibonacci's *Liber Abaci* would be a good place to start. When I got back to London I found a copy at the library, opened the chapter on multiplication and Fibonacci's first suggested method is none other than ... *Vertically and Cross-wise.* I did some more research and discovered that multiplication using

All from 9 and the last from 10 was a favoured technique in several books from sixteenth-century Europe. (In fact, it has been suggested that both methods might have influenced the adoption of × as the multiplication sign. When × made its first appearance as a notation for multiplication in 1631, books had already been published illustrating the two multiplication methods with large ×s drawn as cross-lines.)

Tirthaji's Vedic Mathematics is, it would seem, at least in part, a rediscovery of some very common Renaissance arithmetical tricks. They may or may not have come from India originally, but whatever their provenance, the charm of Vedic Mathematics for me is the way it encourages a childlike joy in numbers and the patterns and symmetries they hold. Arithmetic is essential in daily life and important to do properly, which is why we are taught it so methodically at school. Yet in our focus on practicalities we have lost sight of quite how amazing the Indian number system is. It was a dramatic advance on all previous counting methods and has not been improved upon in a thousand years. We take the decimal place-value system for granted, without realizing how versatile, elegant and efficient it is.

√2

CHAPTER FOUR

Life of Pi

In the early nineteenth century, news of boy wonder George Parker Bidder, the son of a Devonshire stonemason, reached the ears of Queen Charlotte. She had a question for him:

'From the Land's-end, Cornwall, to Farret's-head, in Scotland, is found by measurement to be 838 miles; how long would a snail be creeping that distance, at the rate of 8 feet per day?'

The exchange and the answer – 553,080 days – is mentioned in a popular book of the time, *A short Account of George Bidder, the celebrated Mental Calculator: with a Variety of the most difficult Questions, Proposed to him at the principal Towns in the Kingdom, and his surprising rapid Answers!* The pages list the child's greatest calculations, including such classics as 'What is the square root of 119,550,669,121?' (345,761, answered in half a minute) and 'How many pounds weight of sugar are there in 232 hogsheads, each weighing 12cwt. 1qr. 22lbs?' (323,408lbs, also answered in half a minute.)

Arabic numerals made doing sums easier for everyone, but an unexpected consequence was the discovery that certain people were blessed with truly astonishing arithmetical skills. Often, these prodigies excelled in no other way than their facility with numbers. One of the earliest-known examples was a Derbyshire farmhand, Jedediah Buxton, who amazed locals with his abilities in multiplication despite being barely able to read. He could, for example, calculate the value of a farthing when doubled 140 times. (The answer is 39 digits long, with 2 shillings 8 pence left over.) In 1754, curiosity about Buxton's talent led to him being invited to visit London, where he was examined by members of the Royal Society. He seems to have had some of the symptoms of high-functioning autism, for when he was taken to see Shakespeare's *Richard III* he was left nonplussed by the experience, although he notified his hosts that the actors had taken 5202 steps and spoken 14,445 words.

In the nineteenth century 'lightning calculators' were international stage stars. Some showed aptitude at an extraordinarily young age. Zerah Colburn, from Vermont, was five when he gave his first public demonstration and eight when he sailed to England with dreams of big-time success. (Colburn was born with hexadactyly, but it is not known if his extra fingers gave him an advantage when learning to count.) A contemporary of Colburn's was the Devonshire lad George Parker Bidder. The two prodigies crossed paths in 1818, when Colburn was 14 and Bidder 12, and the encounter, in a London pub, inevitably led to a maths duel.

Colburn was asked how long it would take a balloon to circumnavigate the globe if the balloon were travelling at 3878 feet per minute and the world were 24,912 miles around. It was a suitably international question for the unofficial title of smartest alec on Earth. But after deliberating for nine minutes, Colburn failed to give an answer. A London newspaper gushed that his opponent, on the other hand, took only two minutes before giving the correct reply, '23 days, 13 hours and 18 minutes, [which] was received with marks of great applause. Many other questions were proposed to the American boy, all of which he refused answering; while young Bidder replied to all.' In his charming autobiography, *A memoir of Zerah Colburn, written by himself*, the American gives a different version of the contest. '[Bidder] displayed great strength and power of mind in the higher branches of arithmetic,' he said, before adding dismissively, 'but he was unable to extract the roots, and find the factors of numbers.' The championship was left undecided. Edinburgh University subsequently offered to take over Bidder's education. He went on to become an important engineer, at first in railroads and eventually supervising construction of London's Victoria Docks. Colburn, on the other hand, returned to America, became a preacher and died aged 35.

The ability to calculate rapidly has no great correlation with mathematical insight or creativity. Only a few great mathematicians have demonstrated lightning-calculator skills, and many mathematicians have surprisingly poor arithmetic. Alexander Craig Aitken was a well-known lightning calculator in the first half of the twentieth century, unusual in that he was also a professor of mathematics at Edinburgh University. In 1954 Aitken gave a lecture to the Society of Engineers in London, in which he explained some of the methods in his repertoire, such as algebraic shortcuts and – crucially – the importance of memory. To prove his point he rattled off the decimal expansion for $\frac{1}{97}$, which repeats only after 96 digits.

Aitken ended his talk with the rueful comment that when he acquired his first desk calculator his abilities began to deteriorate. 'Mental calculators may, like the Tasmanian or the Moriori, be doomed to extinction,' he predicted. 'Therefore ... you may be able to feel an almost anthropological interest in surveying a curious specimen, and some of my auditors here may be able to say in the year 2000 CE, "Yes, I knew one such."'

This was one calculation, however, that Aitken got wrong.

'Neurons! On the ready! Go!'

With an impatient snap and swoosh, the contestants in the multiplication round at the Mental Calculation World Cup turned over their papers. The room at Leipzig University was silent as the 17 men and two women contemplated the first question: 29,513,736 × 92,842,033.

Arithmetic is back in vogue. Thirty years after the first cheap electronic calculators precipitated a widespread demise in mental calculation skills, a backlash is under way. Newspapers offer up daily maths brainteasers, popular computer games with arithmetic puzzles sharpen our minds and – at the high end – lightning calculators compete in regular international tournaments. The Mental Calculation World Cup was founded in 2004 by German computer scientist Ralf Laue, and takes place every two years. It was the inevitable culmination of Laue's two hobbies: mental arithmetic and collecting unusual records (such as the Most Grapes Thrown over a Distance of 15ft and Caught in the Mouth in One Minute, which is 55 of them). The internet helped, enabling him to meet kindred spirits – mental arithmeticians are not, in general, extroverts. The global community of human calculators, or 'mathletes', was well represented in Leipzig, with contestants from countries as diverse as Peru, Iran, Algeria and Australia.

How do you measure calculating skills? Laue adopted the categories already chosen by *Guinness World Records* – the multiplication of two eight-digit numbers, the addition of ten ten-digit numbers, extracting the square root of six-figure numbers to eight significant figures, and finding the day of the week of any date between 1600 and 2100. The latter is known as a calendar calculation, and is a flashback to the golden age of lightning calculation, when performers would ask a member of the audience their birth date and then instantly name the day of the week it fell on.

Regulation, and a spirit of competitiveness, have come at the expense of theatrics. The youngest contestant at the World Cup, an 11-year-old boy from India, performed the 'air abacus' – his hands were jerking wildly around rearranging imaginary beads, while all the others were quiet and still, occasionally jotting down their answers. (The rules say that only the final answer can be written down.) After 8 minutes and 25 seconds, Alberto Coto of Spain stuck his hand up like an excited schoolboy. The 38-year-old had completed ten multiplications of two eight-digit numbers in that time, smashing the world record. It was evidently an awesome accomplishment, yet watching him was as compelling as invigilating an exam.

Conspicuously absent from proceedings in Leipzig, however, was perhaps the world's most famous mathlete, the French student Alexis Lemaire, who prefers another yardstick to measure computational power. In 2007 Lemaire, aged 27, made international headlines when, at the Science Museum in London, he took just 70.2 seconds to calculate the thirteenth root of:

85,877,066,894,718,045,602,549,144,850,158,599,202,771,247,
748,960,878,023,151,390,314,284,284,465,842,798,373,290,
242,826,571,823,153,045,030,300,932,591,615,405,929,429,773,
640,895,967,991,430,381,763,526,613,357,308,674,592,650,724,
521,841,103,664,923,661,204,223

Lemaire's achievement was undoubtedly the more spectacular. The number has 200 digits, which can barely be pronounced in 70.2 seconds. But did his feat mean that, as he claims, he is the greatest lightning calculator of all time? This is a matter of deep controversy in the calculation milieu, mirroring the battle almost 200 years ago between Zerah Colburn and George Bidder, both exceptional at their own type of sum.

The term 'thirteenth root of *a*' refers to the number that when multiplied by itself 13 times equals *a*. Only a fixed amount of numbers when multiplied by themselves 13 times equal a 200-digit number. (It is a large fixed amount. The answer is limited to about 400 trillion possibilities, all 16 digits long and beginning with 2.) Because 13 is prime and considered unlucky, Lemaire's calculation was vested with an extra aura of mystery. In fact, 13 brings with it some advantages. For instance when 2 is multiplied by itself 13 times, the answer ends in 2. When 3 is multiplied by itself 13 times the answer ends in 3. The same is true for 4, 5, 6, 7, 8 and 9. In other words, the last digit of the thirteenth root of a number is the same as the last digit of the original number. We get this number for free, without having to do any calculation.

Lemaire has worked out algorithms, which he has not divulged, to calculate the other 14 digits in the final answer. Purists, possibly unfairly, say that his skill is less a feat of calculation and more one of memorizing huge strings of numbers. And they point out that Lemaire cannot find the thirteenth root of *any* 200-digit number given to him. At the Science Museum in London he was presented with several hundred numbers and allowed to select the one that he would do the calculation for.

Still, Lemaire's performance was more in keeping with the tradition of the stage calculators of old. Audiences want to feel the 'wow', rather than understand the process. By contrast, at the Mental Calculation World Cup, Coto had no choice about the problem to be solved and used no hidden techniques when he multiplied 29,513,736 × 92,842,033. He simply used his 1- to 9-times tables. The fastest way to multiply eight digits by eight digits is using the Vedic sutra *Vertically and Cross-Wise*, which breaks the sum into 64 multiplications of single-digit figures. He managed to get the right answer in an average of less than 51 seconds. Knowing what he was doing made it less dazzling, even though it was obviously still a formidable feat.

As I talked with competitors in Leipzig, I discovered that many of them had fallen in love with speed arithmetic thanks to Wim Klein, a Dutch lightning calculator who was a celebrity in the 1970s. Klein was already a veteran of circuses and music halls when, in 1958, he was given a job by Europe's top physics institute – the European Organization for Nuclear Research (CERN) in Geneva, providing calculations for the physicists. He was probably the last human calculator to have been employed as one. As computers developed, his skills became redundant, and in retirement he returned to showbiz, appearing frequently on TV. (Klein, in fact, was one of the first to promote thirteenth-root calculations.)

A century before Klein another human calculator, Johann Zacharias Dase, was also employed by the scientific establishment to do their sums for them. Dase was born in Hamburg and started performing as a lightning calculator in his teens, when he was taken under the wing of two eminent mathematicians. In the days before electronic or mechanical calculators, scientists relied on logarithm tables to do complicated multiplications and divisions. As I will explain in more detail later, every number has its own logarithm, which can be calculated using a laborious procedure of adding fractions. Dase calculated the natural logarithms of the first 1,005,000 numbers, each to seven decimal places. It took him three years, and he said he enjoyed the task. Then, on the recommendation of the mathematician Carl Friedrich Gauss, Dase embarked on another enormous project: compiling a table of factors of all the numbers between 7,000,000 and 10,000,000. This means he looked at every number in the range and calculated its factors, which are the whole numbers that divide that number. For example, 7,877,433 has only two factors: 3 and 2,625,811. When Dase died, aged 37, he had completed a substantial part of it.

Yet it was another calculation for which Dase is best remembered. When still a teenager he calculated pi to 200 places, a record for the time.

Circles are everywhere in the natural world – you see them in the full moon, in the eyes of humans and animals and in the cross-section of an egg. Tie a dog to a post and the path it patrols when the lead is taut is a circle. The circle is the simplest two-dimensional geometrical shape. An Egyptian farmer counting how much of a crop to plant in a round field, or a Roman mechanic measuring the length of wood for a wheel would have needed to make calculations involving circles.

The ancient civilizations realized that the ratio of a circle's circumference to its diameter was always the same no matter how big or small you made the circle. (The circumference is the distance around a circle, and the diameter is the distance across it.) The ratio is known as pi, or π, and it works out as just over three. So, if you take the diameter of a circle and curve it around the circumference, you will find that it fits just over three times.

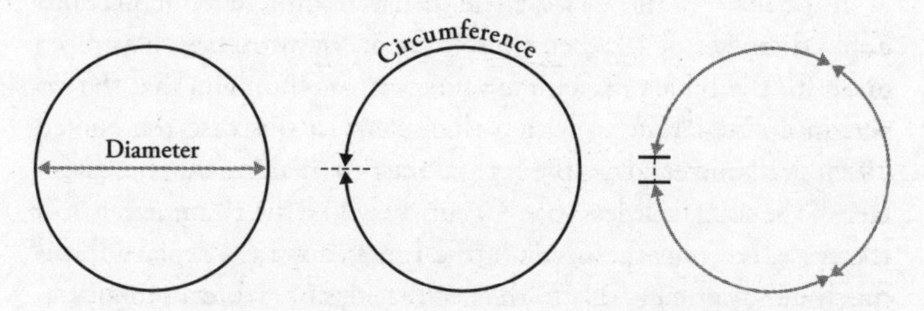

Even though pi is a simple ratio between the basic properties of a circle, the task of finding its exact value has proved to be far from simple. This elusiveness has made pi an object of fascination for thousands of years. It is the only number that is both the name of a song by Kate Bush and a fragrance by Givenchy, whose PR department sent me the following text:

π – Pi
BEYOND INFINITY

Four thousand years have passed and the mystery remains. Although every schoolchild studies π, the familiar symbol still manages to hide an abyss of great complexity.

Why choose π to symbolise the eternal masculine?

It's a matter of signs and directions. If π is the story of the long struggle to achieve the unattainable, it is also a portrait of the fabled conqueror in search of knowledge.

Pi speaks of men, of all men, of their scientific genius, their taste for adventure, their willingness to act, and of their passions to the extreme.

The earliest approximations for pi came from the Babylonians, who used a value of $3\frac{1}{8}$, and the Egyptians, who used $4(\frac{8}{9})^2$, which translate, respectively, into decimals as 3.125 and 3.160. A line in the Bible reveals a situation in which pi is taken as 3: 'Also he made a molten sea of ten cubits from brim to brim, round in compass, and five cubits the height thereof; and a line of thirty cubits did compass it round about' (I Kings 7:23).

If the shape of the sea is a circle with a circumference of 30 cubits and a diameter of 10, then pi is $\frac{30}{10}$, or 3. Many excuses have been given for the Bible's inaccurate value, such as the claim that the sea was in a circular vessel with a thick rim. In this case the quoted 10-cubit diameter covers the sea and the rim (making the true diameter of the sea a little less than 10 cubits), while the circumference of the sea is taken as the inside of the rim. A mystical explanation is much more enticing: due to the peculiarities of Hebrew pronunciation and spelling, the word 'line', or *qwh*, is pronounced *qw*. Totting up the numerological values of the letters gives 111 for *qwh* and 106 for *qw*. Multiplying three by $\frac{111}{106}$ gives 3.1415, which is pi correct to five significant figures.

The first genius whose extreme passion for discovery about pi does justice to the aspirations of Givenchy's aftershave is equally the man who took the most famous bath in the history of science. Archimedes slipped into the tub and noticed that the volume of water he displaced was equal to the volume of his own body under the water. He instantly realized that he could therefore find the volume of any object by submersing it, in particular the crown of the King of Syracuse, and so would be able to ascertain if that piece of royal bling was made of pure gold or not by working out its density. (It wasn't.) As a result, he ran naked into the streets shouting 'Eureka! [I have found it!]', thus displaying – for the citizens of Syracuse, at least – the eternal masculine. Archimedes loved to grapple with problems in the real world, unlike Euclid, who dealt uniquely in abstractions. His many inventions were said to have included a giant catapult and a system of huge mirrors that reflected the sun's rays with such intensity that they set Roman ships ablaze during the Siege of Syracuse. He was also the first person to come up with an apparatus to capture pi.

To do so he first drew a circle, and then he constructed two hexagons – one that he fitted inside the circle, and one that he put outside it, as in the diagrams below. This already tells us that pi must be somewhere between 3 and 3.46, which is determined by calculating the perimeters of the hexagons. If we let the diameter of the circle be 1, then the perimeter of the inner hexagon is 3, which is less than the circumference of the circle, which is pi, which is less than the perimeter of the outer one, which is $2\sqrt{3}$, or 3.46 to two decimal places. (The way that Archimedes calculated this value was by using a method that was essentially a fiddly precursor to trigonometry, and which is too complicated to go into here.)

So, $3 < \text{pi} < 3.46$.

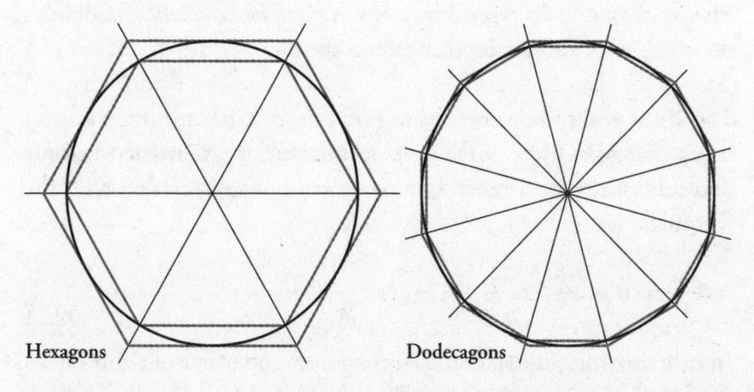

Now if you were to repeat the calculation using two regular polygons with more than six sides, you would get a narrower bound for pi. This is because the more sides that the polygons have, the closer their perimeters are to the circumference, as we can see in the diagram above that uses a 12-sided polygon. The polygons act like walls closing in on pi, squeezing it from above and from below, between narrower and narrower limits. Archimedes started with a hexagon and eventually constructed polygons of 96 sides, allowing him to calculate pi as follows: $3\,\frac{10}{71} < \text{pi} < 3\,\frac{1}{7}$.

This translates into $3.14084 < \text{pi} < 3.14289$, an accuracy of two decimal places.

But pi hunters weren't about to stop there. In order to get closer to the number's true value, all that was required was to create polygons with more sides. Liu Hui, in third-century China, employed a similar method, using the area of a polygon of 3072 sides to pin pi to five decimal places: 3.14159. Two centuries later Tsu Chung-

Chih and his son Tsu Keng-Chih went one digit further, to 3.141592, with a polygon of 12,288 sides.

The Greeks and the Chinese were hampered by cumbersome notation. When mathematicians were eventually able to use Arabic numerals, the record tumbled. In 1596 the Dutch fencing master Ludolph van Ceulen used a souped-up polygon of 60×2^{29} sides to find pi to 20 decimal places. The pamphlet on which he printed his result ended with the words: 'whoever wants to, can come closer', and no one felt the urge as strongly as he did. He went on to calculate pi to 32 and then 35 decimal places – which were engraved on his tombstone. In Germany, *die Ludolphsche Zahl*, Ludolph's number, is still understood as a term for pi.

For 2000 years the only way to pinpoint pi with accuracy was by using polygons. But, in the seventeenth century, Gottfried Leibniz and John Gregory ushered in a new age of pi appreciation with the formula:

$$\frac{\text{Pi}}{4} = 1 - \frac{1}{3} + \frac{1}{5} - \frac{1}{7} + \frac{1}{9} \cdots$$

In other words, a quarter of pi is equal to one minus a third plus a fifth minus a seventh plus a ninth and so on, alternating the addition and subtraction of unit fractions of the odd numbers as they head to infinity. Before this point scientists were only aware of the scattergun randomness of pi's decimal expansion. Yet here was one of the most elegant and uncomplicated equations in maths. Pi, the poster boy of disorder, it turned out, had some kind of order in his DNA.

Leibniz had devised the formula using 'the calculus', a powerful type of mathematics he had discovered, in which a new understanding of infinitesimal amounts was used to calculate areas, curves and gradients. Isaac Newton had also come up with calculus, independently, and the men spent a good deal of time bickering about who had got there first. (For years, Newton was considered to have won the argument, based on the dates of his unpublished manuscripts, but it now appears that a version of calculus was actually first invented in the fourteenth century by the Indian mathematician Madhava.)

The formula that Leibniz found for pi is what is known as an *infinite series*, a sum that goes on and on for ever, and it provides a way to calculate pi. First, we need to multiply both sides of the formula by 4 to get:

$$\text{Pi} = 4 - \frac{4}{3} + \frac{4}{5} - \frac{4}{7} + \frac{4}{9} \cdots$$

Starting with the first term and adding successive terms produces the following progression (converted into decimals):

$$4 \rightarrow 2.667 \rightarrow 3.467 \rightarrow 2.896 \rightarrow 3.340 \rightarrow \ldots$$

The total gets closer and closer to pi in smaller and smaller leaps. Yet this method requires more than 300 terms to get an answer for pi accurate to two decimal places, so it was impracticable for those wanting to use it to find more digits in the decimal expansion.

Eventually, calculus provided other infinite series for pi that were less pretty but more effective for number-crunching. In 1705 the astronomer Abraham Sharp used one to calculate pi to 72 decimal places, smashing van Ceulen's century-old record of 35. While this was quite an achievement, it was also a useless one. There is no practical reason to know pi to 72 digits, or to 35 digits for that matter. Four decimal places is enough for the engineers of precision instruments. Ten decimal places is sufficient to calculate the circumference of the Earth to within a fraction of a centimetre. With 39 decimal places, it is possible to compute the circumference of a circle surrounding the known universe to within an accuracy of a radius of a hydrogen atom. But practicality wasn't the point. Application was not a concern to the pi men of the Enlightenment; digit-hunting was an end in itself, a romantic challenge. A year after Sharp's effort, John Machin reached 100 digits and in 1717 the Frenchman Thomas de Lagny added a further 27. By the turn of the century, the Slovenian Jurij Vega was in the lead with 140.

Zacharias Dase, the German lightning calculator, pushed the record for pi to 200 decimal places in 1844 in an intense two-month burst. Dase used the following series, which looks more convoluted than the pi formula above but is in fact far more user-friendly. This is because, first, it bears down on pi at a respectable rate. An accuracy of two decimal places is reached after the first nine terms. Second, the $\frac{1}{2}$, $\frac{1}{5}$ and $\frac{1}{8}$ that reappear every third term are very convenient to manipulate. If $\frac{1}{5}$ is rewritten as $\frac{2}{10}$ and $\frac{1}{8}$ as $\frac{1}{2} \times \frac{1}{2} \times \frac{1}{2}$, all the multiplications involving those terms can be reduced to combinations of doubling and halving. Dase would have written out a reference table of doubles to help him with his calculations, starting 2, 4, 8, 16, 32, and carried on for as far as he needed – which, since he was calculating pi to 200 places, would be when the final double is 200 digits long. This happens after 667 successive doublings.

Dase used this series:

$$\begin{aligned}\frac{\pi}{4} = \; & \frac{1}{2} - \frac{(1/2)^3}{3} + \frac{(1/2)^5}{5} - \dots \\ & + \frac{1}{5} - \frac{(1/5)^3}{3} + \frac{(1/5)^5}{5} - \dots \\ & + \frac{1}{8} - \frac{(1/8)^3}{3} + \frac{(1/8)^5}{5} - \dots \\ = \; & \frac{1}{2} + \frac{1}{5} + \frac{1}{8} - \left[\frac{(1/2)^3}{3} + \frac{(1/5)^3}{3} + \frac{(1/8)^3}{3}\right] \\ & + \left[\frac{(1/2)^5}{5} + \frac{(1/5)^5}{5} + \frac{(1/8)^5}{5}\right] - \dots\end{aligned}$$

So $\pi = 4\,(0.825 - 0.0449842 + 0.00632 - \dots$

After one term this is 3.3
After two terms this is 3.1200
After three terms this is 3.1452

Dase barely had time to rest on his laurels before the Brits decided to covet his achievement, and within a decade William Rutherford had calculated pi to 440 places. He encouraged his protégé William Shanks, an amateur mathematician who ran a boarding school in County Durham, to go even further. In 1853 he reached 607 digits, and by 1874 he was at 707. His record held for 70 years until D.F. Ferguson, of the Royal Naval College in Chester, found a mistake in Shanks's calculation. He had made an error at the 527th place so all the subsequent numbers were wrong. Ferguson spent the last year of the Second World War calculating pi longhand. One can only assume that he thought the war was already won. By May 1945 he had 530 places, and by July 1946 he reached 620, and no one has ever calculated further using only paper and pen.

Ferguson was the last of the manual digit-hunters and the first of the mechanical. Using a desk calculator he added almost another 200 places in just over a year, so in September 1947 pi was known to 808 decimal places. Computers then changed the race. The first of them to battle pi was the Electronic Numerical Integrator and Computer, or ENIAC, which was built in the final years of the Second World War at the US Army's Ballistic Research Laboratory in Maryland. It was the size of a small house. In September 1949 the ENIAC took 70 hours to calculate pi to 2037 digits – smashing the record by more than a thousand decimal places.

As more and more digits were found in pi, one thing seemed pretty clear: the numbers obeyed no obvious pattern. Yet it was only in 1767 that mathematicians were able to prove that the higgledy-piggledy sequence of digits would never repeat itself. The discovery followed from considering what type of number pi might be.

The most familiar *type* of number is the *natural numbers*. They are the counting numbers starting at one:

1, 2, 3, 4, 5, 6 …

The natural numbers, however, are limited in scope because they expand in only one direction. More useful are the *integers*, which are the natural numbers together with zero and the negatives of the natural numbers:

… –4, –3, –2, –1, 0, 1, 2, 3, 4 …

The integers cover every positive or negative whole number from minus infinity to plus infinity. If there were a hotel with an unlimited amount of floors and an unlimited amount of lower and lower basements, the buttons in the elevator would be the integers.

Another basic type of number is the *fraction*, which are the numbers written $\frac{a}{b}$ when a and b are integers but b is not 0. The top number in a fraction is the *numerator* and the bottom number is the *denominator*. If we have several fractions, the *lowest common denominator* is the lowest number that can be divided by all the denominators without leaving a remainder. So, if we have $\frac{1}{2}$ and $\frac{3}{10}$, the lowest common denominator is 10, since both 2 and 10 divide into 10. What about the lowest common denominator of $\frac{1}{3}$, $\frac{3}{4}$, $\frac{2}{9}$ and $\frac{7}{13}$? In other words, what is the smallest number that 3, 4, 9 and 13 divide into? The answer is surprisingly big: 468! I mention this to make a semantic rather than a mathematical point. The phrase 'lowest common denominator' is often used to describe something basic or unsophisticated. It sounds evocative, but misrepresents the arithmetic. Lowest common denominators can often be big and unconventional: 468 is quite an impressive number! A more arithmetically meaningful phrase for something mainstream and cheap is *highest common factor* – which is the largest number that can be divided into every one of a group of numbers. The highest common factor of 3, 4, 9 and 13, for example, is 1, and you can't get much lower or more unsophisticated than that.

Since fractions are equivalent to ratios between integers, they are also called *rational numbers*, and there is an infinite amount of them. In fact, there is an infinite number of rational numbers between 0 and 1. For example, let's take every fraction where the numerator is 1 and the denominator is a natural number bigger than or equal to 2. This is the set composed of:

$$\frac{1}{2}, \frac{1}{3}, \frac{1}{4}, \frac{1}{5}, \frac{1}{6} \ldots$$

We can go further and prove that there is an infinite number of rational numbers between *any two* rational numbers. Let c and d be any two rational numbers, with c less than d. The point halfway between c and d is a rational number: it is $\frac{(c+d)}{2}$. Call this point e. We can now find a point halfway between c and e. It is $\frac{(c+e)}{2}$. It is rational and also between c and d. We can carry on ad infinitum, always splitting the distance between c and d into smaller and smaller parts. No matter how tiny the distance between c and d is in the first place, there will always be an infinite amount of rational numbers in between them.

Since we can always find an infinite number of rational numbers between any two rational numbers, it might be thought that the rational numbers cover every number. Certainly, this is what Pythagoras had hoped. His metaphysics was based on the belief that the world was made up of numbers and the harmonic proportions between them. The existence of a number that could not be described as a ratio diminished his position, at the very least, if it did not contradict it outright. Yet unfortunately for Pythagoras, there *are* numbers that cannot be expressed in terms of fractions, and – rather embarrassingly for him – it is his own theorem that leads us to one. If you have a square where each side has length 1, then the length of the diagonal is the square root of two, which cannot be written as a fraction. (I have included a proof as an appendix, p. 419.)

Numbers that cannot be written as fractions are called *irrational*. According to legend, their existence was first proved by the Pythagorean disciple Hippasus, which did not endear him to the Brotherhood; he was declared a heretic and was drowned at sea.

When a rational number is written out as a decimal fraction, either it has a finite amount of digits, in the way that $\frac{1}{2}$ can be written 0.5, or the expansion will end up repeating itself, just as $\frac{1}{3}$ is 0.3333… where the 3s go on for ever. Sometimes the recurring loop is more than one digit, as is the case of $\frac{1}{11}$, which is 0.090909… where the digits 09 repeat for ever, or $\frac{1}{19}$, which is 0.0526315789473684210… where 052631578947368421 repeats for ever. By contrast, and this is the crucial point, when a number is irrational its decimal expansion never repeats.

In 1767 the Swiss mathematician Johann Heinrich Lambert proved that pi was indeed irrational. The early pi-men may have hoped after the initial chaos of 3.14159… that the noise would calm and a pattern ensue. Lambert's discovery confirmed that this was impossible. Pi's decimal expansion cavorts towards infinity in a predestined yet apparently indiscriminate way.

Mathematicians interested in the irrationals wanted to categorize them further. In the eighteenth century they started to speculate about a special type of irrational called *transcendental numbers*. These were numbers so mysterious and evasive that finite mathematics could not capture them. The square root of two, $\sqrt{2}$, for example, is irrational but can be described as the solution to the equation $x^2 = 2$. A transcendental number is an irrational that cannot be described by an equation with a finite amount of terms. When the concept of transcendental numbers was first mooted, no one knew if they even existed.

They did exist, although it took about a hundred years before Joseph Liouville, a French mathematician, came up with a few examples. Pi was not among them. Only after another 40 years did the German Ferdinand von Lindemann prove that pi was indeed transcendental. The number existed beyond the realm of finite algebra.

Lindemann's discovery was a milestone for number theory. It also settled, once and for all, what was probably the most celebrated unsolved problem in mathematics: whether or not it was possible to square the circle. In order to explain how it did this, however, I need to introduce the formula that says that the area of a circle is πr^2, where r is the radius. (The radius is the distance from the centre to the side, or half the diameter.) A visual proof of why this is true is an instance where a pie is the best metaphor for pi. Imagine you have two same-sized circular pies, a white one and a grey one, as below in A. The circumference of each pie is pi times the diameter, or pi times twice the radius, or $2\pi r$. When sliced into equal segments the pieces can be rearranged, as in B with quarter segments, or as in C with ten segments. In both cases the length of the side remains $2\pi r$. If we keep on slicing smaller and smaller segments, then the shape would eventually become a rectangle, as in D, with sides r and $2\pi r$. The area of the rectangle – which is the area of the two pies – is therefore $2\pi r^2$, so the area of one pie is πr^2.

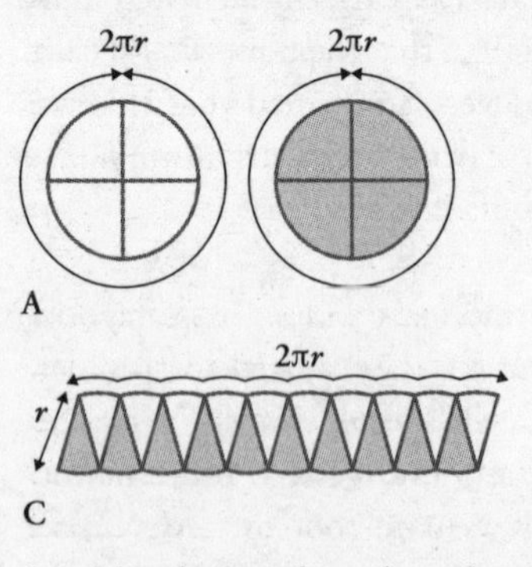

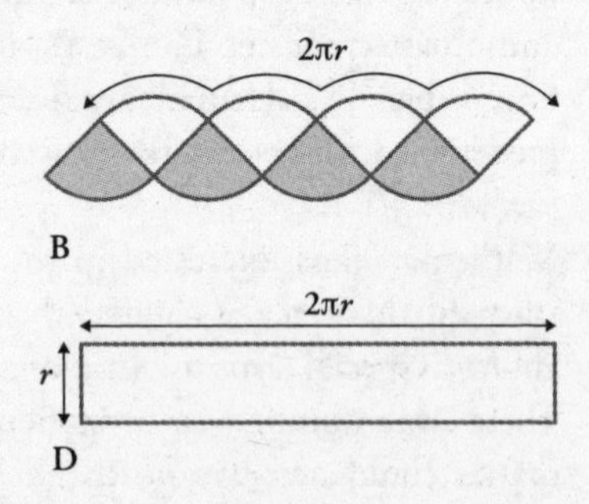

Proof that the area of a circle = πr^2.

To square a circle we must construct (using just a compass and straightedge) a square that has the same area as a given circle. Now we know that a line with length r is the radius of a circle with area πr^2, and we also know that a square with area πr^2 must have a side length $r\sqrt{\pi}$ (since $(r\sqrt{\pi})^2 = r^2(\sqrt{\pi})^2 = r^2\pi = \pi r^2$). So squaring the circle can be reduced to the challenge of constructing the length $r\sqrt{\pi}$ from the length r. Or, taking r as 1 for convenience, the length $\sqrt{\pi}$ from 1.

Using coordinate geometry, which I will cover later, it is possible to express the process of the construction of a line algebraically, as a finite equation. It can be shown that so long as x is the solution to a finite equation, then starting with a line of length 1 we can construct a line with length x. But if x is not the solution to a finite equation – in other words, if x is transcendental – it is impossible to construct a line with length x. Now, the fact that π is transcendental means that $\sqrt{\pi}$ is also transcendental. (You're going to have to trust me on that.) So it is impossible to construct the length $\sqrt{\pi}$. The transcendence of pi proves that the circle cannot be squared.

Lindemann's proof of the transcendence of pi dashed what had been the dream, for thousands of years, of countless mathematicians. Perhaps the most eminent figure ever to have declared that he had squared the circle was Thomas Hobbes, the seventeenth-century English thinker whose book *Leviathan* founded political philosophy. Having become a keen amateur geometer in later life, Hobbes published his solution when he was 67. Even though circle-squaring was still an open question at the time, his proof was received with bemusement by the scientific community. John Wallis, professor at Oxford and the finest British mathematician before Isaac Newton, exposed Hobbes's errors in a pamphlet, thus setting in motion one of the most entertaining – and pointless – feuds in the history of British intellectual life. Hobbes replied to Wallis's comments with an addendum to his book entitled *Six Lessons to the Professors of Mathematics*. Wallis countered with *Due Correction for Mr Hobbes in School Discipline for not saying his Lessons right*. Hobbes followed this with *Marks of the Absurd Geometry, Rural Language, Scottish Church Politics and Barbarisms of John Wallis*. This led to Wallis's *Hobbiani Puncti Dispunctio! or the Undoing of Mr Hobbes's Points*. The quarrelling lasted almost a quarter of a century, until Hobbes's death in 1679. Wallis rather enjoyed the hostilities, since it was a way of casting aspersions on Hobbes's political and religious views, which he despised. And, of course, he was right. In many disputes truths are shared between both sides. Not in Hobbes vs Wallis. Hobbes could not square the circle because it is impossible to do so.

Proof that you cannot square the circle has not put people off trying. In 1897 the Indiana state legislature famously considered a bill containing a proof of squaring the circle by E.J. Goodwin, a country doctor, who offered it 'as a gift to the State of Indiana'. He was, of course, misguided. Since Ferdinand von Lindemann in 1882, the phrase circle-squarer has been mathspeak for crank.

Pi's enigmatic attributes were, in the eighteenth and nineteenth centuries, revealed not only to be at the heart of ancient geometrical problems but also deeply rooted in new fields of science that were not always obviously related to circles. 'This mysterious 3.141592... which comes in at every door and window, and down every chimney,' wrote the British mathematician Augustus De Morgan. For example, the time it takes a pendulum to swing is dependent on pi. The distribution of deaths in a population is a function of pi. If you toss a coin $2n$ times, the probability when n is very large of getting exactly 50 per cent heads and 50 per cent tails is $\frac{1}{\sqrt{(n\pi)}}$.

The man whose name is most associated with offbeat occurrences of pi was the French polymath Georges-Louis Leclerc, the Comte de Buffon (1707–88). Of Buffon's many colourful scientific endeavours, perhaps his most ambitious was the construction of a working version of Archimedes' weapon of mirrors, with which he was said to have set fire to ships. Buffon's contraption was made up of 168 flat mirrors, each six by eight inches, and it was able to ignite a wooden plank at a distance of 150 feet, a good effort, though on a different scale from setting a Roman fleet ablaze.

With regard to pi, Buffon is remembered for having devised an equation that led to a new method for calculating pi, though Buffon did not himself make the connection. Buffon arrived at his equation by studying an eighteenth-century gambling game called 'clean tile', in which you throw a coin on to a tiled surface and bet on whether it will touch the cracks between tiles or rest cleanly. Buffon came up with the following alternative scenario: imagine that a floor is marked with parallel lines spaced evenly apart and that a needle is thrown on it. He then correctly calculated that if the length of the needle is l and the distance between lines is d, then the following equation holds:

$$\text{Probability of the needle touching the line} = \frac{2l}{\pi d}$$

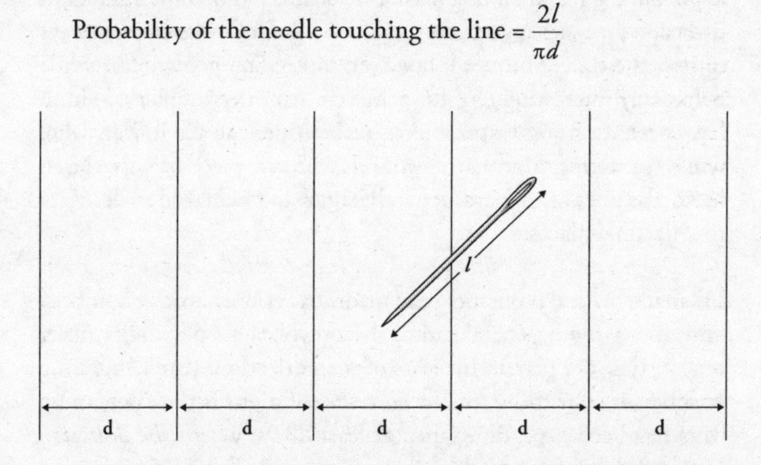

A few years after Buffon died, Pierre Simon Laplace realized that this equation could be used to estimate a value for pi. If you throw lots and lots of needles on the floor, then the ratio of the number of times that the needle hits the line to the total number of throws will be approximately equal to the mathematical probability of the needle touching the line. In other words, after many throws

$$\frac{\text{number of times needle touches a line}}{\text{number of total throws}} \approx \frac{2l}{\pi d}$$

or:

$$\pi \approx \frac{2l\,(\text{number of total throws})}{d\,(\text{number of times needle touches a line})}$$

(The symbol ≈ means 'is approximately equal to'.)

Even though Laplace was the first to write about how pi could be estimated this way, his work followed from Buffon's equation, so Buffon is the person remembered for it. His achievement put him in esteemed company as a member of the club of mathematicians, including Archimedes and Leibniz, who each found a new way to calculate pi.

The more throws of the needle that are taken, the better the approximation, and aiming needles at boards has become a standard diversion for mathematicians unable to think of more creative ways to pass the time. You need, however, to keep on going a fair while before any interesting result is achieved. An early adopter is said to have been a certain Captain Fox in the American Civil War, who, while recovering from battle wounds, threw a piece of wire eleven hundred times on a board of parallel lines and managed to derive pi to 2 decimal places.

Pi's mathematical properties have made it a celebrity among numbers, and also a more general cultural icon. Because pi's digits never repeat, they are perfect for feats of memorization. If remembering numbers is your thing, the *ne plus ultra* of digits is the digits in pi. This has been a pastime since at least 1838, when *The Scotsman* reported that a 12-year-old Dutch boy recited all the 155 digits that were known at the time to an audience of scientists and royals. Akira Haraguchi, a 60-year-old retired engineer, holds the current world record. In 2006 he was filmed in a public hall near Tokyo reciting pi to 100,000 decimal places. The performance took him 16 hours and 28 minutes, including five-minute breaks every two hours to eat rice balls. He explained to a journalist that pi symbolized life since its digits never repeated and followed no pattern. Memorizing pi, he added, was 'the religion of the universe'.

Pi memorization gets a little dull, but pi memorization while juggling, now there's a competitive sport! The record is held by Mats Bergsten, an actuary in Sweden in his late fifties, who has recited 9778 digits while juggling with three balls. He told me, however, that he is proudest of his achievements in the 'Everest test', in which the first 10,000 digits of pi's expansion are divided into 2000 groups of five, beginning with 14159. In the test 50 groups are randomly read out, and the contestant has to say from memory which five numbers precede and succeed each of them. Mats Bergsten is one of only four people in the world who can do this with no errors, and his time, 17min 39secs, is the fastest. It is much more mentally straining, he told me, to recall 10,000 digits randomly than merely remembering them in order.

When Akira Haraguchi recited 100,000 digits of pi by heart, he used a mnemonic technique, assigning syllables to each number from 0 to 9 and then translating pi's decimals into words, which in turn formed sentences. The first fifteen digits sounded like: 'the wife and children have gone abroad; the husband is not scared.' Using words to remember the digits in pi this way is used by schoolchildren in cultures all over the world, but usually this is done not by assigning syllables, but by creating a phrase in which the number of letters in each word represents each consecutive digit in the decimal expansion of pi. A well-known English one is credited to the astrophysicist Sir James Jeans: *How I need a drink, alcoholic in nature, after the heavy lectures involving quantum mechanics. All of thy geometry, Herr Planck, is fairly hard.* 'How' has 3 letters, 'I' has 1, 'need' has 4, and so on.

Among numbers, only pi has inspired this type of fandom. No one wants to memorize the square root of two, which is just as challenging. Pi is also the only number to have inspired its own literary subgenre. Constrained writing is a technique in which some condition is adopted that imposes a pattern or forbids certain things in the text. Entire poems – or 'piems' – have been written under the constraint that the number of letters per word is determined by pi, usually with the convention that a 0 in the expansion requires a ten-letter word. The most ambitious piem is the *Cadaeic Cadenza* by Mike Keith, which follows pi for 3835 digits. It begins as a pastiche of Edgar Allan Poe:

> One; A poem
> A Raven
> Midnights so dreary, tired and weary,
>   Silently pondering volumes extolling all by-now obsolete lore.
> During my rather long nap – the weirdest tap!
>   An ominous vibrating sound disturbing my chamber's antedoor.
> 'This,' I whispered quietly, 'I ignore.'

Keith says that writing with a difficult constraint is an exercise both in discipline and discovery. Since the digits in pi are random, the constraint is, he said, 'like bringing order out of chaos'. When I asked him 'Why pi?' he replied that pi was 'a metaphor for all things infinite, or inscrutable, or unpredictable, or full of endless wonder'.

Pi has gone by this name only since 1706, when the Welshman William Jones introduced the symbol π in his book, the snappily titled *A New Introduction to the Mathematics, for the Use of some Friends who have neither Leisure, Convenience, nor, perhaps, Patience, to search into so many different Authors, and turn over so many tedious Volumes, as is unavoidably required to make but tolerable progress in the Mathematics*. The Greek letter, which was probably an abbreviation for the word periphery, did not immediately catch on, however, becoming standard notation for pi only 30 years later when Leonhard Euler adopted it.

Euler was the most prolific mathematician of all time (he published 886 books), and he is possibly the one who contributed most to an understanding of pi. It was his improved formulae for pi that enabled the eighteenth- and nineteenth-century digit-hunters to peel back more and more decimal places. In the beginning of the twentieth century the Indian mathematician Srinivasa Ramanujan devised many more Euler-style infinite series for pi.

Ramanujan was a largely self-taught mathematician who worked as a clerk in Madras before writing a letter to Cambridge university professor G.H. Hardy. Hardy was flabbergasted to see that Ramanujan had rediscovered results that had taken centuries to achieve, and invited him to England, where the men collaborated before Ramanujan died aged 32. His work showed an extraordinary intuition about the properties of numbers, including pi, and his most famous formula is the following:

$$\frac{1}{\pi} = \frac{2\sqrt{2}}{9801} \sum_{n=0}^{\infty} \frac{(4n)!(1103 + 26390n)}{(n!)^4 396^{4n}}$$

The $\sum_{n=0}^{\infty}$ symbol indicates a series of values all added up, starting with the value when n equals zero, added to the value when n equals one, and so on to infinity. Even without understanding the notation, however, one can appreciate the drama of such an equation. The Ramanujan formula races towards pi with remarkable speed. From the very start, when n is 0 the formula has one term and gives a value of pi accurate to six decimal places. For each increase in the value of n, the formula adds roughly eight new digits to pi. It is an industrial-strength pi-making machine.

Inspired by Ramanujan, in the 1980s the Ukrainian-born mathematicians Gregory and David Chudnovsky devised an even more ferocious formula. Each new term adds roughly 15 digits.

$$\frac{1}{\pi} = \sum_{n=0}^{\infty} (-1)^n \times \frac{(6n)!}{(3n)!\ n!^3} \times \frac{163096908 + 6541681608n}{(262537412640768000)^{n+\frac{1}{2}}}$$

The first time I saw the Chudnovsky formula I was standing on it. Gregory and David are brothers and they share an office at the Polytechnic University in Brooklyn. It consists of an open-plan space with a sofa in the corner, a couple of chairs and a blue floor decorated with dozens of formulae for pi. 'We wanted to put something on the floor and what else can you put on the floor other than stuff which relates to mathematics?' explained Gregory.

In fact, the pi floor pattern was their second choice. The original plan was to lay down a giant reproduction of *Melencolia I* by Albrecht Dürer (reproduced on p. 218). The sixteenth-century woodcut is beloved of mathematicians since it is full of playful references to numbers, geometry and perspective.

'One night, when there was nothing on the surface, we printed 2000 or more pages of [*Melencolia I*] and we laid it on the floor,' said David. 'But if you walked around it, you wanted to throw up! Because your point of view changes extremely abruptly.' David began to study the floors of the cathedrals and castles of Europe in order to work out how he could decorate the office without inducing nausea in anyone walking through. 'I discovered they are mostly laid out in a –'

'Simple geometric style,' interrupted Gregory.

'Black, white, black, white squares …' said David.

'You see, if you really have a complex picture and you try to walk on it, the angle changes so abruptly that your eyes don't like it,' added Gregory. 'So the only way you can do something like that is to –'

'Hang from the ceiling!' David shouted in my left ear, and both men lost themselves in guffaws.

Talking to the Chudnovskys was like wearing stereo headphones with an erratically alternating connection to both ears. They sat me on their sofa and sat on either side of me. Constantly interrupting each other, they finished each other's sentences, speaking in a highly melodic English thick with Slavic tones. The brothers were born in Kiev, when it was in the Soviet Republic of Ukraine, although they have lived in the United States since the late 1970s and are American citizens. They have collaborated on so many papers and books together that they encourage you to think of them as one mathematician, not two.

For all their genetic, conversational and professional homogeneity, however, the men look very different. This is mostly because Gregory, who is 56, suffers from *myasthenia gravis*, an auto-immune disorder of the muscles. He is so thin and frail that he spends most of his time lying down. I never saw him get up off the sofa. Still, the energy his limbs lacked was compensated for by a brilliantly expressive face that burst into life as soon as he talked about maths. He has pointed features, large brown eyes, a white beard and wispy unkempt hair. David, who has blue eyes, is five years older, rounder of body and fuller of face. He was clean-shaven and his short hair was hidden under an olive-green baseball cap.

The Chudnovskys are arguably the mathematicians who have done the most to popularize pi in recent years. In the early 1990s they built a supercomputer in Gregory's Manhattan apartment out of mail-order parts that, using their own formula, calculated the number to more than two billion decimal places – a record at the time.

This amazing achievement was chronicled in a *New Yorker* article, which in turn inspired the 1998 film *Pi*. The main protagonist was an unruly-haired maths genius looking for hidden patterns in stock-market data on a homemade supercomputer. I was curious to discover if the Chudnovskys had seen the film, which garnered favourable reviews and has become a reference for low-budget, black-and-white psycho-mathematical thrillers. 'No, no, we haven't seen it,' said Gregory.

'You have to realize, usually movie-makers repeat their internal state,' added David sarcastically.

I tell them I thought they might have been flattered by the attention.

'No, no,' grinned Gregory.

'Let me tell you another thing,' David cut in. 'Two years ago I came back from France. A couple of days before I left, there was a huge book fair. I stopped at a stand where there was a book that had a detective story on it. It was written by an engineer. It was a murder mystery, you know. A lot of dead bodies, mostly mistress in hotel, and the source that determined everything he did was pi.'

Gregory was smiling from ear to ear and said under his breath: 'OK, I am not going to read this book, *zat's* for sure.'

David carried on: 'So I talked to the guy. He is a very educated man.' He paused, shrugged his shoulders and raised his pitch by an octave: 'As I say, I bear no responsibility!'

David said that he was taken aback the first time that he saw billboards advertising the Givenchy perfume. 'All the way down the street was pi ... pi ... pi ...' He was wailing now: '*Pi ... pi ... pi!* Do I bear any responsibility?'

Gregory glanced at me and said: 'For some reason, the general public is fascinated with this stuff. They get kind of the wrong inference.' There are many professional mathematicians, he said, who study pi. He added wryly: 'Usually these people are not allowed to see the light of day.'

In the 1950s and 1960s advances in computer technology were reflected in the number of new digits found in pi. By the end of the 1970s the record had been broken nine times and stood at just over a million decimal places. In the 1980s, however, a combination of even faster computers and brand-new algorithms led to a frenetic new era of digit-hunting. Yasumasa Kanada, a young computer scientist at Tokyo University, was first off the block in what became a two-way pi race between Japan and the United States. In 1981 he used an NEC computer to calculate pi to 2 million digits in 137 hours. Three years later he was up to 16 million. William Gosper, a mathematician in California, then nudged into the lead with 17.5 million, before David H. Bailey, at NASA, bettered him with 29 million. In 1986 Kanada overtook them both with 33 million and broke his own record three times in the next two years to reach 201 million with a new machine, the S-820, which did the calculation in just under six hours.

Away from the digit-hunting spotlight, the Chudnovskys were also beavering away on pi. Using a new method of communication called the internet, Gregory connected the computer at his bedside to two IBM supercomputers at different sites in the US. The brothers then devised a program to calculate pi based on the new superfast pi formula they had discovered. They were allowed on to the computers only when no one else was using them, at nights and weekends.

'It was a great thing,' remembered Gregory nostalgically. In those days there was no computer capacity to store the numbers that the brothers were calculating. 'They kept the pi on magnetic tape,' he said.

'Mini-tape. And you had to call the guy and ask ...' added David.

'And say tape number such and such,' continued Gregory. 'And sometimes if somebody else is more important your tapes are dismounted in the middle of computation.' His eyes rolled as if to throw his hands in the air.

Despite the obstacles, the Chudnovskys kept on going, pushing beyond a billion digits. Kanada then nudged ahead of them briefly, before the Chudnovskys retook the lead with 1.13 billion. David and Gregory then decided that if they were serious about calculating pi they needed their own machine.

The Chudnovskian supercomputer lived in a room in Gregory's apartment. Made up of processors linked by cables, the whole thing cost, according to their estimates, about $70,000. It was a steal, compared to the millions of dollars that would have bought a machine of similar capacity; although it came with its own complications to their lifestyles. The computer, which they called *m zero*, was switched on at all times, just in case switching it off was irreversible, and it needed 25 fans in the room to keep it cool. The brothers were careful not to switch on too many lights in the apartment just in case the added demand blew the wiring.

In 1991 David and Gregory's homemade contraption calculated pi to more than two billion places. Then they got distracted by other problems. By 1995 Kanada was ahead once again, and he reached 1.2 trillion digits in 2002, a record that lasted only until 2008, when compatriots at the University of Tsukuba revealed 2.6 trillion. In December 2009 the Frenchman Fabrice Bellard claimed a new record using the Chudnovsky formula: almost 2.7 trillion places. The calculation had taken 131 days on his desktop PC.

If you wrote a trillion digits in small type, the distance would cover from here to the sun. If you put 5000 digits on a page (which is very small type) and stacked the pages on top of each other, the pi in the sky would be 10km high. What is the point of calculating pi to such absurd lengths? One reason is very human: records exist to be broken.

But there is another, more important motivation. Finding new digits in pi is ideal for testing the processing capacity and reliability of computers. 'I have no interest as a hobby for extending the known value of pi itself,' Kanada once said. 'I have a major interest for improving the performance of the computation.' Pi calculation is now essential for quality-testing supercomputers because it is a 'high-duty job which requires large main memory, operates huge number-crunching and gives [an] easy [way] to check [the] correct answer. Mathematical constants like the square root of two, *e** [and] gamma are some of the candidates, but pi is most effective.'

The story of pi has wonderful circularity. It is the simplest and most ancient ratio in maths, which has been reinvented as a massively important tool on the frontline of computer technology.

In fact, the Chudnovskys' interest in pi came primarily from their desire to build supercomputers, a passion that still burns brightly. The brothers are currently designing a chip that they claim will be the fastest in the world, only 2.7cm wide but containing 160,000 smaller chips and 1.75km of wire.

On discussion of their new chip, Gregory became very high-spirited: 'Computers double their power every 18 months, not because they are faster but because they can pack more stuff in. But there is a catch,' he said. The mathematical challenge was how to partition the smaller pieces so that they can talk to each other in the most efficient way. His laptop showed the chip's circuitry. 'I'd say the problem with this chip is that it is a capitalist chip!' he exclaimed. 'The problem is that most of the stuff here is not doing anything. There are not too much proles here.' He pointed at one section. 'This is just management of the stores inside the chip,' he lamented. 'The majority of these guys just do warehousing and accounting. This is awful! Where is the manufacturing sector?'

---

* The mathematical constant *e* is an irrational number beginning 2.718281828, which Gregory Chudnovsky calls 'twice Tolstoy' since the Russian novelist was born in 1828. It has no relation to Einstein's equation $E=mc^2$, where E means energy.

In Carl Sagan's bestselling book *Contact*, an extraterrestrial informs a woman on Earth that after a certain amount of digits the randomness in pi stops and there is a message written in 0s and 1s. This message occurs after 10^{20} decimal places – which is the number described by 1 followed by 20 zeros. Since we currently know pi to 'only' 2.7 trillion places (27 followed by 11 zeros), we have a little way to go just to check that he was making it up. Actually, we have further to go since the message is apparently written in base 11.

The idea that there is a pattern in pi is an exhilarating one. Mathematicians have been looking for signs of order in the decimal expansion of pi for as long as there have been decimal expansions. The irrationality of pi means that the numbers keep on spewing out with no repeated pattern, but this does not eliminate the possibility of patches of order – such as a message written in 0s and 1s. So far, however, no one has found anything significant. The number, does, though, have its quirks. The first 0 comes in position 32, which is much later than expected if the digits are randomly distributed. The first time a digit is repeated six times consecutively is 999999 at the 762nd decimal place. The likelihood of six 9s occurring so early if they occur randomly is less than 0.1 per cent. This sequence is known as the Feynman point, since physicist Richard Feynman once remarked that he wanted to memorize pi until that point and finish by saying 'nine, nine, nine, nine, nine, nine, and so on'. The next time pi throws up six consecutive identical digits happens in position 193,034 and they are also 9s. Is this a message from beyond, and if so what is it saying?

A number is considered *normal* if each of the digits 0 to 9 occurs equally often in the number's decimal expansion. Is pi normal? Kanada looked at the first 200 billion digits of pi and found that the digits occurred with the following frequencies:

0	20,000,030,841	5	19,999,917,053
1	19,999,914,711	6	19,999,881,515
2	20,000,136,978	7	19,999,967,594
3	20,000,069,393	8	20,000,291,044
4	19,999,921,691	9	19,999,869,180

Only the digit 8 seems a little overabundant, yet the difference is statistically insignificant. It would appear that pi is normal, yet no one has been able to prove it. And neither has anyone been able to prove that such a proof is impossible. There is a chance, therefore, that pi is *not* normal. Maybe after 10^{20} there really are just 0s and 1s?

A different, but related issue, is the positioning of the numbers. Are they distributed randomly? Stan Wagon analysed the first ten million digits of pi with a 'poker test': take five consecutive digits and consider them as if they are a poker hand.

Type of hand	Actual occurrence	Expected occurrence
All digits different	604,976	604,800
One pair, three different	1,007,151	1,008,000
Two pairs	216,520	216,000
Three of a kind	144,375	144,000
Full house	17,891	18,000
Four of a kind	8887	9000
Five of a kind	200	200

The right column is how many times we would expect to see the poker hands if pi was normal and each decimal place had an equal chance of being occupied by any digit. The results are well within the boundaries of what we would expect. Each pattern of numbers seems to appear with the frequency it would had each decimal place been randomly generated.

There are websites that can find you the first occurrence in pi of the numbers of your birthday. The first time the sequence 0123456789 occurs is at the 17,387,594,880th place – which was discovered only when Kanada got that far in 1997.

I asked Gregory if he ever believed an order would be found in pi. 'There is no order,' he replied dismissively. 'And if there would be an order, it would be weird and not right. So there is no point wasting the time.'

Instead of focusing on patterns in pi, some see its very randomness as a tremendous expression of mathematical beauty. Pi is predetermined but it seems to mimic randomness extraordinarily well.

'It is a very good random number,' agreed Gregory.

Shortly after the Chudnovskys started computing pi they got a call from the United States government. David squealed an impersonation of the voice at the end of the line: 'Could you please send pi?'

Random numbers are needed in industry and commerce. For example, just say a market-research company needs to poll a representative sample of a thousand people from a population of a million. The company will use a random-number generator to select the sample group. The better the generator is at providing random numbers, the more representative the sample will be – and the poll will be more accurate. Likewise, streams of random numbers are needed to simulate unpredictable scenarios when testing computer models. The more random the numbers, the more robust the testing will be. In fact, projects can fail if the random numbers used to test them are not random enough. 'You are only as good as your random number,' remarked David. 'If you use terrible random numbers, you will end up in a terrible condition,' concluded Gregory. Of all sets of available random numbers, the decimal expansion of pi is the best.

Yet there is a philosophical paradox. Pi is self-evidently not random. Its digits may behave like random digits, but they are fixed. For example, if the digits in pi were truly random then there would be only a 10 per cent chance that the first digit after the decimal point is a 1, yet we know that it is 1 with absolute certainty. Pi exhibits randomness non-randomly – which is fascinating, and weird.

Pi is a mathematical concept that has been studied for thousands of years, yet it holds many secrets. There have been no great advances in understanding its nature since transcendence was proved almost a century and a half ago.

'We actually don't know about most of this stuff,' said Gregory.

I asked if there would ever be a new advance in understanding pi.

'Of course, of course,' said Gregory, 'There are always advances. Mathematics moves forward.'

'It will be more miraculous but it won't be nice,' David said.

Nineteen sixty-eight was a year of counter-cultural uprisings around the world, and Britain was not immune to such generational upheaval. In May the Treasury announced the introduction of a revolutionary new coin.

The 50p piece was designed to replace the old ten-shilling note as part of the switch from imperial to decimal currency. Yet what set the coin apart was not its denomination, but its unorthodox shape.

'It's no ordinary coin,' exhorted the *Daily Mirror*. 'Why, the Decimal Currency Board even go so far as to call it a "multilateral-curve heptagon".' Never before had a country introduced a seven-sided coin. And never before had a nation been so outraged over the aesthetics of a geometric shape. Leading the foment was retired army colonel Essex Moorcroft, of Rosset, Derbyshire, who formed the Anti-Heptagonists. 'We take our motto from Cromwell's heartfelt cry, "Take away this bauble". I have founded the society because I believe our Queen is insulted by this heptagonal monstrosity,' he said. 'It is an ugly coin and an insult to our Sovereign, whose image it bears.'

Nonetheless, the 50p entered circulation in October 1969, and Colonel Moorcroft did not take to the barricades. Indeed, by January 1970 *The Times* reported that 'the curvaceous heptagon seems to have won for herself some affection'. Today the 50p piece is considered a distinctive and cherished part of British heritage. When a 20p piece was introduced in 1982, it too was heptagonal.

The 50p and 20p pieces are, in fact, design classics. Their seven-sided shape means that they are easily distinguished from circular coins, helping in particular the blind and partially sighted. They are also the most thought-provoking coins in circulation. The circle is not the only interesting round shape in mathematics.

A circle can be defined as a curve for which every point is equidistant from a fixed point, the centre. This property has many practical applications. The wheel – generally trumpeted as humanity's first great invention – is the most obvious. An axle attached to the centre of a wheel will stay at a fixed point above the ground when the wheel rotates smoothly along a surface, which is why carts, cars and trains run smoothly without bobbing up and down.

For the transportation of very heavy loads, however, an axle might not take the weight. One alternative is to use rollers. A roller is a long tube with a circular cross-section, laid out on the ground. If a heavy object with a flat base (such as a giant cuboidal piece of stone intended for a pyramid) is put on several rollers, then it can be pushed smoothly over the rollers, with new rollers being put in front as it inches forward.

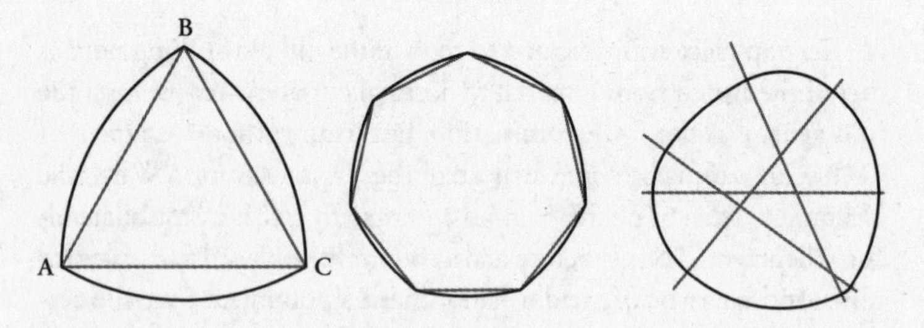

Curves of constant width include the Reuleaux triangle (left) and the multilateral-curve heptagon, better known as the 50p piece (centre).

The crucial feature of a roller is that the distance between the ground and the top of the roller is always the same. This is obviously the case with a circular cross-section, since the width of a circle (the diameter) is always the same.

Do all rollers have to have a circular cross-section? Are there any other shapes that will work? It may appear counter-intuitive, but there are, in fact, a variety of shapes that will make perfect rollers. An example is the 50p piece.

If you welded coins together to form rollers that had a 50p piece as a cross-section and then placed this book on these rollers, the book would not wobble up and down as you pushed it along. It would ride as smoothly as if on circular cylinders.

The reason this happens is because the 50p piece is a *curve of constant width*. Wherever you measure it around its perimeter, the 50p has the same breadth. So, when a 50p rolls along the floor, the distance from the floor to the top of the coin is always equal. A book lying atop a set of 50p rollers therefore stays at the same height.

Surprisingly, there are many, many curves of constant width. The simplest is the Reuleaux triangle. It is constructed from an equilateral triangle by putting the point of a compass at each vertex and drawing the arc that goes between the other two vertices. In the diagram above, put the compass at A and move the pencil from B to C, and then repeat with the other vertices. The multilateral-curve heptagon is constructed the same way. Curves of constant width do not need to be symmetrical. It is possible to construct them from any number of lines crossing, as shown above right. The sections of perimeter are always arcs of a circle centred in the opposing vertex.

The curvilinear triangle owes its name to Franz Reuleaux, a German engineer who first wrote about its applications in his 1876 book *Kinematics of Machinery*. This was read many years later by H.G. Conway, a former president of the Institution of Mechanical Engineers, who sat on the British Treasury's Decimal Currency Board. Conway suggested a non-circular curve of constant breadth for the 50p piece since this property made it suitable for use in coin-operated machines. Machines distinguish coins by measuring diameter, and the 50p piece has the same width whichever position it is in. (A square coin, even with rounded sides, can never have constant width, which is why there are no four-sided coins.) Seven sides were chosen since it was considered the most aesthetically pleasing.

While the Reuleaux triangle reinvented the roller, it didn't reinvent the wheel. Wheels cannot be made of Reuleaux triangles because non-circular curves of constant width do not have a 'centre' – a fixed point that is equidistant from every point on the perimeter. If you put an axle on a Reuleaux triangle and rolled it, the height of the shape would stay the same but the axle would judder around.

One useful property of a Reuleaux triangle is that it can be rotated inside a square so that it touches all four sides of the square at all times. This property was exploited by Harry James Watts, an English engineer living in Pennsylvania, in 1914, when he designed one of the most bizarre tools in existence: a drill that can drill square holes. (The corners are rounded, rather than sharp, so, strictly speaking, the hole is a modified square.)

The cross-section of Watts's invention is simply a Reuleaux triangle, with three portions removed to make a cutting edge. It comes with a special chuck to compensate for the wobble of the drill bit's centre as it rotates. The Watts square-hole drill is still used today.

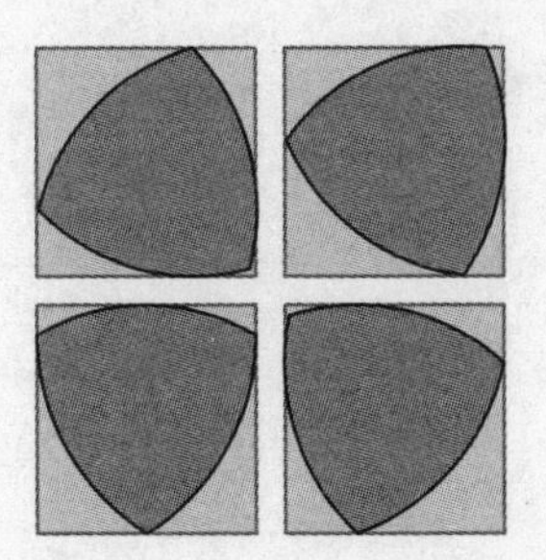

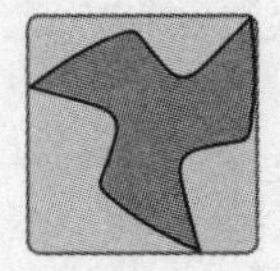

Left: Reuleaux triangle rotating in square.
Above: Cross-section of Watts's square drill.

5663.17715

CHAPTER FIVE

The *x*-factor

Mathematicians tend to like magic tricks. They can be fun and often conceal interesting theory. Here's a classic trick that's also a neat way of appreciating the virtues of algebra. Start by choosing any three-digit number in which the first and last digits differ by at least two – for example, 753. Now, reverse this number to get 357. Subtract the smaller from the larger: 753 – 357 = 396. Finally, add this number to its reverse: 396 + 693. The sum you get is 1089.

Try it again, with a different number, say 421.

421 – 124 = 297
297 + 792 = 1089

The answer is the same. In fact, no matter what three-digit number you start with, you always end up with 1089. As if by magic, 1089 is conjured out of nowhere, a rock in the shifting sands of randomly chosen numbers. Though it is perplexing to think that we get to the same result from any starting point after only a few simple operations, there is an explanation, and we'll get to it shortly. The mystery of the recurrent 1089 is almost instantly unravelled when the problem is written in symbols rather than in numbers.

While using numbers for the purposes of entertainment has been a consistent theme of mathematical discovery, maths only properly got started as a tool for solving practical problems. The Rhind Papyrus, which dates back to around 1600 BC, is the most comprehensive surviving mathematical document from ancient Egypt. It contains 84 problems covering areas such as surveying, accounting, and how to divide a certain number of loaves among a certain number of men.

The Egyptians stated their problems rhetorically. Problem 30 of the Rhind Papyrus asks, 'If the scribe says, What is the heap of which $\frac{2}{3} + \frac{1}{10}$ will make ten, let him hear.' The 'heap' is the Egyptian term for the *unknown quantity*, which we now refer to as x, the fundamental and essential symbol of modern algebra. Nowadays we would say that Problem 30 is asking: What is the value of x such that $\frac{2}{3} + \frac{1}{10}$ multiplied by x is 10. To put it more concisely: What is x such that $(\frac{2}{3} + \frac{1}{10})\, x = 10$?

Because the Egyptians didn't have the symbolic tools that we have now, such as brackets, equal signs, or xs, they solved the above question using trial and error. Estimating for the heap, they then worked out the answer. The method is called the 'rule of false position' and is rather like playing golf. Once you are on the green, it's easier to see how to get the ball into the hole. Similarly, once you have an answer, even a wrong answer, you know how to get closer to the right one. By comparison, the modern method of solution is to combine the fractions in the equation with the variable x, so that:

$$(\frac{2}{3} + \frac{1}{10})\, x = 10$$

Which is the same as:

$$(\frac{20}{30} + \frac{3}{30})\, x = 10$$

Or:

$$(\frac{23}{30})\, x = 10$$

Which further reduces to:

$$x = 10\,(\frac{30}{23})$$

And finally:

$$x = \frac{300}{23}$$

Symbolic notation makes life so much easier.

The Egyptian hieroglyph for addition was 𓂻, a pair of legs walking from right to left. Subtraction was 𓂽, a pair of legs walking from left to right. As number symbols evolved from tally notches to numerals, so the symbols for arithmetical operations also evolved.

Still, the Egyptians had no symbol for the unknown quantity, and neither did Pythagoras or Euclid. For them, maths was geometric, tied to what was constructible. The unknown quantity required a further level of abstraction. The first Greek mathematician to introduce a symbol for the unknown was Diophantus, who used the Greek letter sigma, ς. For the square of the unknown number he used Δ^Y, and for the cube he used K^Y. While his notation was a breakthrough in its time, since it meant problems could be expressed more concisely, it was also confusing since, unlike the case with x, x^2 and x^3, there was no obvious visual connection between ς and its powers Δ^Y and K^Y. Despite his symbols' shortcomings, however, he is nevertheless remembered as the father of algebra.

Diophantus lived in Alexandria sometime between the first and the third centuries CE. Nothing else is known of his personal life except for the following riddle, which appeared in a Greek collection of puzzles and is said to have been inscribed on his tomb:

> God vouchsafed that he should be a boy for the sixth part of his life; when a twelfth was added, his cheeks acquired a beard; He kindled for him the light of marriage after a seventh, and in the fifth year after his marriage He granted him a son. Alas! Late-begotten and miserable child, when he had reached the measure of half his father's life, the chill grave took him. After consoling his grief by the science of numbers for four years he reached the end of his life.

The words are perhaps less an accurate description of Diophantus's family circumstances than they are a tribute to the man whose innovative notation presented new methods for solving problems like the one above. The ability to express mathematical sentences clearly, devoid of confusing verbiage, opened the doors to new techniques. Before I show how to solve the epitaph, let's look at some of them.

Algebra is the generic term for the maths of equations, in which numbers and operations are written as symbols. The word itself has a curious history. In medieval Spain, barbershops displayed signs saying *Algebrista y Sangrador*. The phrase means 'Bonesetter and Bloodletter', two trades that used to be part of a barber's repertoire.

(This is why a barber's pole has red and white stripes – the red symbolizes blood, and the white symbolizes the bandage.)

The root of *algebrista* is the Arabic *al-jabr*, which, in addition to referring to crude surgical techniques, also means restoration or reunion. In ninth-century Baghdad, Muhammad ibn Musa al-Khwarizmi wrote a maths primer entitled *Hisab al-jabr w'al-muqabala*, or *Calculation by Restoration and Reduction*. In it, he explains two techniques for solving arithmetical problems. Al-Khwarizmi wrote out his problems rhetorically, but here, for ease of understanding, they are expressed in modern symbols and terminology.

Consider the equation $A = B - C$.

Al-Khwarizmi described *al-jabr*, or restoration, as the process by which the equation becomes $A + C = B$. In other words, a negative term can be made positive by resetting it on the other side of the equal sign.

Now, consider the equation $A = B + C$.

Reduction is the process that turns the equation into $A - C = B$.

Thanks to modern notation, we can now see that both restoration and reduction are examples of the general rule that *whatever* you do to one side in an equation, you must do to the other as well. In the first equation we *added* C to both sides. In the second equation we *subtracted* C from both sides. Because by definition the expressions on either side of an equation are equal, they must continue to be equal when another term is simultaneously added to or subtracted from either side. It follows that if we multiply one side by an amount, we must multiply the other by the same amount, and the same applies for division and other operations.

The equals sign is like a picket fence separating the gardens of two very competitive families. Whatever the Joneses do to their garden, the Smiths next door will do exactly the same.

Al-Khwarizmi wasn't the first person to use restoration and reduction – these operations could also be found in Diophantus; but when Al-Khwarizmi's book was translated into Latin, the *al-jabr* in the title became *algebra*. Al-Khwarizmi's algebra book, together with another one he wrote on the Indian decimal system, became so widespread in Europe that his name was immortalized as a scientific term: Al-Khwarizmi became Alchoarismi, Algorismi and, eventually, algorithm.

Between the fifteenth and seventeenth centuries mathematical sentences moved from rhetorical to symbolic expression. Slowly, words were replaced with letters. Diophantus might have started letter symbolism with his introduction of ς for the unknown quantity, but the first person to effectively popularize the habit was François Viète in sixteenth-century France. Viète suggested that upper-case vowels – A, E, I, O, U – and Y be used for unknown quantities, and that the consonants B, C, D, etc., be used for known quantities.

Within a few decades of Viète's death, René Descartes published his *Discourse on Method*. In it, he applied mathematical reasoning to human thought. He started by doubting all of his beliefs and, after stripping everything away, was left with only certainty that he existed. The argument that one cannot doubt one's own existence, since the process of thinking requires the existence of a thinker, was summed up in the *Discourse* as *I think, therefore I am*. The statement is one of the most famous quotations of all time, and the book is considered a cornerstone of Western philosophy. Descartes had originally intended it as an introduction to three appendices of his other scientific works. One of them, *La Géométrie*, was equally a landmark in the history of maths.

In *La Géométrie* Descartes introduces what has become standard algebraic notation. It is the first book that looks like a modern maths book, full of *a*s, *b*s and *c*s and *x*s, *y*s and *z*s. It was Descartes's decision to use lower-case letters from the beginning of the alphabet for known quantities, and lower-case letters from the end of the alphabet for the unknowns. When the book was being printed, however, the printer started to run out of letters. He enquired if it mattered if *x*, *y* or *z* was used. Descartes replied not, so the printer chose to concentrate on *x* since it is used less frequently in French than *y* or *z*. As a result, *x* became fixed in maths – and the wider culture – as the symbol for the unknown quantity. That is why paranormal happenings are classified in the X-Files and why Wilhelm Röntgen came up with the term X-ray. Were it not for issues of limited printing stock, the Y-factor could have become a phrase to describe intangible star quality and the African-American political leader might have gone by the name Malcolm Z.

With Descartes' symbology, all traces of rhetorical expression had been expunged.

The equation that Luca Pacioli in 1494 would have expressed as:

4 Census p 3 de 5 rebus ae 0

and Viète would have written in 1591 as:

4 in A quad – 5 in A plano + 3 aequatur 0

in 1637 Descartes had nailed as:

$4x^2 - 5x + 3 = 0$

Replacing words with letters and symbols was more than convenient shorthand. The symbol x may have started as an abbreviation for 'unknown quantity', but once invented, it became a powerful tool for thought. A word or an abbreviation cannot be subjected to mathematical operations in the way that a symbol such as x can. Numbers made counting possible, but letter symbols took mathematics into a domain far beyond language.

When problems were expressed rhetorically, as in Egypt, mathematicians used ingenious, but rather haphazard, methods to solve them. These early problem-solvers were like explorers stuck in a fog with few tricks to help them move about. When a problem was expressed using symbols, however, it was as though the fog lifted to reveal a precisely defined world.

The marvel of algebra is that once a problem is restated in symbolic terms, often it is almost solved.

For example, let's re-examine Diaphantus's epitaph. How old was he when he died? Translating that statement, using the letter D to symbolize his age when he died, the epitaph says that for $\frac{D}{6}$ years he was a boy, that another $\frac{D}{12}$ years passed before he sprouted facial hair, and that he wed after another $\frac{D}{7}$. Five years after that he had a son, who lived for $\frac{D}{2}$ years, and four years later Diophantus himself breathed his last. The sum of all these time intervals adds up to D, since D is the number of years Diophantus lived. So:

$$\frac{D}{6} + \frac{D}{12} + \frac{D}{7} + 5 + \frac{D}{2} + 4 = D$$

The lowest common denominator of the fractions is 84, so this becomes:

$$\frac{14D}{84} + \frac{7D}{84} + \frac{12D}{84} + 5 + \frac{42D}{84} + 4 = D$$

Which can be rearranged as:

$$D(\tfrac{14+7+12+42}{84}) + 9 = D$$

Or:

$$D(\tfrac{75}{84}) + 9 = D$$

Which is:

$$D(\tfrac{25}{28}) + 9 = D$$

Moving the Ds to the same side:

$$D - D(\tfrac{25}{28}) = 9$$

$$D(\tfrac{28}{28}) - D(\tfrac{25}{28}) = 9$$

$$D(\tfrac{3}{28}) = 9$$

Multiplying out:

$$D = 9 \times \tfrac{28}{3} = 84$$

The father of algebra died aged 84.

We can now return to the trick at the start of the chapter. I asked you to name a three-digit number for which the first and last digits differed by at least two. I then asked you to reverse that number to give you a second number. After that, I asked you to subtract the smaller number from the larger number. So, if you chose 614, the reverse is 416. Then, 614 – 416 = 198. I then asked you to add this intermediary result to its reverse. In the above case, this is 198 + 891.

As before, the answer is 1089. It always will be, and algebra tells you why. First, though, we need to find a way of describing our protagonist, the three-digit number in which the first and last digits differ by at least two.

Consider the number 614. This is equal to 600 + 10 + 4. In fact, any three-digit number written abc can be written $100a + 10b + c$ (note: abc in this case is **not** $a \times b \times c$). So, let's call our initial number abc, where a, b and c are single digits. For the sake of convenience, make a bigger than c.

The reverse of abc is cba which can be expanded as $100c + 10b + a$.

We are required to subtract cba from abc to give an intermediary result. So $abc - cba$ is:

$(100a + 10b + c) - (100c + 10b + a)$

The two b terms cancel each other out, leaving an intermediary result of:

$99a - 99c$, or
$99(a - c)$

At a basic level algebra doesn't involve any special insight, but rather the application of certain rules. The aim is to apply these rules until the expression is as simple as possible.

The term $99(a - c)$ is as neatly arranged as it can be.

Since the first and last digits in *abc* differ by at least 2, then $a - c$ is either 2, 3, 4, 5, 6, 7 or 8.

So, $99(a - c)$ is one of the following: 198, 297, 396, 495, 594, 693 or 792. Whatever three-figure number we started with, once we have subtracted it from its reverse, we have an intermediary result that is one of the above eight numbers.

The final stage is to add this intermediary number to its reverse.

Let's repeat what we did before and apply it to the intermediary number. We'll call our intermediary number *def*, which is $100d + 10e + f$. We want to add *def* to *fed*, its reverse. Looking closely at the list of possible intermediary numbers above, we see that the middle number, e, is always 9. And also that the first and third numbers always add up to 9, in other words $d + f = 9$. So, *def* + *fed* is:

$100d + 10e + f + 100f + 10e + d$

Or:
$100(d + f) + 20e + d + f$

Which is:
$(100 \times 9) + (20 \times 9) + 9$

Or:
$900 + 180 + 9$

Hey presto! The total is 1089, and the riddle is laid bare.

The surprise of the 1089 trick is that from a randomly chosen number we can always produce a fixed number. Algebra lets us see beyond the legerdemain, providing a way to go from the concrete to the abstract – from tracking the behaviour of a specific number to tracking the behaviour of *any* number. It is an indispensable tool, and not just for maths. The rest of science also relies on the language of equations.

In 1621, a Latin translation of Diophantus's masterpiece *Arithmetica* was published in France. The new edition rekindled interest in ancient problem-solving techniques, which, combined with better numerical and symbolic notation, ushered in a new era of mathematical thought. Less convoluted notation allowed greater clarity in describing problems. Pierre de Fermat, a civil servant and judge living in Toulouse, was an enthusiastic amateur mathematician who filled his own copy of *Arithmetica* with numerical musings. Next to a section dealing with Pythagorean triples – any set of natural numbers a, b and c such that $a^2 + b^2 = c^2$, for example 3, 4 and 5 – Fermat scribbled some notes in the margin. He had noticed that it was impossible to find values for a, b and c such that $a^3 + b^3 = c^3$. He was also unable to find values for a, b and c such that $a^4 + b^4 = c^4$. Fermat wrote in his *Arithmetica* that for any number n greater than 2, there were no possible values a, b and c that satisfied the equation $a^n + b^n = c^n$. 'I have a truly marvellous demonstration of this proposition which this margin is too narrow to contain,' he wrote.

Fermat never produced a proof – marvellous or otherwise – of his proposition even when unconstrained by narrow margins. His jottings in *Arithmetica* may have been an indication that he had a proof, or he may have believed he had a proof, or he may have been trying to be provocative. In any case, his cheeky sentence was fantastic bait to generations of mathematicians. The proposition became known as Fermat's Last Theorem and was the most famous unsolved problem in maths until the Briton Andrew Wiles cracked it in 1995. Algebra can be very humbling in this way – ease in stating a problem has no correlation with ease in solving it. Wiles's proof is so complicated that it is probably understood by no more than a couple of hundred people.

Improvements in mathematical notation enabled the discovery of new concepts. The *logarithm* was a massively important invention in the early seventeenth century, thought up by the Scottish mathematician John Napier, the Laird of Merchiston, who was, in fact, much more famous in his lifetime for his work on theology. Napier wrote a best-selling Protestant polemic in which he claimed that the Pope was the Antichrist and predicted that the Day of Judgement would come between 1688 and 1700. In the evening he liked to wear a long robe and pace outside his tower chamber, which added to his reputation as a necromancer. He also experimented with fertilizers on his vast estate near Edinburgh, and came up with ideas for military hardware, such as a chariot with a 'moving mouth of mettle' that would 'scatter destruction on all sides' and a machine for 'sayling under water, with divers and other strategems for harming of the enemyes' – precursors of the tank and the submarine. As a mathematician, he popularized the use of the decimal point, as well as coming up with the idea of logarithms, coining the term from the Greek *logos*, ratio, and *arithmos*, number.

Don't be put off by the following definition: *the logarithm, or log, of a number is the exponent when that number is expressed as a power of 10.* Logarithms are more easily understood when expressed algebraically: if $a = 10^b$, then the log of a is b.

So, log 10 = 1 (because $10 = 10^1$)
log 100 = 2 (because $100 = 10^2$)
log 1000 = 3 (because $1000 = 10^3$)
log 10,000 = 4 (because $10{,}000 = 10^4$)

Finding the log of a number is self-evident if the number is a power of 10. But what if you're trying to find the log of a number that isn't a power of 10? For example, what is the logarithm of 6? The log of 6 is the number a such that when 10 is multiplied by itself a times you get 6. However, it seems completely nonsensical to say that you can multiply 10 by itself a certain number of times to get 6. How can you multiply 10 by itself a fraction of times? Of course, the concept *is* nonsensical when we imagine what this might mean in the real world, but the power and beauty of mathematics is that we do not need to be concerned with any meaning beyond the algebraic definition.

The log of 6 is 0.778 to three decimal places. In other words, when we multiply 10 by itself 0.778 times, we get 6.

Here is a list of the logarithms of the numbers from 1 to 10, each to three decimal places.

log 1 = 0	log 6 = 0.778
log 2 = 0.301	log 7 = 0.845
log 3 = 0.477	log 8 = 0.903
log 4 = 0.602	log 9 = 0.954
log 5 = 0.699	log 10 = 1

So, what's the point of logarithms? Logarithms turn the more difficult operation of multiplication into the simpler process of addition. More precisely, the multiplication of two numbers is equivalent to the addition of their logs. If $X \times Y = Z$, then $\log X + \log Y = \log Z$.

We can check this equation using the table above.

$3 \times 3 = 9$
$\log 3 + \log 3 = \log 9$
$0.477 + 0.477 = 0.954$

Again,
$2 \times 4 = 8$
$\log 2 + \log 4 = \log 8$
$0.301 + 0.602 = 0.903$

The following method can therefore be used in order to multiply two numbers together: convert them into logs, add them to get a third log, and then convert this log back into a number. For example, what is 2×3? We find the logs of 2 and 3, which are 0.301 and 0.477, and add them, which is 0.778. From the list above, 0.778 is log 6. So, the answer is 6.

Now, let's multiply 89 by 62.

First, we need to find their logs, which we can do by putting the number into a calculator or Google. Until the late twentieth century, however, the only way of doing this was done by consulting log tables. The log of 89 is 1.949 to three decimal places. The log of 62 is 1.792.

So, the sum of the logs is 1.949 + 1.792 = 3.741.

The number whose log is 3.741 is 5518. This is again found by using the log tables.

So, $89 \times 62 = 5518$.

Significantly, the only piece of calculation we have done to work out this multiplication was a fairly simple addition.

Logarithms, wrote Napier, were able to free mathematicians from the 'tedious expense of time' and the 'slippery errors' involved in the 'multiplications, divisions, square and cubical extractions of great numbers'. Using Napier's invention, not only could multiplication be made into the addition of logs, but division was made into the subtraction of logs; calculating of square roots was made into the division of logs by two; and calculating cube roots, into the division of logs by three.

The convenience that logarithms brought made them the most significant mathematical invention of Napier's time. Science, commerce and industry benefited massively. The German astronomer Johannes Kepler, for example, used logs almost immediately to calculate the orbit of Mars. It has recently been suggested that he might not have discovered his three laws of celestial mechanics without the ease of calculation offered by Napier's new numbers.

In his 1614 book *A Description of the Admirable Table of Logarithmes*, Napier used a slightly different version of logarithms than those used in modern mathematics. Logarithms can be expressed as a power of any number, which is called the base. Napier's system used an unnecessarily complicated base of $1 - 10^{-7}$ (which he then multiplied by 10^7). Henry Briggs, England's top mathematician in Napier's day, visited Edinburgh to congratulate the Scot on his discoveries. Briggs went on to simplify the system by introducing base-ten logarithms – which are also known as Briggsian logarithms, or common logarithms, because ten has been the most popular base ever since. In 1617 Briggs published a table of the logs of all numbers from 1 to 1000 to eight decimal places. By 1628 Briggs and the Dutch mathematician Adriaan Vlacq had extended the log table to 100,000, to ten decimal places. Their calculations involved laborious number-crunching – although, once the sums were done correctly, they never needed to be done again.

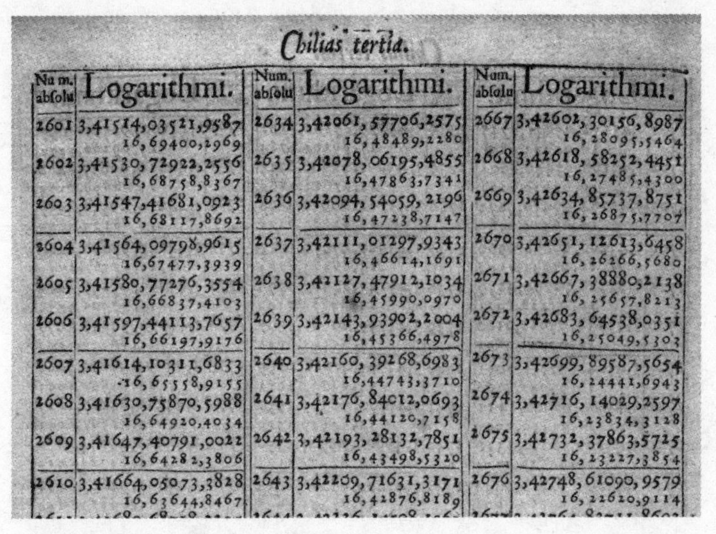

Chilias tertia.

Num. absolu	Logarithmi.	Num. absolu	Logarithmi.	Num. absolu	Logarithmi.
2601	3,41514,03521,9587	2634	3,42061,57706,2575	2667	3,42602,30156,8987
	16,69400,2969		16,48489,2280		16,28095,5464
2602	3,41530,72922,2556	2635	3,42078,06195,4855	2668	3,42618,58252,4451
	16,68758,8367		16,47863,7341		16,27485,4300
2603	3,41547,41681,0923	2636	3,42094,54059,2196	2669	3,42634,85737,8751
	16,68117,8692		16,47238,7147		16,26875,7707
2604	3,41564,09798,9615	2637	3,42111,01297,9343	2670	3,42651,12613,6458
	16,67477,3939		16,46614,1691		16,26266,5680
2605	3,41580,77276,3554	2638	3,42127,47912,1034	2671	3,42667,38880,2138
	16,66837,4103		16,45990,0970		16,25657,8213
2606	3,41597,44113,7657	2639	3,42143,93902,2004	2672	3,42683,64538,0351
	16,66197,9176		16,45366,4978		16,25049,5303
2607	3,41614,10311,6833	2640	3,42160,39268,6983	2673	3,42699,89587,5654
	16,65558,9155		16,44743,3710		16,24441,6943
2608	3,41630,75870,5988	2641	3,42176,84012,0693	2674	3,42716,14029,2597
	16,64920,4034		16,44120,7158		16,23834,3128
2609	3,41647,40791,0022	2642	3,42193,28132,7851	2675	3,42732,37863,5725
	16,64282,3806		16,43498,5320		16,23227,3854
2610	3,41664,05073,3828	2643	3,42209,71631,3171	2676	3,42748,61090,9579
	16,63644,8467		16,42876,8189		16,22620,9114

Page of Briggs's log tables from 1624.

That is, until 1792, when the young French republic decided to commission ambitious new tables – the log of every number to 100,000 to 19 decimal places, and from 100,000 to 200,000 to 24 decimal places. Gaspard de Prony, the man who headed the project, claimed that he could 'manufacture logarithms as easily as one manufactures pins'. He had a staff of nearly 90 human calculators, many of whom were former servants or wig dressers whose pre-revolutionary skills had become redundant (if not treasonous) in the new regime. Most of the calculations were finished by 1796, but by then the government had lost interest, and de Prony's gigantic manuscript was never published. Today it is housed in the Paris Observatory.

Briggs's and Vlacq's tables remained the basis for all log tables for 300 years, until the Englishman Alexander J. Thompson in 1924 began work manually on a new set accurate to 20 places. Yet instead of giving an old concept a modern sheen, Thompson's work was already outdated when he finished it, in 1949. By then computers could generate the tables easily.

When you plot the digits 1 to 10 on a ruler positioned to their log values, you get the following pattern:

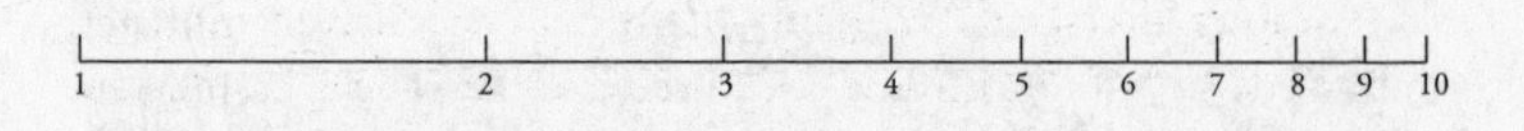

We can carry on like this, say, up to 100.

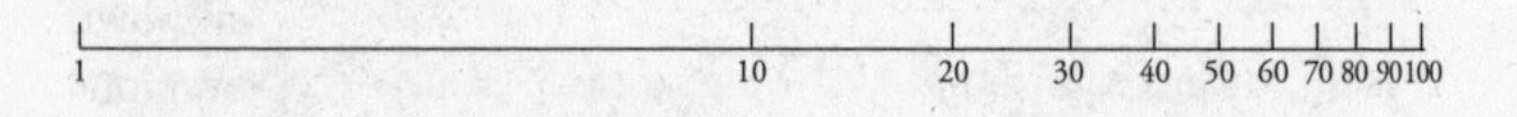

This is what is known as a logarithmic scale. In the scale, numbers get progressively closer together the higher they are.

Some scales of measurement are logarithmic, which means that for every unit you go up on the scale, it represents a tenfold change in what it is measuring. (In the second scale above, the distance between 1 and 10 is equal to the distance between 10 and 100.) The Richter scale, for example, which measures the amplitude of waves recorded by seismographs, is the most commonly used logarithmic scale. An earthquake that registers 7 on the Richter scale triggers an amplitude that is ten times more than an earthquake that registers a 6.

In 1620 the English mathematician Edmund Gunter was the first person to mark the logarithmic scale on a ruler. He noticed that he could multiply by adding lengths of this ruler. If a compass was placed with the left spike at 1, and the right at *a*, then when the left spike was moved to *b*, the right spike pointed to $a \times b$. The diagram below shows the compass set to 2 and then positioned with the left spike at 3, putting the right spike at $2 \times 3 = 6$.

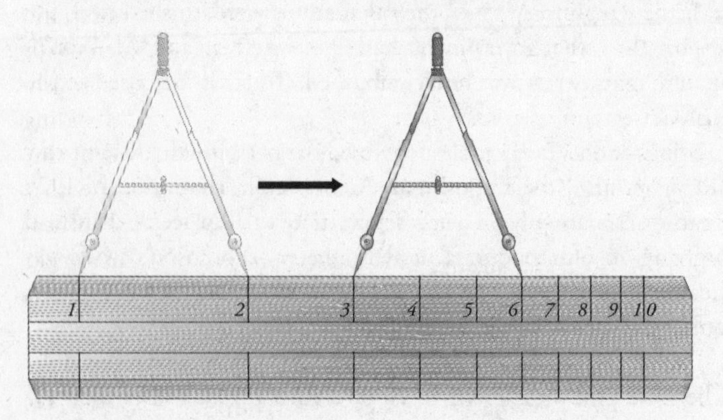

Gunter multiplication.

Not long afterwards William Oughtred, an Anglican minister, improved on Gunter's idea. He dispensed with the compasses, instead placing two wooden log scales next to each other to create a device known as the slide-rule. Oughtred devised two styles of slide-rule. One version used two straight rulers and the other used a circular disc with two cursors. But Oughtred, for unknown reasons, didn't publish news of his invention. In 1630, however, one of his students, Richard Delamain, did. Oughtred was outraged, accusing Delamain of being a 'pickpurse', and the feud over the slide-rule's origins continued until Delamain's death. 'This scandall,' complained Oughtred at the end of his life, 'hath wrought me much prejudice and disadvantage.'

The slide-rule was a calculating machine of fantastic ingenuity, and while it may now be obsolete it still has fanatical devotees. I visited one of them, Peter Hopp, in Braintree, Essex. 'Between the 1700s and 1975 every single technological innovation was invented using a slide-rule,' he told me when he picked me up at the station. Hopp, a retired electrical engineer, is an extremely affable man with wispy eyebrows, blue eyes, and luxurious jowls. He was taking me to see his slide-rule collection, one of the world's largest, which contains more than a thousand of these forgotten heroes of our scientific heritage. On the drive to his home we chatted about collecting. Hopp said the best stuff was auctioned directly on the internet, where competition inevitably pushed prices higher. A rare slide-rule, he said, can easily cost hundreds of pounds.

When we arrived at his house, his wife made us a cup of tea and we retired to his study, where he presented me with a wooden 1970s Faber-Castell slide-rule with a magnolia-coloured plastic finish. The rule was the size of a normal 30cm ruler and had a sliding middle section. On it, several different scales were marked in tiny writing. It also had a transparent movable cursor marked with a hairline. The shape and feel of the Faber-Castell were deeply evocative of a kind of post-war, pre-computer-age nerdiness – when geeks had shirts, ties and pocket protectors rather than T-shirts, sneakers and iPods.

I went to secondary school in the 1980s, by which time slide-rules were no longer used, so Hopp gave me a quick tutorial. He recommended that as a beginner I should use the log scale from

1 to 100 on the main ruler and adjacent log scale from 1 to 100 on the sliding middle section.

Multiplication of two numbers using a slide-rule – which also used to be called a slipstick in the US – is performed by lining up the first number marked out on one scale with the second number marked out on the other scale. You don't even need to understand what logs are – you just need to slide the middle ruler to the correct position and read the scale.

For example, say I want to multiply 4.5 by 6.2. I need to add the length that is 4.5 on one ruler to the length that is 6.2 on the other. This is done by sliding the 1 on the middle ruler to the point where 4.5 is on the main ruler. The answer to the multiplication is the point on the main ruler adjacent to where 6.2 is on the middle ruler. The diagram below makes this clear:

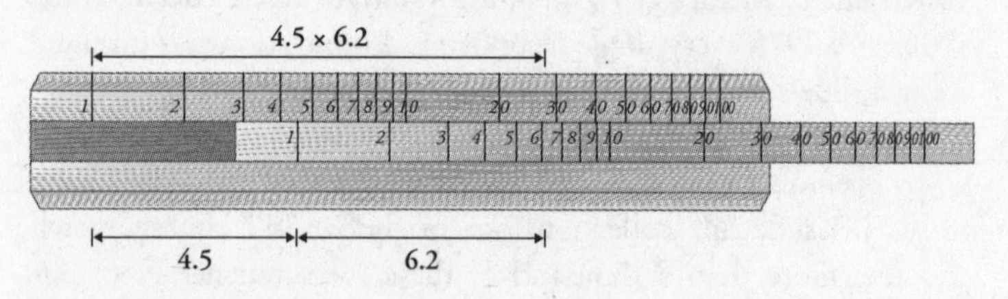

How to multiply with a slide-rule.

Using the hairline cursor, it is easy to see where one scale meets the other. Moving up from 6.2 on the middle ruler, I can see that it crosses the main ruler at *just under* 28, which is a correct answer. Slide-rules are not precise machines. Or rather, we are imprecise in our use of them. In reading a slide-rule, we are estimating where a number is on an analogue scale, rather than finding a clear result. Yet despite their inherent imprecision, Hopp said that – for his purposes as an engineer, at least – slide-rules were accurate enough for most uses.

The log scale on the slide-rule I used went from 1 to 100. There are also scales that go from 1 to 10, which are used for greater accuracy because there is more space between the numbers. For this reason, whenever you use a slide-rule it's always best to convert the original sum into numbers between 1 and 10 by moving the decimal point. For example, in order to multiply 4576 by 6231,

I would turn this into the multiplication of 4.576 by 6.231. Once I have the answer, I will move the decimal point six places back to the right. When I put in 4.576 and line it up with 6.231 I get around 28.5, meaning that the answer to 4576 × 6231 is about 28,500,000. The precise answer, as calculated above, using logs, is 28,513,056. Not a bad estimate. Usually, a slide-rule like the Faber-Castell will give you accuracy to three significant figures – which is often all that is required. What I lost in accuracy, however, I gained in speed – this sum took me under five seconds to do. Using log tables would have taken me ten times longer.

The oldest item in Peter Hopp's collection was a wooden slide-rule from the early eighteenth century, used by taxmen for making calculations on alcohol volume. Before meeting Hopp, I had been sceptical as to how interesting slide-rule collecting could be as a pastime. At least stamps and fossils can be pretty! Slide-rules, on the other hand, are proudly functional tools of convenience. Hopp's antique slide-rule, however, was beautiful, with elegantly crafted numbers on fine wood.

Hopp's vast collection reflected the small improvements that were made over the centuries. In the nineteenth century new scales

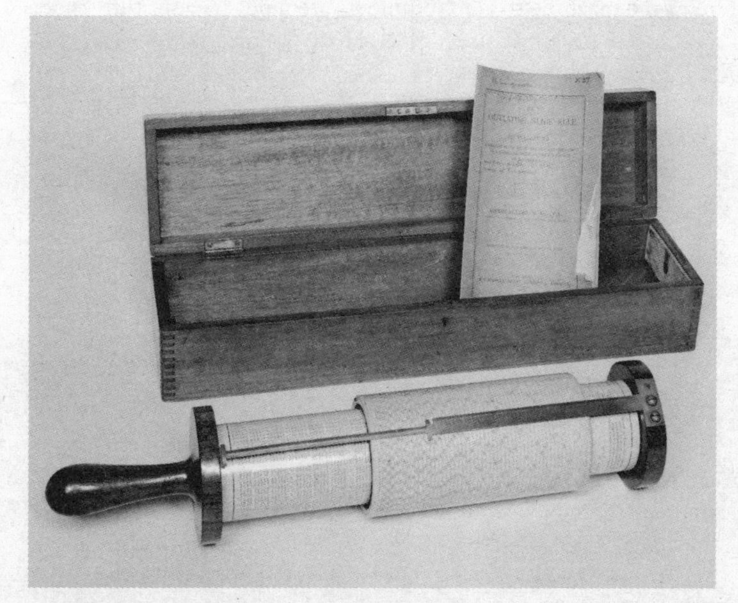

Professor Fuller's Calculator.

were added. Peter Roget – whose compulsive list-making (as a coping mechanism for mental illness) resulted in his timeless, classic, definitive *Thesaurus* – invented the log-log scale, which enabled calculation of fractions of powers, such as $3^{2.5}$, and square roots. As manufacturing techniques improved, new devices of increasing ingenuity, precision and splendour were designed. For instance, Thacher's Calculating Instrument looks like a rolling pin on a metal mount, and Professor Fuller's Calculator has three concentric, hollow brass cylinders and a mahogany handle. A 41ft-helix spirals around the cylinder, giving an accuracy of five significant figures. The Halden Calculex, on the other hand, looks like a timepiece and is made of glass and chromed steel. Slide-rules, I decided, are indeed objects of surprising appeal.

Among these others, I spotted a contraption on Hopp's shelf that looked like a pepper-grinder, and enquired what it was. He said it was a Curta. The Curta is a black, palm-sized cylinder with a crank handle on the top, and was a unique invention – the only mechanical pocket calculator ever produced. Demonstrating how it worked, Hopp cranked the handle round for one rotation, which reset the machine to zero. Numbers are input by adjusting knobs positioned on the Curta's side. Hopp set the numbers to 346 and turned the handle once. He then reset the knobs to 217. When he turned the handle again, the sum of both numbers – 563 – was displayed on the top of the machine. Hopp said that the Curta could also subtract, multiply, divide and perform other mathematical operations. It used to be very popular with sports-car enthusiasts, he added. Navigators were able to calculate driving times by cranking it without taking their eyes off the road for too long. It was easier to read than a slide-rule, and less susceptible to bumps in the road.

Although the Curta is not a slide-rule, its ingeniousness at calculating has endeared it to collectors of mathematical instruments. Immediately upon using it, it was my favourite item in Hopp's collection. For a start, it was a literal take on number-crunching – in went the numbers, and with a crank of the handle, the result appeared. The notion of grinding out an answer, however, was too crude a way to describe a gadget made up of 600 mechanical parts that moved with the precision of a Swiss watch.

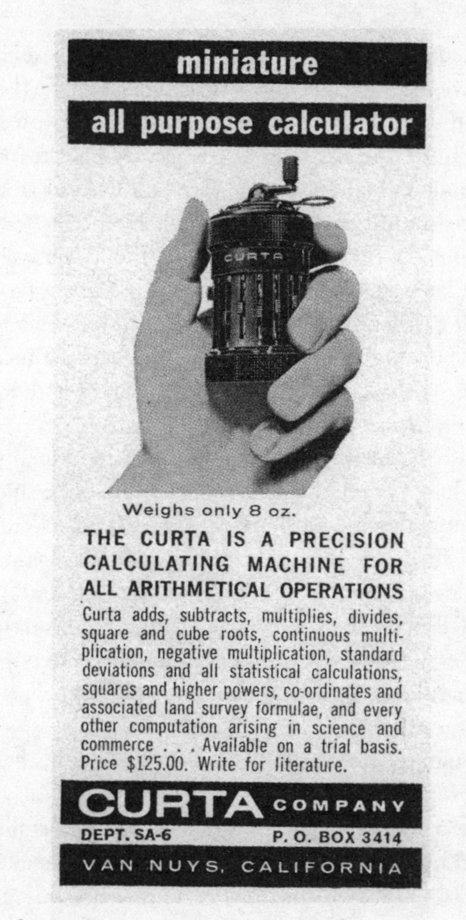

Curta advert from 1971.

Even more intriguing, the Curta has a particularly dramatic history. Its inventor, Curt Herzstark designed the prototype for the device while a prisoner at the Buchenwald concentration camp during the final years of the Second World War. Herzstark, an Austrian whose father was Jewish, was given special dispensation to work on his calculating machine because he was known to the camp authorities as an engineering genius. Herzstark was told that if it worked, it would be given to Adolf Hitler as a present – after which he would be declared Aryan, and his life would be spared. When the end of the war came, and Herzstark was freed, he left with his nearly finished plans folded inside his pocket. After several attempts to find an investor, he eventually managed to convince the Prince of Liechtenstein – where the first Curta was manufactured in 1948. From then until the early 1970s, a factory in the principality produced about 150,000 of them. Herzstark lived in an apartment in Liechtenstein until he died, aged 86, in 1988.

Throughout the 1950s and the 1960s the Curta was the only pocket calculator in existence that could produce exact answers. But both the Curta and the slide-rule were all but rendered extinct by an event in the history of arithmetical paraphernalia as cataclysmic as the meteorite that is said to have annihilated the dinosaurs: the birth of the electronic pocket calculator.

It is hard to think of an object that has disappeared so quickly after such a long period of dominance than the slide-rule. For 300 years it reigned supreme until, in 1972, Hewlett-Packard launched the HP-35. The device was promoted as a 'high precision portable electronic slide-rule', but it wasn't like a slide-rule at all. It was the size of a small book with a red LED display, 35 buttons, and an on-off switch. Within a few years it was impossible to buy a general-purpose slide-rule except second-hand, and the only people interested in them were collectors.

Even though the electronic calculator killed off his beloved slip-stick, Peter Hopp bears no grudge. He likes to collect early electronic calculators too. When our conversation moved to them, he showed me his HP-35, and started reminiscing about the time he first saw one in the early 1970s. At the time, Hopp was beginning his career at Marconi, the electrical communications firm. One of his colleagues had bought an HP-35, which had cost him £365 – at the time, about half the annual salary of a junior engineer. 'It was so valuable he kept it locked in his desk and never let anyone use it,' Hopp said. The colleague, however, had another reason for his secrecy. He believed he had found a way of using the calculator that could save the company 1 per cent of its expenditure. 'He had top-secret meetings with the bosses. It was all hush, hush,' said Hopp. In fact, though, his colleague had made a mistake. Calculators aren't perfect instruments – type in 10 and divide by 3. You get 3.3333333. Multiply the result by 3, however, and you do not get back to where you started; rather, you get 9.9999999. Hopp's colleague had used what was an anomaly in digital calculators to create something from nothing. Hopp recalled the incident with a smile: 'When the plan was peer-reviewed by someone who used a slide-rule, the improvements were judged illusory.'

The story demonstrates why Hopp laments the demise of the slide-rule. The device provided the user with a visual understanding of numbers, which meant that even before he had worked out the answer he had a rough idea of what it would be. Nowadays, Hopp said, people plug numbers into a calculator without any intuitive sense of whether the answer is correct.

Still, the digital electronic calculator was an improvement on the analogue slide-rule. The pocket calculator was easier to use, gave precise answers and by 1978 was priced under £5, making it accessible to the general public.

It is now more than three decades since the slipstick slipped away, which means it is surprising to discover that there is, in fact, one situation in the modern world where they are still commonly used. Pilots use them to fly planes. A pilot's slide-rule is circular, called a 'whizz wheel', and measures speed, distance, time, fuel consumption, temperature and air density. In order to qualify as a pilot, you must be proficient with a whizz wheel, which seems utterly strange, bearing in mind the high-end computer technology now used in cockpits. The slide-rule requirement is because pilots must also be able to fly small planes without onboard computers. Yet often pilots flying the most modern jets prefer to use their whizz wheels. Having a slide-rule at hand means you can work out estimates very quickly, and also have a more visual understanding of the numerical parameters of the flight. Flying jets is safer because of pilots' dexterity with an early seventeenth-century calculating machine.

The astronomically high prices of the early electronic calculators made them luxury business products. The inventor Clive Sinclair called his first product the Executive. One marketing idea involved using geishas to target high-rolling businessmen in Japan. After a night of entertaining, the geisha would whip out a Sinclair Executive from under her kimono so that the host could add up the bill. He would then feel obliged to buy it.

As prices dropped, calculators were seen not only as arithmetical aides but also as versatile toys. *The Pocket Calculator Game Book*, published in 1975, suggested many recreational activities for the high-tech electronic marvel. 'Pocket calculators are new to our lives. Unknown five years ago, they are becoming as popular as televisions or hi-fi sets,' it said. 'Yet they are different in that they are not a passive entertainment but require intelligent input and definite intention for their use. We are not so much interested in what the pocket calculator can do as we are in what you can do with your pocket calculator.' In 1977 the bestselling *Fun & Games with Your Electronic Calculator* included a dictionary of words that can be made using only the letters O, I, Z, E, h, S, g, L and B, which are the LED digits 0, 1, 2, 3, 4, 5, 6, 7 and 8 when turned upside-down. The longest words are:

Seven letters:

OBELIZE
ELEgIZE
LIBELEE
OBLIgEE
gLOBOSE
SESSILE
LEgIBLE
BESIEgE
BIggISh
LOOBIES
LEgLESS
ZOOgEOg

Eight letters:

ISOgLOSS
hEELLESS
EggShELL

Nine letters:

gEOLOgIZE
ILLEgIBLE
EISEgESIS

Surprisingly, the list does not include 'BOOBLESS' – the word whose use by teenage boys to their flat-chested female classmates is probably responsible for turning a generation of girls off mathematics. Still, *Fun & Games with Your Electronic Calculator* is probably the only numbers book that improves your English more than your arithmetic.

Enthusiasm for playing with one's calculator was quickly extinguished as more enjoyable electronic games were introduced to the market. It soon became clear that instead of inspiring love of numbers, calculators would have the opposite effect – bringing about a decline in mental arithmetic skills.

Whereas the logarithm was a brand-new invention made possible by advances in notation, the *quadratic equation* was an ancient mathematical staple that was spruced up by new symbology. In modern notation, we say that a quadratic equation is one that looks like this:

> $ax^2 + bx + c = 0$, where x is the unknown and a, b and c are any constants.
> For example, $3x^2 + 2x - 4 = 0$.

Quadratics, in other words, are equations with an x and an x^2. They occur most basically in calculations involving area. Consider the following problem from a Babylonian clay tablet: a rectangular field with area 60 units has one side that is 7 units bigger than the other. How big are the sides of the field? To find the answer we need to sketch the problem, as in the diagram below. The problem reduces to solving the quadratic equation $x^2 + 7x - 60 = 0$.

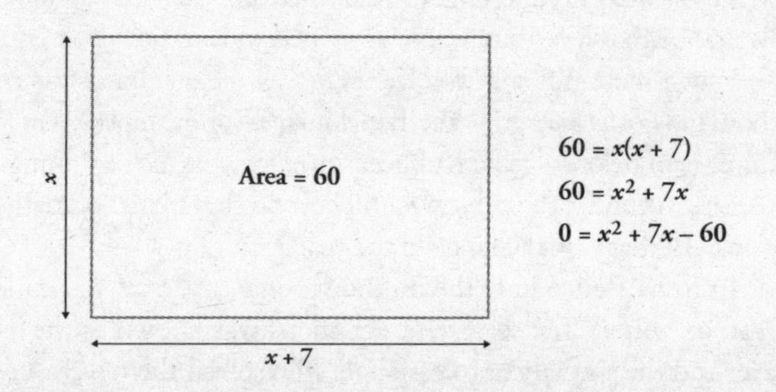

A convenient feature of quadratic equations is that they can be solved by substituting the values for a, b and c in this one-size-fits-all formula:

$$x = \frac{-b \pm \sqrt{b^2 - 4ac}}{2a}$$

The ± means that there are two solutions, one for the formula with a + and one with a –. In the Babylonian problem, $a = 1$, $b = 7$ and $c = -60$, which gives the two solutions 5 and –12. The negative solution is meaningless when describing area, so the answer is 5.

Quadratics are used in calculations other than those analysing area. Physics was essentially born with Galileo Galilei's theory of falling bodies, which he supposedly discovered by dropping cannonballs from the Leaning Tower of Pisa. The formula he derived to describe the distance of a falling object was a quadratic equation. Since then, quadratics have become so crucial to the understanding of the world that it is no exaggeration to say that they underpin modern science.

Even so, not every problem can be reduced to equations in x^2. Some require the next power of x up the scale: x^3. These are called *cubic* equations, and are of the form:

> $ax^3 + bx^2 + cx + d = 0$, where x is the unknown and a, b, c and d are any constants.
> For example, $2x^3 - x^2 + 5x + 1 = 0$.

Cubic equations often emerge with calculations involving volumes, in which one may need to multiply the three dimensions of a solid object. Even though they are just one level up from quadratics, cubic equations are much more difficult to solve. Whereas the quadratic was cracked thousands of years ago – the Babylonians, for example, were able to solve them before algebra had been invented – at the beginning of the sixteenth century, the cubic was still beyond the abilities of mathematicians. All that would change in the year 1535.

In Renaissance Italy the unknown quantity, or x, was called the *cosa*, or 'thing'. The science of equations was known as the 'cossick art', and the specialist professionals who solved them were 'cossists', literally, 'thingists'. The thingists were not just ivory-tower academics, but also tradesmen who hired out their mathematical skills to a burgeoning commercial class that needed help with sums. Dealing in unknowns was a competitive business, and like master craftsmen, cossists kept their best techniques close to their chests.

Despite their secrecy, however, in 1535 a rumour circulated through Bologna that two cossists had discovered how to solve the cubic equation. For the cossick community, the news was pretty damn exciting. Conquering the cubic would elevate an equation professional above his peers and allow him to charge higher rates.

In modern academia the announcement of the proof of a famous unsolved problem would be presented through the publication of a paper, perhaps at a press conference, but back in the Renaissance the thingists agreed to a public mathematical duel.

On 13 February crowds gathered at the University of Bologna to see Niccolò Tartaglia and Antonio Fiore fight it out. The rules of the contest were that each man would challenge the other with 30 cubic equations. For each equation solved correctly, the solver would win a banquet paid for by his opponent.

The contest ended in a knockout victory for Tartaglia. (His name translates as 'Stammerer', and was a nickname he gained thanks to a sabre wound that had left him facially disfigured and with a severe speech impediment.) Tartaglia solved all of Fiore's problems in two hours, while Fiore was unable to solve even one of Tartaglia's. As the first person to discover a method of solving cubic equations, Tartaglia was the envy of mathematicians across Europe, but he wouldn't tell anyone how he did it. In particular, he resisted the entreaties of Girolamo Cardano, who was probably history's most colourful mathematician of significance.

Cardano was a doctor by profession, and he was internationally known for his cures – he once travelled to Scotland to treat the asthma of the country's archbishop. He was also a prolific writer. In his autobiography he lists 131 printed books, 111 unprinted books, and 170 manuscripts he rejected himself as not good enough. *Consolation*, his compendium of advice to the sorrowful, was a bestseller throughout Europe, and is understood by literary scholars to be the book that Hamlet has in his hands during his 'to be or not to be' soliloquy. He was also a professional astrologer and claimed to have invented 'metoposcopy', the reading of character from the irregularities of one's face. In mathematics, Cardano's major contribution was the invention of probability, which I will return to later.

Cardano was desperate to know how Tartaglia had solved the cubic equation, so wrote to him, asking if he could include his cubic solution in a book he was writing. When Tartaglia refused, Cardano asked again, this time promising that he wouldn't tell anyone else. Again, Tartaglia refused.

Extracting the cubic formula from Tartaglia became an obsession of Cardano's. He eventually thought up a ruse – he invited Tartaglia to Milan on the pretext of setting up an introduction to a potential benefactor, the governor of Lombardy. Tartaglia accepted the offer, but once he arrived, discovered that the governor was out of town. Instead, he encountered only Cardano. Worn out from Cardano's incessant pestering, Tartaglia relented, telling Cardano that if he could keep the formula to himself, he would reveal it to him. But when he passed over the information to Cardano, the crafty Tartaglia wrote the solution in a deliberately abstruse way: as a bizarre poem of 25 lines.

Despite this impediment, the multi-talented Cardano deciphered the method, and he almost kept his promise. He told the solution to only one person, his personal secretary, a young boy named Lodovico Ferrari. This turned out to be problematic, not because Ferrari was indiscreet, but because he improved on Tartaglia's method to find a way to solve *quartic equations*. These are equations that require the power of x^4. For example, $5x^4 - 2x^3 - 8x^2 + 6x + 3 = 0$. A quartic may arise when multiplying one quadratic with another.

Cardano was in a fix – he couldn't publish Ferrari's discovery without betraying Tartaglia's word, but neither could he deny Ferrari the public acclamation that he deserved. Cardano, however, managed to find a clever way out. It turned out that Antonio Fiore, the man who lost the cubic duel against Tartaglia, did in fact know how to solve the cubic, and he had learned the method from an older mathematician, Scipione del Ferro, who had told Fiore from his deathbed. Cardano discovered this after approaching del Ferro's family and going through the late mathematician's unpublished notes. Cardano thus felt morally justified in publishing the result, crediting del Ferro as the original inventor, and Tartaglia as the reinventor. The method was included in Cardano's *Ars Magna*, the most important book on algebra of the sixteenth century.

Tartaglia never forgave Cardano, and died an angry and bitter man. Cardano, however, lived until he was nearly 75. He died on 21 September 1576, the date he had predicted when casting his horoscope years before. Some maths historians claim he was in perfect health and drank poison just to ensure his prediction would come true.

Rather than just looking at equations with higher and higher powers of x, we can also increase complexity by adding a second unknown number, y. The school algebra favourite that is known as *simultaneous equations* is usually the task of solving two equations that each have two variables. For example:

$y = x$
$y = 3x - 2$

To solve the two equations, we substitute the value of the variable in one equation with the value from the other. In this case, since $y = x$, then:

$x = 3x - 2$
Which reduces to $2x = 2$
So $x = 1$, and $y = 1$

It's also possible to understand any equation in two variables visually. Draw a horizontal line and a vertical line that intersect. Define the horizontal line as the x-axis, and the vertical line as the y-axis. The axes intersect at 0. The position of any point in the plane can be determined by referencing a point on both axes. The position (a,b) is defined as the intersection of a vertical line through a on the x-axis and a horizontal line through b on the y-axis.

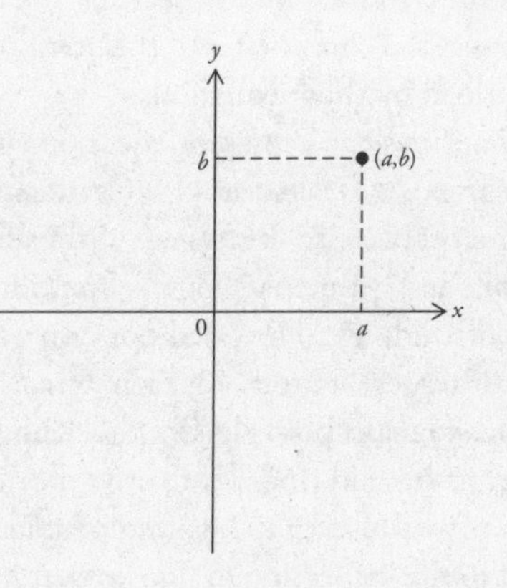

For any equation in x and y, the points where (x,y) has values for x and y that satisfy the equation describe a line on a graph. For example, the points (0,0), (1,1), (2,2) and (3,3) all satisfy our first equation above, $y = x$. If we mark, or plot, these points on a graph, it becomes clear that the equation $y = x$ generates a straight line, as in the figure below. Likewise, we can draw the second equation, $y = 3x - 2$. By assigning x a value and then working out what y is, we can establish that the points (0,–2), (1,1), (2,4) and (3,7) are on the line described by this equation. It is also a straight line, which crosses the y-axis at –2, below right:

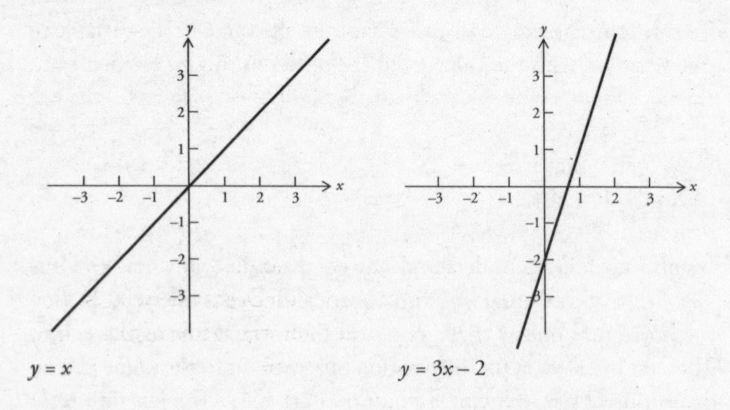

If we superimpose one of these lines over the other, we see that they cross at the point (1,1). So, we can see that the solution of simultaneous equations is the coordinates of the point of intersection of the two lines described by those equations.

The idea that lines can represent equations was the major innovation of Descartes' *La Géométrie*. His 'Cartesian' coordinate system was revolutionary because it forged a thus-far uncharted path between algebra and geometry. For the first time, two separate and distinct areas of study were revealed not only to be linked but also to be alternate representations of each other. One of Descartes' motivations was to make both algebra and geometry easier to understand because, as he said, independently 'they extend to only very abstract matters which seem to be of no practical use, [geometry] is always so tied to the inspection of figures that it cannot exercise the understanding without greatly tiring the imagination, while … [algebra] is so subjected to certain rules and numbers that it has become a confused and obscure art which oppresses the mind instead of being a science which cultivates it'. Descartes was no fan of overexertion. He was one of history's late risers, famously staying in bed until midday whenever he could.

The Cartesian marriage of algebra and geometry is a powerful example of the interplay between abstract ideas and spatial imagery, a recurring theme in mathematics. Many of the most impressive proofs in algebra – such as the proof of Fermat's Last Theorem – rely on geometry. Likewise, now that they could be described algebraically, 2000-year-old geometrical problems were given a new lease of life. One of the most exciting characteristics of maths is how seemingly different topics are interrelated, and how this in itself leads to vibrant new discoveries.

In 1649 Descartes moved to Stockholm to be personal tutor to Queen Christina of Sweden. She was an early bird. Unaccustomed to both the Scandinavian winter and to waking up at 5 a.m., he caught pneumonia shortly after arriving and died.

One of the most obvious corollaries of Descartes' insight that equations in x and y can be written as lines was the recognition that different types of equation produce different types of line. We can start classifying them here:

Equations such as $y = x$ and $y = 3x - 2$, in which the only terms are x and y, always produce straight lines.

By contrast, equations with quadratic terms – ones that include values for x^2 and/or y^2 – always produce one of the following four types of curve: circle, ellipse, parabola or hyperbola.

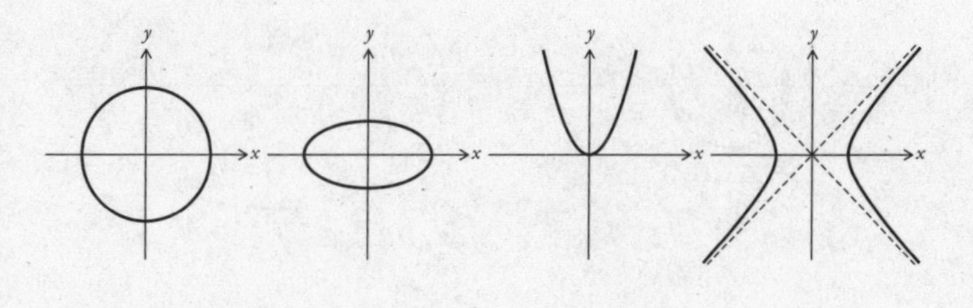

Circle *Ellipse* *Parabola* *Hyperbola*

The fact that every circle, ellipse, parabola and hyperbola that can be drawn can be described by a quadratic equation in *x*s and *y*s is helpful for science because all these curves are found in the real world. The parabola is the shape that describes the trajectory of an object flying through the air (ignoring air resistance and assuming a uniform gravitational field). When a soccer player kicks a ball, for example, it traces a parabola. The ellipse is the curve that describes how planets orbit around the sun, and the path followed by the shadow of the tip of a sundial during a day is a hyperbola.

Consider the following quadratic equation, which is like a machine for drawing circles and ellipses:

$$\frac{x^2}{a^2} + \frac{y^2}{b^2} = 1, \text{ where } a \text{ and } b \text{ are constants}$$

The machine has two knobs, one for a and one for b. By adjusting the values of a and b we can create any circle or ellipse with centre 0 that we want.

For example, when a is the same as b the equation is a circle with radius a. When $a = b = 1$, the equation is $x^2 + y^2 = 1$ and produces a circle with radius 1, also called the 'unit circle', as shown below left. And when $a = b = 4$, the equation is $\frac{x^2}{16} + \frac{y^2}{16} = 1$ and this is the circle with radius 4. If, on the other hand, a and b are different numbers, then the equation is an ellipse that crosses the x-axis at a, and the y-axis at b. For example, the curve below right is the ellipse when $a = 3$ and $b = 2$.

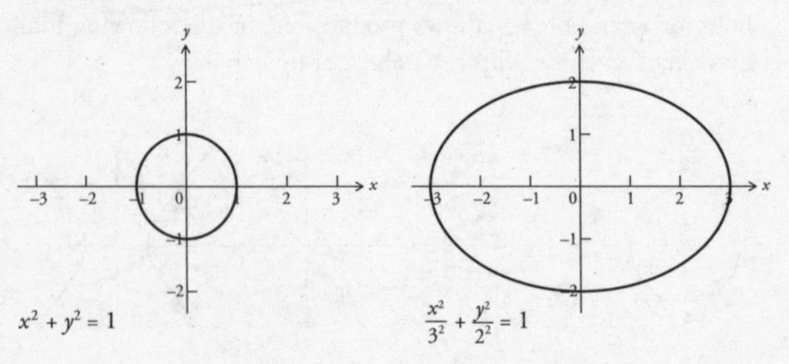

In 1818 the French mathematician Gabriel Lamé started to play around with the formula for the circle and ellipse. He wondered what would happen if he started to tweak the exponent, or power, rather than the values of *a* and *b*.

The effect of this adjustment was fascinating. For example, consider the equation $x^n + y^n = 1$. When $n = 2$, this creates the unit circle. Here are the curves produced by $n = 2$, $n = 4$ and $n = 8$:

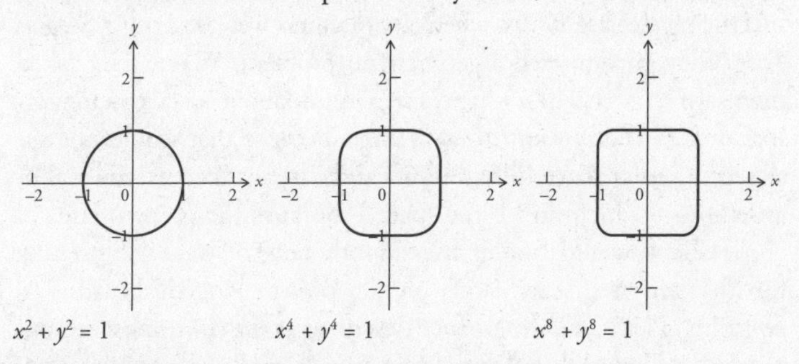

$x^2 + y^2 = 1$ $\quad$ $x^4 + y^4 = 1$ $\quad$ $x^8 + y^8 = 1$

When *n* is 4, the curve looks like an aerial view of a Babybel cheese squashed in a box. Its sides have become flattened and there are four rounded corners. It's as if the circle is trying to become a square. When *n* is 8, the curve is even more like a square.

In fact, the higher you push *n*, the closer the curve is to a square. In the limit, when $x^\infty + y^\infty = 1$, the equation *is* a square. (If anything deserves to be called the squaring of the circle, surely this is it.)

The same thing happens to an ellipse. If we take the ellipse described by $(\frac{x}{3})^n + (\frac{y}{2})^n = 1$, then by increasing the values of *n*, the ellipse will eventually turn into a rectangle.

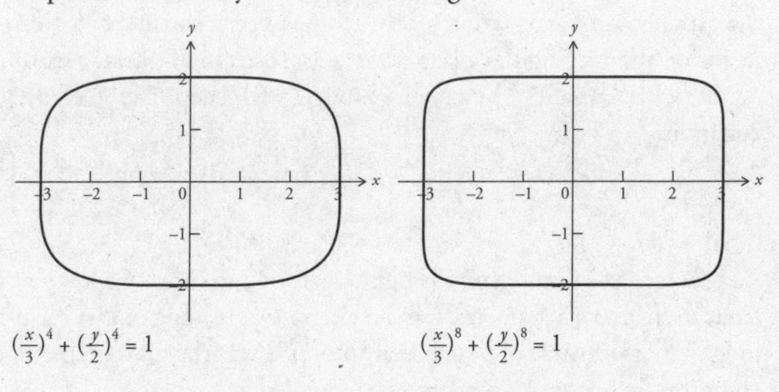

$(\frac{x}{3})^4 + (\frac{y}{2})^4 = 1$ $\quad$ $(\frac{x}{3})^8 + (\frac{y}{2})^8 = 1$

In downtown Stockholm there is a main public plaza called Sergels Torg. It's a large rectangular space, with a pedestrian lower level and a traffic circle on top. It's the place activists choose to hold political rallies, and where sports fans congregate when Sweden's national teams win a major event. The plaza's dominant feature is a central section with a sturdy 1960s sculpture that locals love to hate – a 37m-high glass and steel obelisk that lights up at night.

During the late 1950s, when city planners were designing Sergels Torg, they encountered a geometrical problem. What, they asked themselves, is the best shape for a roundabout in a rectangular space? They didn't want to use a circle because that would not use the rectangular space fully. But nor did they want to use an oval or an ellipse – which do fill the space – because the pointed ends of either shape would hinder the smooth flow of traffic. Searching for an answer, the architects on the project looked abroad, and consulted Piet Hein, a man once described as the third-most famous person in Denmark (after the physicist Niels Bohr and the writer Karen Blixen). Piet Hein was the inventor of the grook, a style of short aphoristic poem that he published in Denmark during the Second World War as a form of passive resistance against Nazi occupation. He was also a painter and mathematician, so possessed the right combination of artistic sensibilities, lateral thinking and scientific understanding to give fresh ideas to Scandinavian planning problems.

Piet Hein's solution was to find a shape that was halfway between an ellipse and a rectangle, using simple mathematics. To achieve this, he used the method described on the previous page. He adjusted the exponent in the equation for the ellipse to get a shape that would fit inside the rectangular plaza at Sergels Torg. In algebraic terms, he did what Lamé did by playing around with the n in the ellipse equation:

$$\left(\frac{x}{a}\right)^n + \left(\frac{y}{b}\right)^n = 1$$

As I showed previously, increasing the n from 2 to infinity takes you from a circle to a square, or from an ellipse to a rectangle. Piet Hein judged that the value of n such that the curve was the most aesthetic compromise between round and right-angled was when $n = 2.5$. He

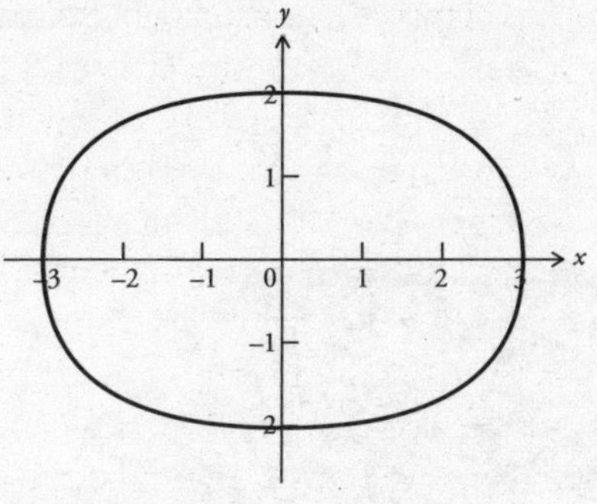

$(\frac{x}{3})^{2.5} + (\frac{y}{2})^{2.5} = 1$

could have called his new shape a 'squircle'. Instead, he called it a superellipse.

More than just an elegant piece of maths, Piet Hein's superellipse touched on a deeper human theme – the ever-present conflict in our surroundings between circles and straight lines. As he wrote, 'In the whole pattern of civilization there have been two tendencies, one toward straight lines and rectangular patterns and one toward circular lines.' His piece continued, 'There are reasons, mechanical and psychological, for both tendencies. Things made with straight lines fit well together and save space. And we can move easily – physically or mentally – around things made with round lines. But we are in a straitjacket, having to accept one or the other, when often some intermediate form would be better. The superellipse solved the problem. It is neither round nor rectangular, but in between. Yet it is fixed, it is definite – it has a unity.'

Stockholm's superelliptical roundabout was copied by other architects, most notably in the design for the Azteca stadium in Mexico City – which held the World Cup final in 1970 and 1986. In fact, Piet Hein's curve spread to fashion, becoming a feature of 1970s Scandinavian furniture design. It's still possible to buy superelliptic plates, trays and door-handles from the company run by Piet Hein's son.

Piet Hein's playful mind did not stop at the superellipse, however. For his next project, he wondered what a three-dimensional version of the shape would look like. The result was halfway between

Supereggstatic: Uri Geller.

a sphere and a box. He could have called it a 'sphox'. Instead, he called it a 'superegg'.

One unexpected feature of the superegg was that it could stand on its end without falling. In the 1970s Piet Hein marketed super-eggs made out of stainless steel as a 'sculpture, novelty or a charm'. They are beautiful, curious objects. I have one on my mantelpiece. Uri Geller also has one. John Lennon gave it to him, explaining that he had received the egg from aliens who visited him in his New York apartment. 'Keep it,' Lennon told Geller. 'It's too weird for me. If it's my ticket to another planet, I don't want to go there.'

CHAPTER SIX

Playtime

Maki Kaji runs a Japanese magazine that specializes in number puzzles. Kaji considers himself an entertainer who uses numbers as the tools of his trade. 'I feel more like a film or theatre director than a mathematician,' he explained. I met Kaji at his office in Tokyo. He was neither geeky nor formal, two characteristics one might expect to find in a numbers guy-turned-successful businessman. Kaji was wearing a black T-shirt under a trendy beige cardigan, and a pair of John Lennon glasses. Aged 57, he has a trim white goatee and side-burns, and often grins effervescently. Kaji keenly told me about his other hobbies, besides number puzzles. For instance, he collects rubber bands, and on a recent trip to London found what to him amounted to a glorious cache – a 25g pack of branded rubber bands from WH Smith and a 100g pack from an independent stationer. He also amuses himself by photographing arithmetically gratifying car licence plates. In Japan, licence plates consist of two numbers followed by another two numbers. Kaji carries a small camera at all times and snaps every registration he sees for which the first pair multiplied by each other equals the second pair.

Assuming that no Japanese car has a 00 for the second pair of digits, each plate Kaji photographs is a line in the times tables of the digits 1 to 9. For instance, 11 01 can be thought of as $1 \times 1 = 1$. Likewise, 12 02 is $1 \times 2 = 2$. We can carry on the list, and work out that there are 81 possible combinations. Kaji has already collected more than 50. Once he has gathered the full set of times tables, he plans to exhibit them in a gallery.

The idea that numbers can entertain is as old as maths itself. The ancient Egyptian Rhind Papyrus, for example, contains the following list as part of the answer to problem 79. Unlike the other problems in the papyrus, this one has no apparent practical application.

Kaji snaps 3 × 5 = 15 in a Tokyo car park.

Houses	7
Cats	49
Mice	343
Spelt	2401*
Hekat†	16,807
Total	19,607

The list is the inventory of seven houses, each of which had seven cats, each of which ate seven mice, each of which ate seven grains of spelt, each of which came from a separate *hekat*. The numbers form a *geometric progression* – which is a sequence where each term is calculated by multiplying the previous term by a fixed number, in this case, seven. There are seven times more cats than houses, seven times more mice than cats, seven times more grains of spelt than mice, and seven times more *hekats* than grains of spelt. We could rewrite the total number of items as $7 + 7^2 + 7^3 + 7^4 + 7^5$.

It wasn't just the Egyptians, however, who found such a sequence irresistible. Almost exactly the same sum reappeared in the early nineteenth century in a Mother Goose nursery rhyme:

---

* In the original, the spelt figure is mistakenly written 2301.

† Egyptian unit of volume.

As I was going to St Ives,
I met a man with seven wives,
Every wife had seven sacks,
Every sack had seven cats,
Every cat had seven kits.
Kits, cats, sacks, wives,
How many were going to St Ives?

The verse is the most famous trick question in English literature since, presumably, the man and his phalanx of females and confined felines were coming *from* St Ives. Irrespective of the direction of travel, however, the total number of kits, cats, sacks and wives is $7 + 7^2 + 7^3 + 7^4$, which is 2800.

Another, less well-known, appearance of the riddle is as a problem in Leonardo Fibonacci's *Liber Abaci*, from the thirteenth century. This version involved seven women on their way to Rome with increasing numbers of mules, sacks, loaves, knives and sheaths. The extra 7^6 brings the series to 137,256.

What is the allure of rising powers of seven that they have appeared in such different ages and contexts? Each case demonstrates the turbo-charged acceleration of geometric progressions. The rhyme is a poetic way of showing how quickly small numbers can lead to big ones. On first hearing, you think there might be a fair amount of kits, cats, sacks and wives – but not almost 3000 of them! Likewise, the playful problems set in the Rhind Papyrus and the *Liber Abaci* express the same mathematical insight. And the number 7, though it would seem like it should have some special quality to make it so common across these problems, is rather irrelevant. When you multiply any number by itself a few times, the sum quickly reaches a counter-intuitively high amount.

Even when multiplying the lowest number possible, 2, by itself, the sum swirls to the heavens at a dizzying pace. Place one grain of wheat on the corner square of a chessboard. Place two grains on the adjacent square, and then start filling up the rest of the board by doubling the grains of wheat per square. How much wheat would you need to fill the final square? A few truckloads, or a container, maybe? There are 64 squares on a chessboard, so we have doubled up 63 times, meaning that the number is 2 multiplied by itself 63 times, or 2^{63}. In grains, this number is about 100 times more than the world's current annual wheat production. Or, to consider it another way, if you started counting a grain of wheat per second at the very moment of the Big Bang 13 billion or so years ago, then you would not even have counted up to a tenth of 2^{63} by now.

Mathematical riddles, rhymes and games are now collectively known as *recreational maths*. It is a wide-ranging and vibrant field, an essential feature of which is that the topics are accessible to the dedicated layperson, even though they might touch on impossibly complicated theory. Or they might not even involve theory at all, but rather merely kindle an appreciation of the wonder of numbers – such as the thrill of collecting pictures of licence plates.

A landmark event in the history of recreational maths is said to have taken place by the banks of the Yellow River in China around 2000 BC. According to legend, Emperor Yu saw a turtle creep out of the water. It was a divine turtle, with black and white dots on its underbelly. The dots denoted the first nine numbers and formed a grid on the turtle's belly that (if the dots were instead written as Arabic numerals) looked like A:

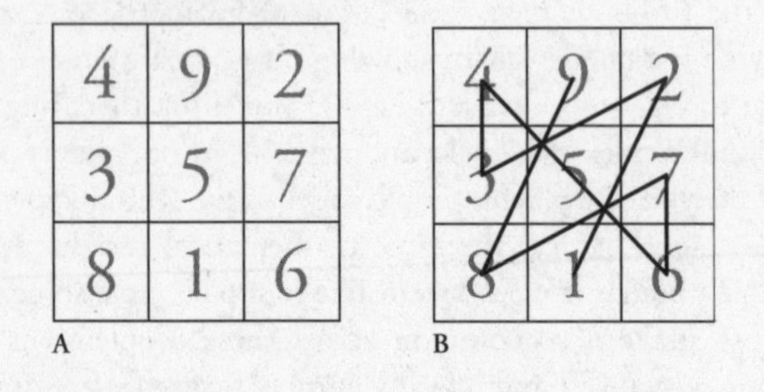

A square like this one, that contains all the consecutive numbers starting from 1 and arranged so that all the rows, columns and corner-to-corner diagonals add up to the same total, is known as a *magic square*. The Chinese called this square the *lo shu*. (Its rows, columns and diagonals all add up to 15.) The Chinese believed that the *lo shu* symbolized the inner harmonies of the universe and used it for divination and worship. For example, if you start at 1 and draw a line between the numbers of the square in order, you map out the pattern that can be seen in B and the figure opposite, which

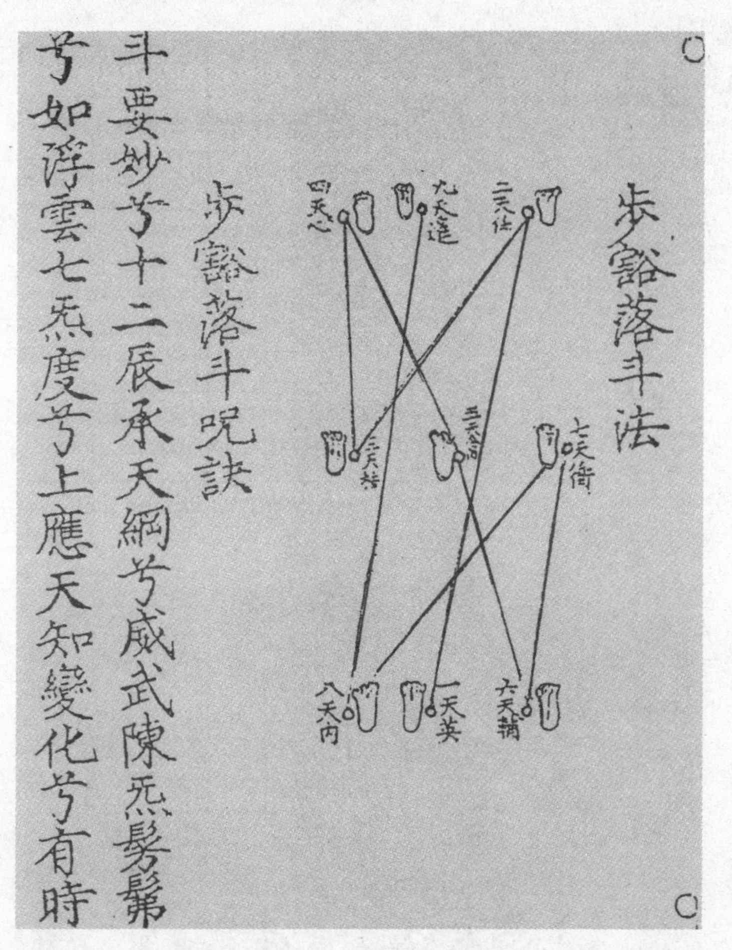

Taoist woodblock print with the yubu.

shows the instructions for the movement of Taoist priests through a temple. The pattern, which is called the *yubu*, also underlies some of the rules of *feng shui*, the Chinese philosophy of aesthetics.

China was not the only culture to see the mystical side of the *lo shu*. Magic squares have been objects of spiritual importance for Hindus, Muslims, Jews and Christians. Islamic culture found the most creative uses. In Turkey and India virgins were required to embroider magic squares on the shirts of warriors. And if a magic square were

Melencolia I: *Dürer's famous woodcut shows an angel lost in thought surrounded by mathematical and scientific objects, such as a compass, a sphere, a set of scales, an hourglass and a magic square. Art historians, especially those with a mystical bent, have long pondered the symbolism of the geometrical object in the middle left of the image, which is known as 'Dürer's solid'; mathematicians have long pondered the mystery of how on earth to construct it.*

placed over the womb of a woman in labour, it was believed that the birth would be an easier one. Hindus wore amulets with magic squares as protective charms, and Renaissance astrologers associated them with the planets in our solar system. It is easy to mock a predisposition for the occult in our ancestors, yet modern man can understand their fascination with magic squares. Both simple and yet subtly complex, a magic square is like a numerical mantra, an object of endless contemplation and a self-contained expression of order in a disordered world.

One of the pleasures of magic squares is that they are not restricted to 3 × 3 grids. A famous example of a 4 × 4 square comes from the work of Albrecht Dürer. In *Melencolia I* (shown opposite), Dürer included a 4 × 4 square that is best known for containing the year in which he engraved it: 1514.

Dürer's square, in fact, is übermagic. Not only do the rows, columns and diagonals add up to 34, but so do the combinations of four numbers marked by dots and linked in the squares below.

The patterns Dürer's square produces are amazing, and the more you look, the more you find. For example, the sum of the squares of the numbers on the first and second rows adds up to 748. You get the same total by adding up the squares of the numbers in rows 3

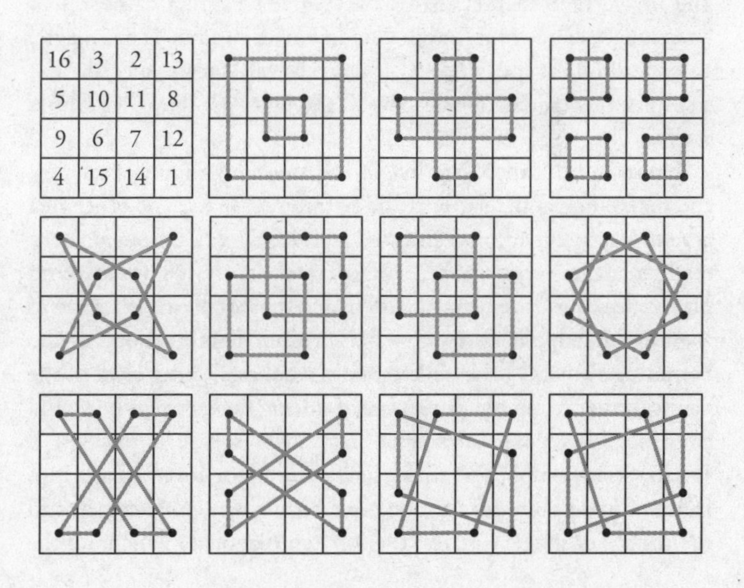

and 4, or the squares of the numbers in rows 1 and 3, or the squares of the numbers in rows 2 and 4, or the squares of the numbers in both diagonals. Wow!

For further amazement, rotate Dürer's square by 180 degrees, then subtract 1 from the squares containing 11, 12, 15 and 16. The result is the following:

The image is from the side of the Sagrada Família cathedral in Barcelona, designed by Antoni Gaudí. Gaudí's square is not magic, as two numbers are repeated, but it is still pretty special. The columns, rows and diagonals now all add up to 33: the age of Christ at his death.

Hours of fun can be had by playing around with magic squares, and marvelling at the patterns and harmonies. In fact, no other area of non-practical maths has attracted as much attention from amateur mathematicians over such a long period. In the eighteenth and nineteenth centuries, literature on magic squares flourished. One of the most notable enthusiasts was Benjamin Franklin, one of the Founding Fathers of the United States who, as a young clerk of the Pennsylvania Assembly, got so bored during debates that he would construct his own squares. His best-known square is the 8 × 8 variation shown opposite, which he is said to have invented as a boy. In this square Franklin included one of his own enhancements to the theory of magic squares: the 'broken diagonal', which are the

Fig. III. *Page* 351.

52	61	4	13	20	29	36	45
14	3	62	51	46	35	30	19
53	60	5	12	21	28	37	44
11	6	59	54	43	38	27	22
55	58	7	10	23	26	39	42
9	8	57	56	41	40	25	24
50	63	2	15	18	31	34	47
16	1	64	49	48	33	32	17

In a letter published in 1769, Benjamin Franklin commented about a book of magic squares: 'In my younger days … I had amused myself in making these kind of magic squares and, at length, had acquired such a knack at it, that I could fill the cells of any magic square, of reasonable size, with a series of numbers as fast as I could write them, disposed in such a manner, as that the sums of every row, horizontal, perpendicular, or diagonal, should be equal; but not being satisfied with these, which I looked on as common and easy things, I had imposed on myself more difficult tasks, and succeeded in making other magic squares, with a variety of properties, and much more curious.' He then introduced the square above, printed in his Experiments and Observations on Electricity, made at Philadelphia in America*, in 1769.*

numbers in the black squares and grey squares, shown in A and B below. While his square isn't a proper magic square because the full diagonals don't add up to the number 260, his newly invented broken diagonals do. The sums of the black squares in C and D and E, and the sum of the grey squares in E and, of course, the sum of every row and column, also add up to 260.

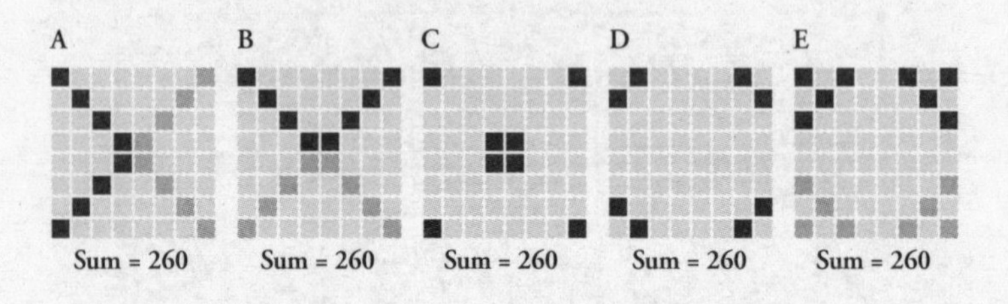

Franklin's square contains even more beguiling symmetries. The sum of the numbers in every 2 × 2 subsquare is 130, as is the sum of any four numbers that are arranged equidistant from the centre. Franklin is also said to have invented another square in his forties. Over the course of a single evening, he composed an incredible 16 × 16 square that he claimed was 'the most magically magical of any magic square ever made by any magician'. (It is in the appendices, p. 420.)

One of the reasons for the enduring popularity of constructing magic squares is that there is a surprising number of them. Let's count them, starting from the smallest: there is just one magic square in a 1 × 1 grid: the number 1. There are no magic squares with four numbers in a 2 × 2 grid. There are eight ways to arrange the digits 1 to 9 so that the resulting 3 × 3 square is magic, but each of these eight squares is really the same square either rotated or reflected, so it's conventional to say that there is only one true 3 × 3 magic square. The figure on the opposite page shows how to generate each possibility, starting with the *lo shu*.

Amazingly, after three, the number of magic squares that can be made grows staggeringly fast. Even after reducing the number by ignoring rotations and reflections, it is possible to make 880 magic squares in a 4 × 4 grid. In a 5 × 5 grid the number of magic squares

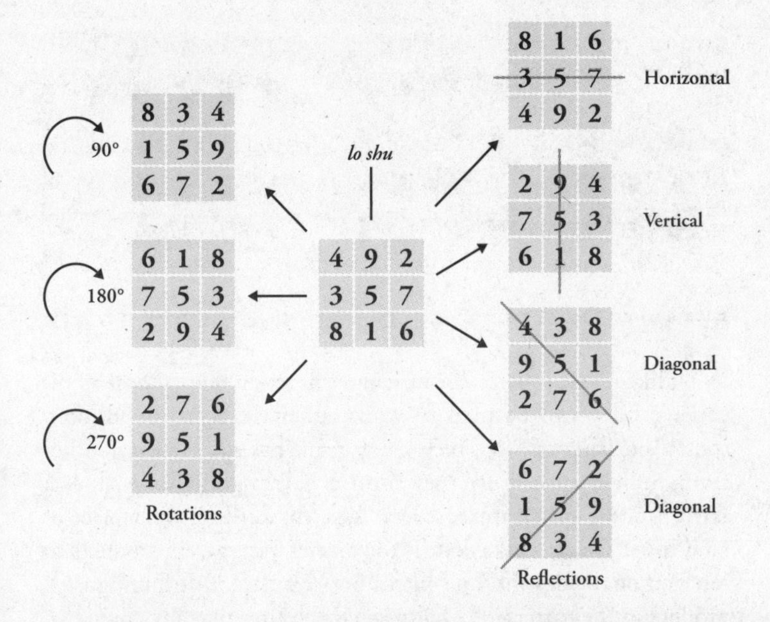

is 275,305,224, a result calculated in 1973, and only through the use of a computer. And though this number seems astronomically high, it is, in fact, tiny compared to the number of all possible arrangements of the digits 1 to 25 in a 5 × 5 square. The total number of arrangements is calculated by multiplying 25 by 24 by 23 and so on until 1, which is about 1.5 followed by 25 zeros, or 15 septillions.

The number of 6 × 6 magic squares is not even known, although it is likely to be in the order of 1 followed by 19 zeros. The number is so huge that it exceeds even the total number of grains of wheat in the chessboard example on p. 215.

Magic squares have not been just the province of amateurs. At the end of his life, the eighteenth-century Swiss mathematician Leonhard Euler became curious about them. (He was almost totally blind by this time, which makes his research into what is an essentially spatial application of numbers especially awe-inspiring.) In particular, his work included the study of a modified version in which each number or symbol in the grid appears exactly once in each row and column. He called it a Latin square.

●	▲	■
▲	■	●
■	●	▲

£	¥	$	€
€	$	¥	£
$	£	€	¥
¥	€	£	$

2	3	4	0	1
0	1	2	3	4
3	4	0	1	2
1	2	3	4	0
4	0	1	2	3

Latin squares.

Unlike magic squares, Latin squares have several practical applications. They can be used to work out brackets in round-robin sports tournaments, in which every team has to play every other team, and in agriculture they form a handy grid that enables a farmer to test, for example, several different fertilizers on a piece of land to see which works best. If the farmer has, say, six products to test and he divides the land into a 6 × 6 square, distributing each product in the pattern of a Latin square ensures that any change in soil conditions affects each treatment equally.

Maki Kaji, the Japanese puzzle-maker I introduced at the start of the chapter, ushered in a new era of number-square fascination. The idea came to him as he was browsing an American puzzle magazine. As a non-English-speaker, he scanned pages of incomprehensible word games before stopping when he came across an intriguing-looking grid of numbers. The puzzle, entitled 'Number Place', was a partially completed 9 × 9 Latin square that used the digits 1 to 9. Based on the rules that each number was allowed to appear only once per row and per column, the would-be solver needed to figure out how to fill in the missing gaps using a process of logical deduction. Solvers were aided by a further condition: the square was divided into nine 3 × 3 subsquares, each of which was marked in bold. Each number 1–9 was allowed only once per subsquare. Kaji solved Number Place and got excited – this was precisely the sort of puzzle he wanted to put in his new magazine.

Number Place, which had made its first appearance in 1979, was the creation of Howard Garns, a retired architect and puzzle enthusiast from Indiana. Though he enjoyed solving Garns's puzzle, Kaji

Soroban champion Yuzan Araki, aged eight, with trophy and medals. 'I like to calculate fast,' he said.

Yuji Miyamoto, the inventor of Flash Anzan, and his youngest class of soroban pupils. *(See pp. 68–75.)*

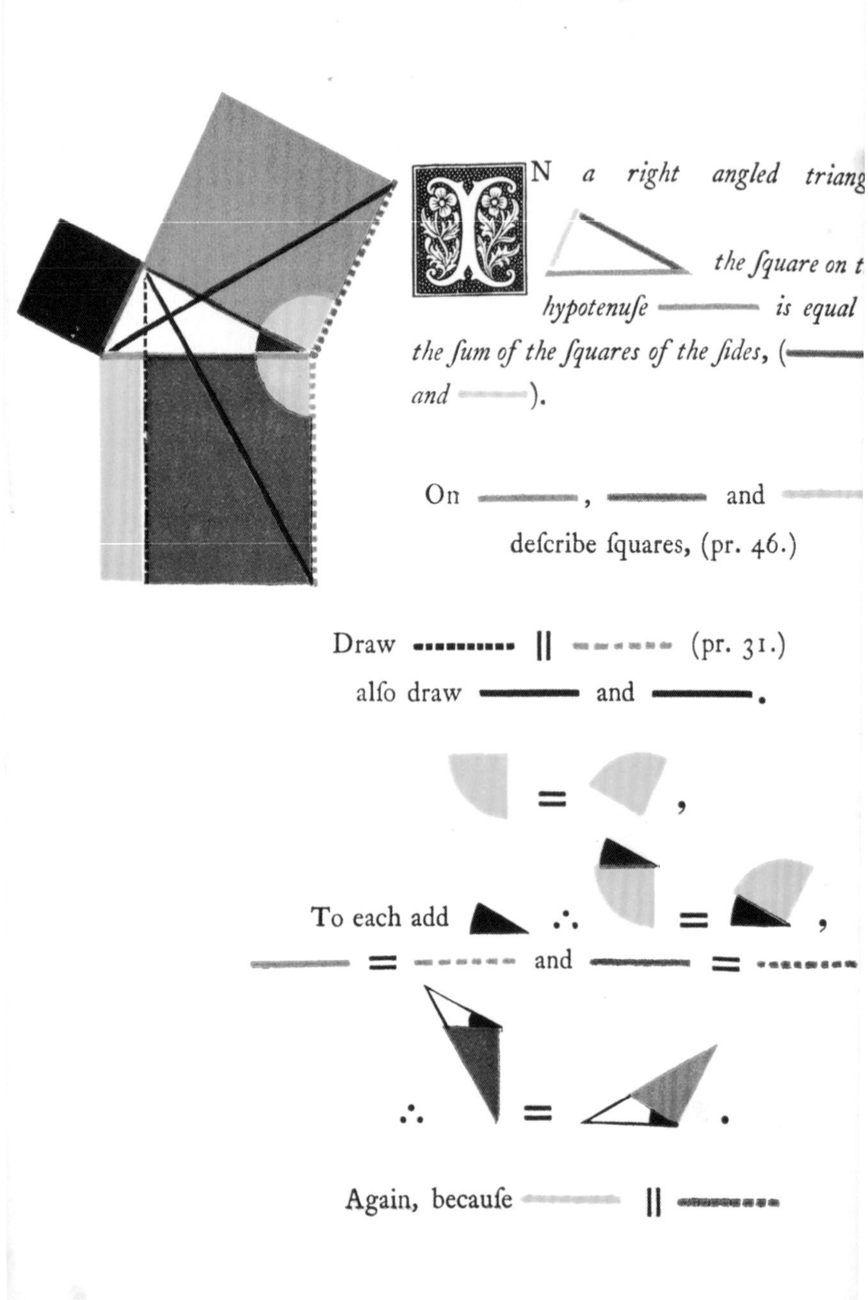
N a right angled triang
the ſquare on t
hypotenuſe is equal
the ſum of the ſquares of the ſides, (
and).
On , and
deſcribe ſquares, (pr. 46.)
Draw || (pr. 31.)
alſo draw and .
= ,
To each add ∴ = ,
= and =
∴ = .
Again, becauſe ||

BOOK I. PROP. XLVII. THEOR. 49

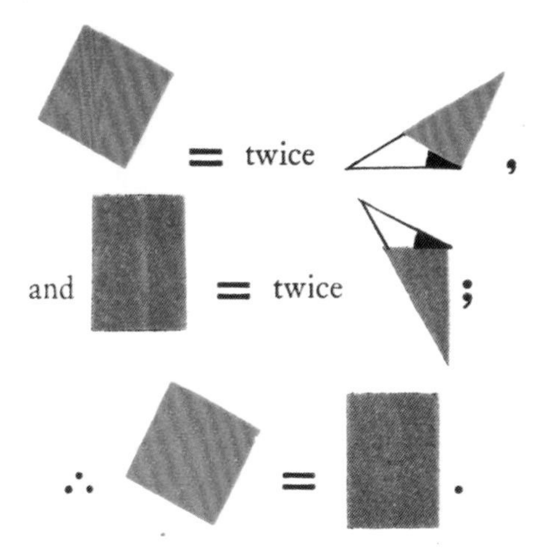

In the ſame manner it may be ſhown

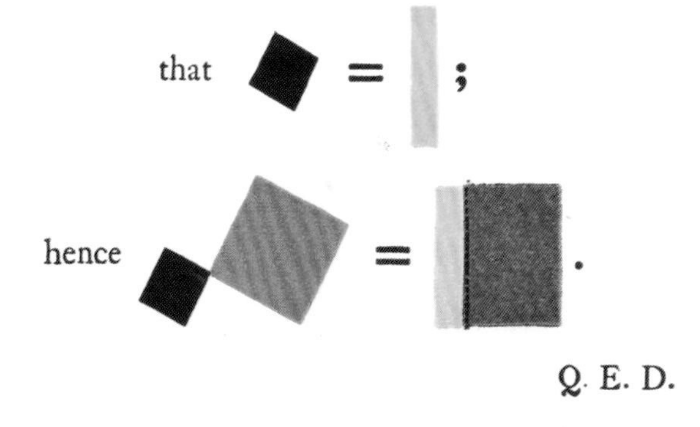

Pythagoras's theorem from Oliver Byrne's remarkable 1847 version of Euclid's The Elements, *in which the propositions are expressed using blocks of colour.* *(See p. 97.)*

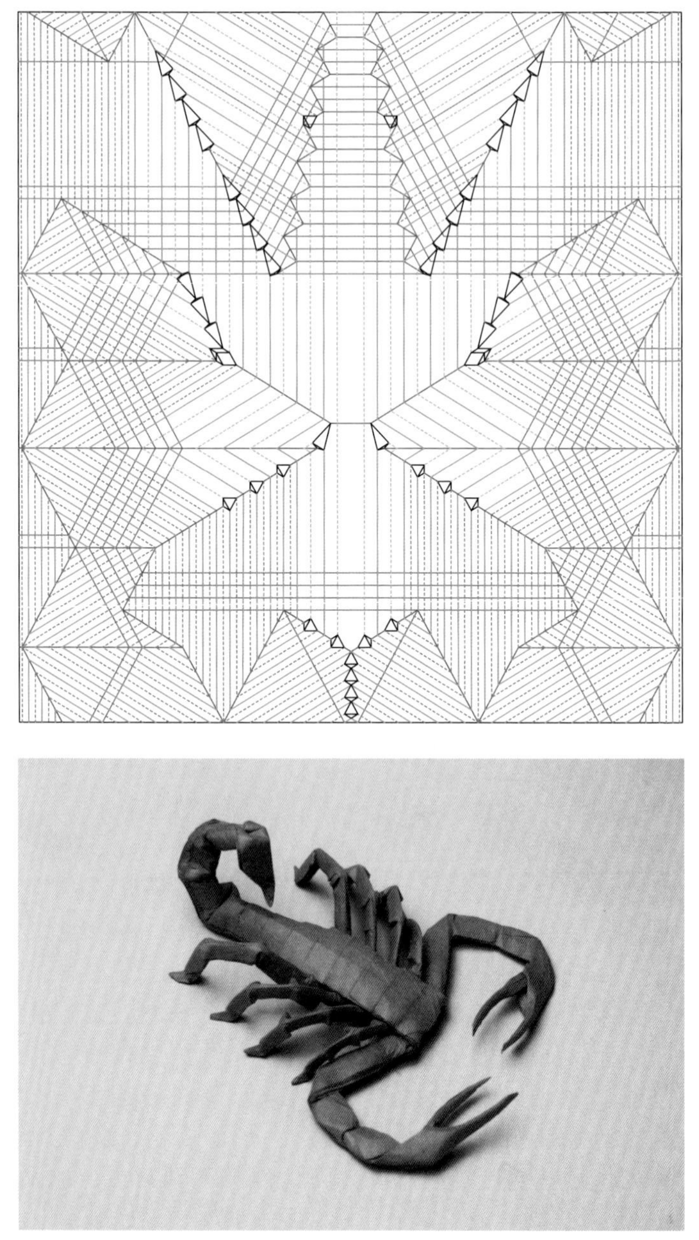

An origami scorpion and its crease pattern, created by former NASA physicist Robert Lang. *(See p. 108.)*

Islamic geometers covered their holy sites with sophisticated tile patterns, like this mosaic from the Alhambra palace in Granada. The ever-repeating patterns were representations of how God reveals himself through mathematical forms. (See pp. 98–9.)

A Latin square at Rothamsted Research Farm in Harpenden. Six chemical treatments are arranged so that each row and column has one square of each treatment. (See p. 224.)

Modern-day Pythagorean Jerome Carter and his wife Pamela, outside their dream home in Scottsdale, Arizona (see pp. 77–9). He has advised hip-hop royalty on the numbers behind their names.

The Shankaracharya of Puri, right, sitting on his throne, next to his chief disciple and interpreter. A picture of Shankara is on the wall behind. (See pp. 133–9.)

Contestants limbering up at the Mental Calculation World Cup (see p. 145). German Jan van Koningsveld, in the foreground, won the square root and calendar calculation rounds.

Pi men: brothers Gregory and David Chudnovsky, left and right, with their collaborator Tom Morgan, centre. They built a supercomputer in Gregory's New York apartment that calculated pi to more than two billion decimal places. (See pp. 165–9.)

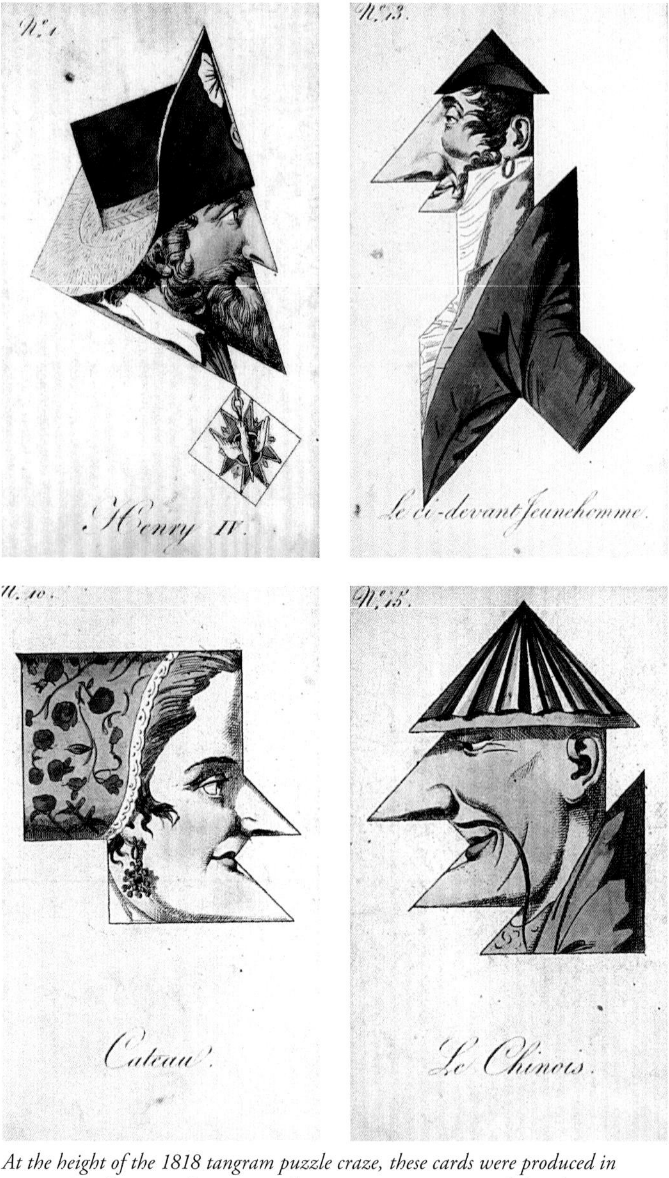

At the height of the 1818 tangram puzzle craze, these cards were produced in France. The figures – of Henry IV, the Young man, Cateau and the Chinese man – were to be assembled from the seven geometrical pieces.

OPPOSITE *The tangram-style trench patience game, played (and partly designed) by German soldiers in the First World War.* *(See pp. 230–2.)*

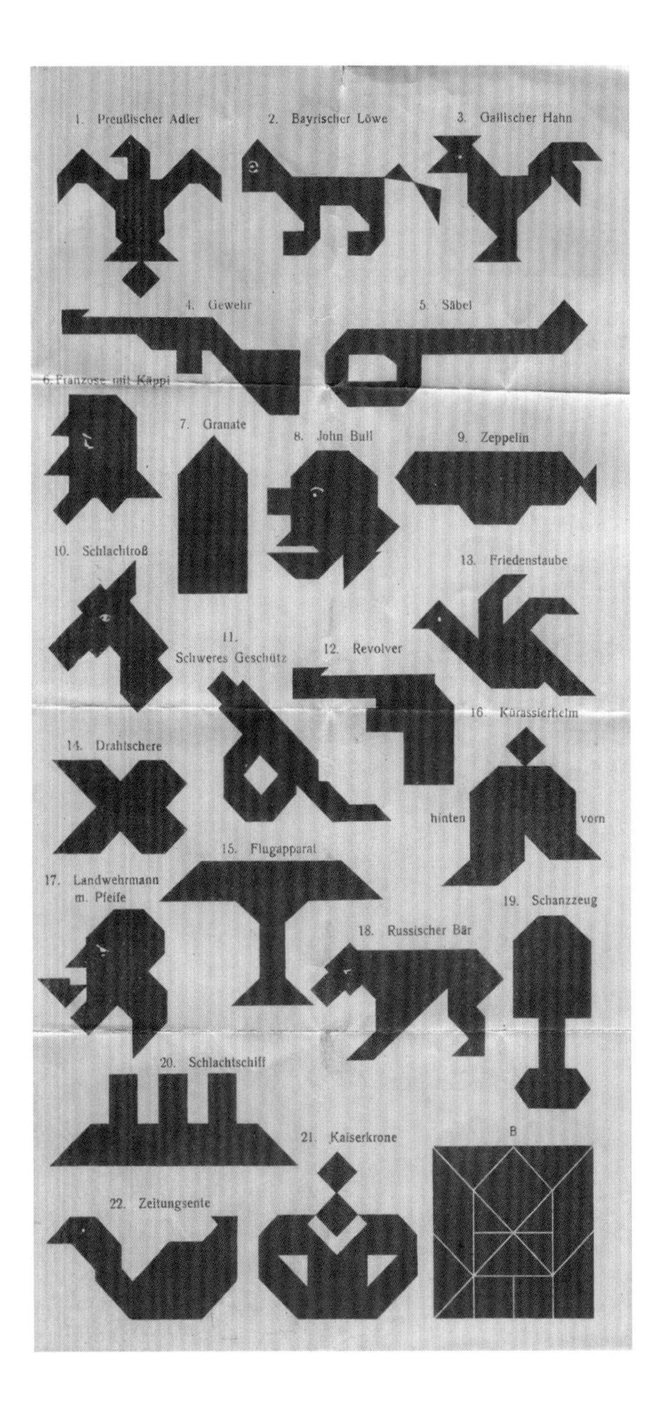

1. Preußischer Adler
2. Bayrischer Löwe
3. Gallischer Hahn
4. Gewehr
5. Säbel
6. Franzose mit Käppi
7. Granate
8. John Bull
9. Zeppelin
10. Schlachtroß
13. Friedenstaube
11. Schweres Geschütz
12. Revolver
16. Kürassierhelm
14. Drahtschere
hinten
vorn
15. Flugapparat
17. Landwehrmann m. Pfeife
19. Schanzzeug
18. Russischer Bär
20. Schlachtschiff
21. Kaiserkrone
B
22. Zeitungsente

Erik Demaine, computer scientist (see pp. 108 and 242–3).

Ivan Moscovich, puzzlist (see pp. 245–6).

Raymond Smullyan, logician (see pp. 86 and 245).

Neil Sloane, sequence supremo (see pp. 255–65).

The four men above were all guests at the 2008 Gathering for Gardner in Atlanta, an event that celebrates the work of Martin Gardner, the king of recreational mathematics (see pp.243–53). Gardner, pictured right at his home in Oklahoma, still works standing up at his wooden desk.

MARTIN
One Simple Solution

Claw blimey! Eddy Levin with his golden mean gauge in his garden in north London. He found phi in a flower, and also a peacock feather. *(See pp. 283–301.)*

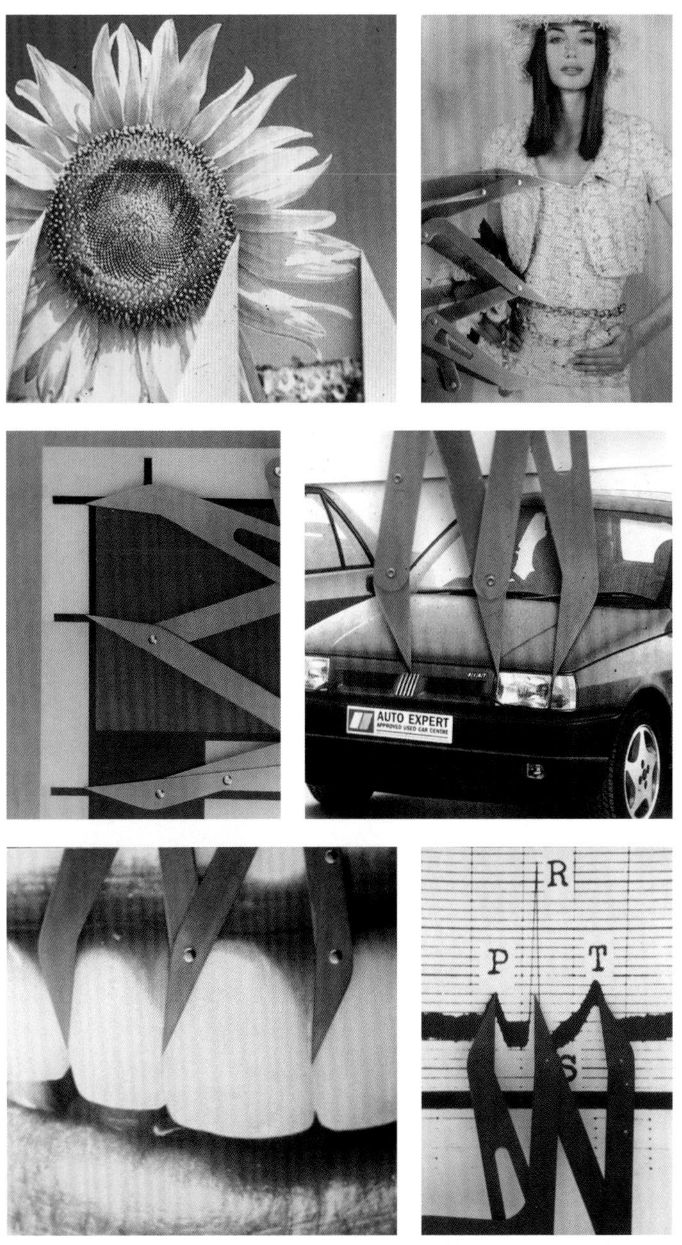

Levin says that wherever there is beauty, there is the golden ratio: in a sunflower, a dress, a Mondrian painting, a Fiat car, a set of perfect teeth and the graph of a heartbeat.

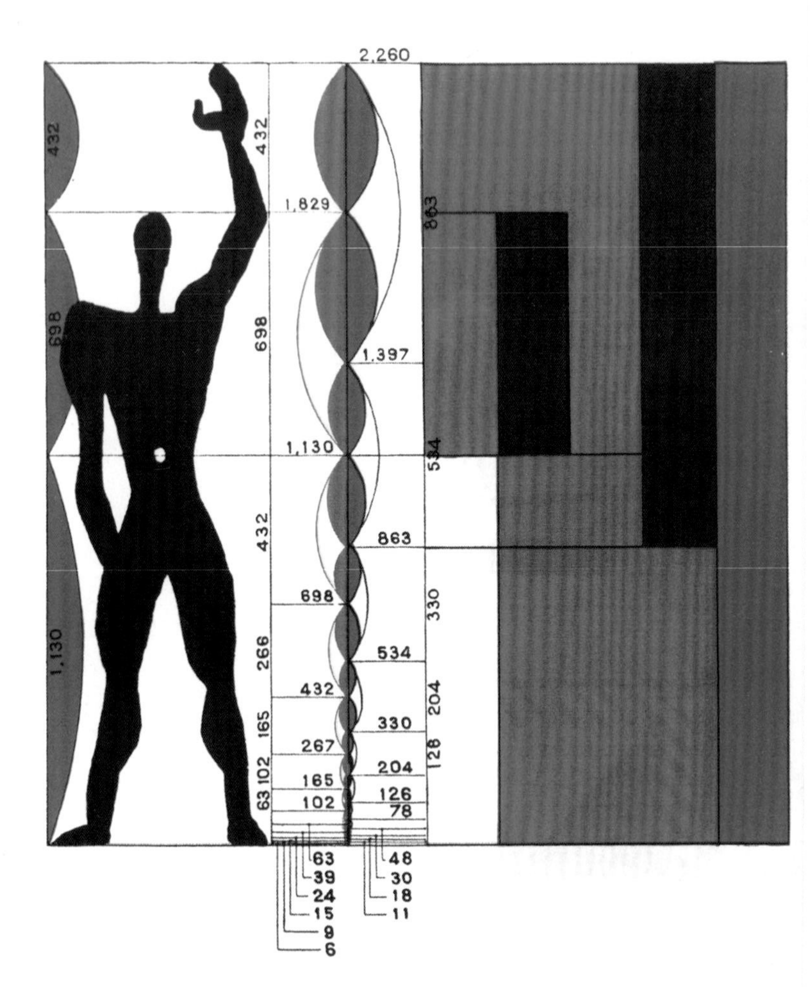

In 1948, the French architect Le Corbusier published Le Modulor, a system of proportions based on the golden ratio (see p. 299). He said the Modulor was a 'range of harmonious measurements to suit the human scale, universally applicable to architecture and to mechanical things'.

Anthony Baerlocher, seen here in his Reno office, sets the odds for more than half of the world's slot machines (see pp. 316–23). Slots are now the most profitable, and addictive, casino game.

Ed Thorp conquered two other casino games: he made the first wearable computer to win at roulette and invented card-counting to win at blackjack. He then used maths to make vast sums in the financial markets. (See pp. 337–47.)

Use your loaf! Baguettes from Greggs mount up in my kitchen in an experiment to discover the maths behind measurement error. *(See p. 349.)*

Daina Taimina took six months and 5.5km of pink yarn to make the world's largest hyperbolic crochet model, here next to Mango the cat. *(See pp. 383–4 and 395–8.)*

<table>
<tr><td></td><td>1</td><td>5</td><td></td><td>4</td><td></td><td>8</td><td></td><td></td></tr>
<tr><td></td><td>8</td><td></td><td>3</td><td></td><td></td><td></td><td>5</td><td></td></tr>
<tr><td>6</td><td></td><td></td><td></td><td></td><td>5</td><td></td><td>2</td><td></td></tr>
<tr><td>8</td><td></td><td></td><td></td><td>6</td><td></td><td>5</td><td></td><td></td></tr>
<tr><td></td><td></td><td>4</td><td>8</td><td></td><td>3</td><td>6</td><td></td><td></td></tr>
<tr><td></td><td></td><td>6</td><td></td><td>9</td><td></td><td></td><td></td><td>1</td></tr>
<tr><td></td><td>2</td><td></td><td>1</td><td></td><td></td><td></td><td></td><td>8</td></tr>
<tr><td></td><td>9</td><td></td><td></td><td></td><td>8</td><td></td><td>7</td><td></td></tr>
<tr><td></td><td></td><td>8</td><td></td><td>7</td><td></td><td>2</td><td>1</td><td></td></tr>
</table>

decided to redesign it so that the numbers provided were distributed in a symmetrical pattern around the grid, matching the format used for crosswords. He called his version Sudoku, the Japanese for 'the number must appear only once'.

Sudoku appeared in the early issues of Kaji's puzzle magazine, which launched in 1980, but Kaji said it attracted no attention. It was only once the puzzle travelled overseas that it spread like wildfire.

Just as a Japanese-speaker with no English could understand Number Place, so an English-speaker with no Japanese could understand Sudoku. In 1997 a New Zealander named Wayne Gould walked into a bookstore in Tokyo. Although he was initially disoriented by the fact that everything was in Japanese, his eyes eventually landed on something familiar. He saw a book cover with what looked like a crossword grid with numbers on it, and though the image was obviously some kind of puzzle, he didn't instantly understand the rules. Still, he bought the book, thinking he'd figure it out later. On a holiday in southern Italy he worked backwards until he cracked the puzzle. Gould had just retired from being a judge in Hong Kong and was teaching himself how to program computers, so he decided he would try to write a program that generated Sudokus. A top programmer might take a couple of days for this task. It took Gould six years.

But the effort was worth it, and in September 2004 he persuaded New Hampshire's *Conway Daily Sun* to publish one of his puzzles. It was an immediate success. The following month he decided to approach the British national press. Gould thought the most effective way to pitch his idea was to present a mock-up of that day's paper with the Sudoku already in it. He knew enough about forgery from covering trials in Hong Kong to make a convincing fake of *The Times*'s second section, and took it to the paper's head office. After waiting a few hours in reception, Gould showed them his dummy paper, and they seemed to like it. In fact, immediately after he left, a *Times* executive sent Gould an email asking him not to show the Sudoku puzzles to anyone else. Two weeks later the puzzle first appeared, and three days after that, the *Daily Mail* introduced its own version. In January 2005 the *Daily Telegraph* joined the game, and not long afterwards every British newspaper had to have a daily puzzle to keep up with the competition. That same year the *Independent* reported a 700 per cent rise in UK pencil sales, and attributed it to the craze. By summer, shelves of Sudoku books appeared in bookshops, newsagents and airports, and not just in the UK but around the world. At one point in 2005 six of the top 50 books on *USA Today*'s bestseller list were Sudoku titles. By the end of the year the puzzle had spread to 30 countries, and *Time* magazine named Wayne Gould one of the 100 people who most shaped the world that year, along with Bill Gates, Oprah Winfrey and George Clooney. By the end of 2006, Sudokus were being published in 60 countries; and by the end of 2007, in 90. According to Maki Kaji, the number of regular Sudoku players now exceeds 100 million people.

Completing any puzzle is immensely gratifying to the ego, but part of the extra allure of completing a Sudoku is the inner beauty and balance of the perfect Latin square that gives it its form. Sudoku's success is testament to an age-old, cross-cultural fetish for number squares. And unlike many other puzzles, its success is also a remarkable victory for mathematics. The puzzle is maths by stealth. Although Sudoku contains no arithmetic, it does require abstract thought, pattern recognition, logical deduction and the generation of algorithms. The puzzle also encourages an aggressive attitude towards problem-solving, and fosters an appreciation of mathematical elegance.

For instance, as soon as you understand the rules of Sudoku, the concept of a *unique solution* is remarkably clear. For every pattern of numbers in the grid at the start, there is only one possible final arrangement for the numbers in the empty spaces. However, it is not the case that every partially filled grid has a unique solution. It is perfectly possible that a 9 × 9 square with some numbers filled in has no solutions, just as it is perfectly possible that it has many solutions. When Sky TV launched a Sudoku show, they drew what they claimed was the world's largest Sudoku by cutting a 275 × 275ft grid out of a chalk hillside in the English countryside. However, the given numbers were placed in such a way that there were 1905 valid ways to complete the square. The proclaimed largest Sudoku did not have a unique solution, and therefore was not a Sudoku at all.

The branch of mathematics that involves the counting of combinations, such as all the 1905 solutions to Sky TV's faux Sudoku, is called *combinatorics*. It is the study of permutations and combinations of things, such as grids of numbers, but also, famously, the schedules of travelling salesmen. Let's say, for instance, that I'm a travelling salesman and I have 20 shops to visit. In what order should I visit them so that my total distance is the shortest? The solution requires me to consider all the permutations of paths between all the shops, and is a classic (and extremely difficult) combinatorial problem. Similar problems arise throughout business and industry, for example in scheduling flight departure times at airports or having an efficient postal sorting system.

Combinatorics is the branch of maths that most consistently deals with extremely high numbers. As we saw with magic squares, a small set of numbers can be rearranged in an astonishingly large number of ways. Though they both share square grids, there are fewer Latin squares than magic squares for the same size of grid, though the number of Latin squares is still colossal. The number, for example, of 9 × 9 Latin squares is 28 digits long.

How many possible Sudokus are there? If a 9 × 9 Latin square is to qualify as a finished Sudoku grid, the 9 subsquares must also include every digit, and this reduces the total of Sudoku-ready 9 × 9 squares to 6,670,903,752,021,072,963,960. Many of these grids, however, are different versions of the same square when reflected or rotated (as I showed with the 3 × 3 magic square on p. 223). Eliminating squares that are rotations and reflections, the number of distinct possible finished Sudoku grids is about 5.5 billion.

Still, this is not the total number of possible Sudokus, which is much larger since each finished grid will be the solution to many Sudokus. For example, a Sudoku in a newspaper has one unique solution. Once you fill in one of the squares, however, you are creating a new grid with a new set of givens, in other words a new Sudoku with the same unique solution, and so on for each square you fill in. So if a Sudoku has, say, 30 given numbers, then we will be able to create another 50 Sudokus with the same unique solution until we complete the grid. (That's one new Sudoku for each extra number, until there are 80 givens in the 81-square grid.) Finding the total number of Sudokus is not that interesting since what we will find is that most of these Sudokus have grids with very few blank spaces in them, which is not in the spirit of the puzzle. Instead, mathematicians get much more excited about the number of digits you can leave in the square. The number one combinatorics question about Sudoku is what is the least amount of numbers you can leave in the square so that there is only one way to fill in the grid?

The Sudokus published in newspapers usually include around 25 given numbers. To date, no one has found a Sudoku that has a unique solution with fewer givens than 17. In fact, 17-clue Sudokus inspire something of a combinatorics cult. Gordon Royle, of the University of Western Australia, maintains a database of 17-clue Sudokus, and receives three or four new ones every day from puzzle-makers around the world. So far, he has collected almost 50,000. But even though he is the world's expert on 17-clue puzzles, he says he doesn't know how close he is to finding the total number of possible puzzles. 'A while ago I would have said that we were close to the end, but then an anonymous contributor sent in nearly 5000 new ones,' he said. 'We never really worked out how "anon17" could do it, but it clearly involved a clever algorithm.'

In Royle's opinion, no one has found a 16-clue Sudoku because, as he said, 'Either we are not clever enough or our computers are not powerful enough.' Most likely, anon17 didn't reveal his method because he was using someone else's very large computer when he wasn't supposed to. Answering combinatorics problems often relies on giving the hard work of number-crunching over to a computer. 'The total possible space of 16-clue possible puzzles is far too vast for us to explore more than a tiny proportion of it without some new theoretical ideas,' Royle claimed. But he has a gut feeling that no 16-clue Sudokus will ever be found, adding: 'We have so many 17-clue puzzles now it would really be a bit strange if there was a 16-clue puzzle that we had not stumbled across.'

Maki Kaji's business card has the words *Godfather of Sudoku*. Wayne Gould describes himself as the Stepfather of Sudoku. I finally met up with Gould over coffee in a West London deli. He was wearing a New Zealand rugby top and had a typically antipodean, easy-going manner. Gould has a gap between his front teeth, which together with his thick glasses, short silvery hair and youthful enthusiasm reminded me of a young university lecturer rather than a former judge. Sudoku has transformed Gould's life. He has been busier in retirement than he ever was before. He supplies puzzles for free to more than 700 newspapers in 81 countries, earning money from selling his program and books, which he says gives him only about 2 per cent of the global Sudoku market. Still, Sudoku has earned him a seven-figure fortune. And he is a celebrity. When I asked him how his wife felt about his unexpected fame, he paused. 'We separated last year,' he stuttered. 'After 32 years of marriage. Maybe it was having all that money. Maybe it gave her a freedom she never knew she had.' Through the silence the message I heard was a heartbreaking one: he might have launched a global craze, but the adventure had come at too high a personal cost.

I've always thought that one reason for Sudoku's success is its exotic name, which resonates with the romance of superior Oriental wisdom, despite the fact that the American Howard Garns came up with the idea in Indiana. In fact, there is a tradition of puzzles coming from the East. The very first international puzzle craze dates from the early nineteenth century, when European and American sailors returning from China brought home sets of geometrical shapes, typically made of wood or ivory, that had seven pieces – two large triangles, two small triangles, a medium-sized triangle, a rhomboid and a square. Put together, the pieces made up a larger square. Accompanying the sets were booklets with dozens of outlines of geometrical shapes, human figures and other objects. The aim of the puzzle was to use all seven of the pieces to create each printed silhouette.

The puzzle had originated with the Chinese tradition of arranging tables of different shapes for banquets. One Chinese book, from the twelfth century, showed 76 banquet placements, many of which were made to look like objects, such as a fluttering flag, a range of mountains and flowers. At the turn of the nineteenth century a Chinese writer with the playful nickname Dim-Witted Recluse adapted this ceremonial choreography for finger-sized geometrical blocks and put the figures in a book, *Pictures Using Seven Clever Pieces.*

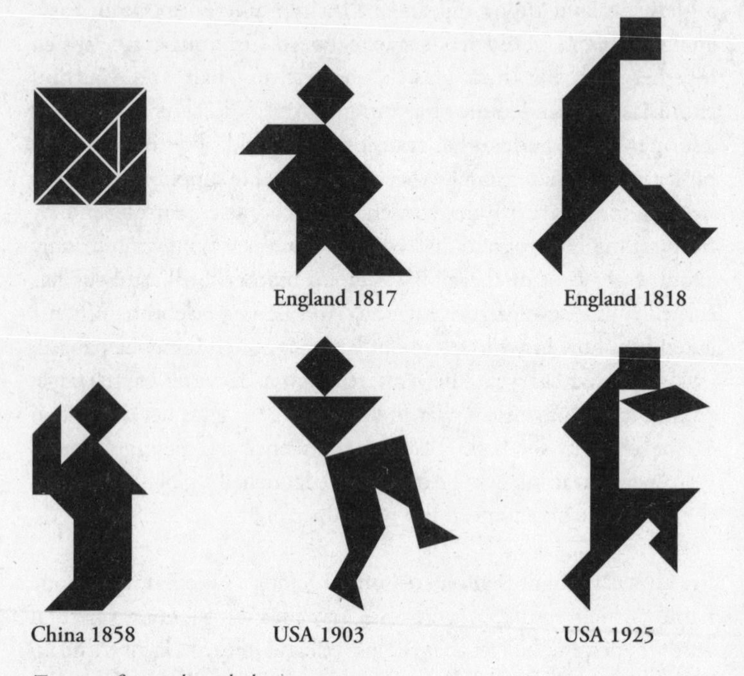

Tangram figures through the ages.

Originally called the Chinese Puzzle, the sets later gained the name 'tangram'. The first book of tangram puzzles to be published outside China was printed in London in 1817. Immediately, the book started a fad. Between 1817 and 1818 dozens of tangram books came out in France, Germany, Italy, the Netherlands and Scandinavia. Cartoonists of the day captured the craze by portraying men unwilling to go to bed with their wives, chefs unable to do the cooking and doctors refusing to attend patients because they were too busy rearranging triangles. The craze was more pronounced in France, perhaps because one of the books claimed that the puzzle was Napoleon's favourite amusement during his exile on the South Atlantic island of St Helena. The former emperor was an early adopter since ships stopped off there on their way back from Asia.

I love the tangram. Men, women and animals magically come to life. By a slight repositioning of just one piece the personality of the figure changes entirely. With their angular and often grotesque contours, the figures are wonderfully suggestive. The French took this personification to an extreme by actually painting images within the silhouettes.

It is hard to believe how engrossing the puzzle is until you have tried it out. In fact, though it looks easy, solving tangram problems can be surprisingly difficult. The shapes can easily deceive, as when two similar-looking silhouettes have totally different underlying structures. The tangram can serve as a warning against complacency, reminding you that the essence of objects may not always be what you first see. Take a look at the following tangram figures. It looks as if a small triangle has been removed from the first to make the second. In fact, both figures use all the pieces and they are arranged in completely different ways.

In the mid nineteenth century, tangrams were embraced by schools, though they still remained an adult pastime. The German company Richter rebranded the tangram as the *Kopfzerbrecher*, or brain-buster, and, due to the product's success, introduced more than a dozen similar rearrangement puzzles with different shapes cut into different pieces. During the First World War, Richter puzzles became a much-loved diversion for troops stuck in the trenches. Demand was so great that another 18 puzzles were launched. One of them was called the *Schützengraben Geduldspiel* – the trench patience game – which contained military shapes such as a Zeppelin,

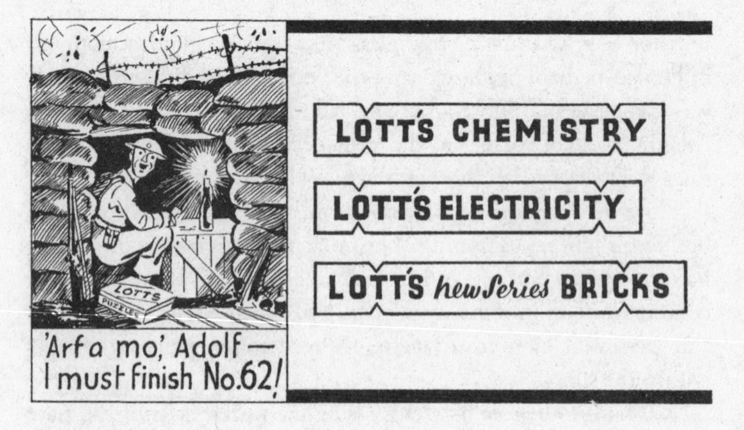

Second World War advert for Lott's Bricks' puzzles.

a revolver and a grenade. Some of the figures were devised by soldiers, who had sent the ideas in from the front.

Richter's puzzles were widely sold beyond Germany's borders before the First World War. When the UK banned German imports during the war, Lott's Bricks Ltd of Watford patriotically stepped in and produced British-made copies until the 1940s.

As each generation has created new figures, the tangram has never really gone out of fashion in almost 200 years. You can still buy the puzzle in toyshops and bookstores. The stock of published outlines is now more than 5900.

Despite the association of the tangram with puzzles of this kind, it was not the world's first rearrangement puzzle. In ancient Greece the similar *stomachion* divided a square into 14 pieces. (*Stomachi* means 'stomach', and the name is thought to be a result of the bellyache that the puzzle induced, although not from ingesting the pieces.) Archimedes wrote a treatise on the *stomachion*, only a fraction of which survives. Based on this fragment, it has been suggested that the treatise was an attempt to calculate the number of different ways the pieces of the *stomachion* could be positioned to make a perfect square. Only recently was this particular ancient problem solved. In 2003 the computer scientist Bill Cutler found that there are 536 ways (excluding solutions that are identical under rotation or reflection).

The stomachion, *which is also known as the 'loculus of Archimedes'.*

Since the time of Archimedes a keen interest in recreational puzzles has been a trait shared by many mathematicians. 'Man is never more ingenious than in the invention of games,' said Gottfried Leibniz, for instance, whose love of peg solitaire resonated with his obsession with binary numbers: a hole either has a peg in it or it doesn't, it is either a 1 or a 0. However, the most playful of the great mathematicians was Leonhard Euler, who, in order to crack an eighteenth-century brainteaser, invented a whole new branch of mathematics.

In Königsberg, the former Prussian capital that is now the Russian city of Kaliningrad, there used to be seven bridges that crossed the River Pregel. Locals wanted to know if it would be possible to make a journey across all seven bridges without crossing any bridge more than once.

To come up with a proof that a circuit of this kind was impossible, Euler created a graph in which each landmass was represented by a dot, or node, and each bridge by a line, or link. He worked out a theorem that related the number of links touching each node to whether it was possible to make a circuit of the graph, and in this case it was impossible.

The conceptual leap Euler made was to realize that what was important to solve the problem was not information about the exact position of the bridges, but how they were connected. The London Underground map borrows this idea: it is not geographically accurate, but is faithful to how the Tube lines are linked together. Euler's theorem launched graph theory, and presaged the development of topology, a very rich area of maths that studies the

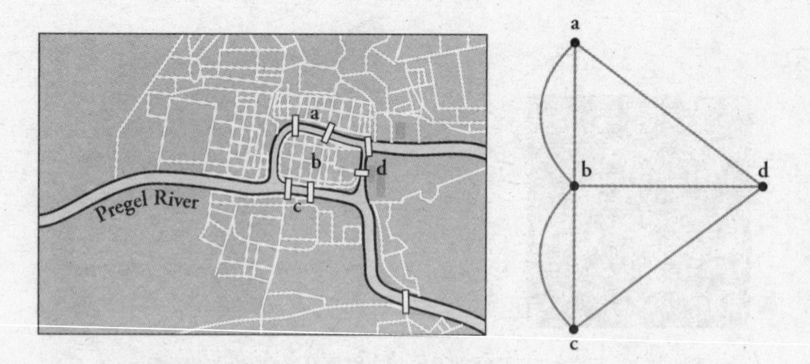

Königsberg in the eighteenth century: as a map and as a graph.

properties of objects that do not change when the object is squeezed, twisted or stretched.

Fascination in 1817 with tangrams was nothing compared with the extraordinary levels of excitement generated by the world's second international puzzle craze. From the day in December 1879 when the Fifteen puzzle launched in a Boston toyshop, manufacturers couldn't meet the demand. 'Neither the wrinkled front of age nor the cherubic brow of childhood is proof against the contagion,' declared the *Boston Post*.

The Fifteen puzzle consisted of 15 square wooden blocks placed in a square carton so that the blocks made a 4 × 4 square with one square missing. The blocks were numbered 1 to 15 and put in the box randomly. The aim of the puzzle was to slide the blocks around the 4 × 4 square using the empty space and finish with them positioned in numerical order. Playing with the Fifteen puzzle was so addictive and fun that the fad soon spread from Massachusetts to New York, and then across the United States. 'It has swept over the land from East to West with the violence of the sirocco, scorching men's brains as it passed, and apparently making them temporarily insane,' quivered the *Chicago Tribune*. According to the *New York Times*, no pestilence 'has ever visited this or any other country which has spread with [such] awful celerity'.

The puzzle soon went overseas, with one shop in London reportedly selling nothing else. Within six months it had reached the other side of the world. 'Not a few have already been driven

The Fifteen puzzle was initially known as the Gem puzzle.

insane,' claimed a letter in the *Otago Witness*, New Zealand, on 1 May 1880.

The Fifteen puzzle was the creation of Noyes Chapman, a postmaster in upstate New York, who almost two decades previously had been trying to make a physical model of a 4 × 4 magic square. He made small wooden squares for each of the 16 numbers and fitted them snugly into a square box. When he realized that leaving out one block provided a space into which any adjacent block could be slid, he saw that trying to rearrange the numbers would make a particularly fun game. Chapman made a few versions for family and friends, but never capitalized on his invention. It was only when a savvy Boston carpenter decided to commercialize the puzzle that it finally took off.

The Fifteen puzzle was especially tormenting to those who attempted it because sometimes it was solvable and sometimes it was not. Once the blocks were randomly put in there seemed to be only two outcomes: either they could be rearranged in numerical order, or they could be rearranged so that the first three rows were in order but the last line went 13-15-14. The craze was fuelled, in part, by a desire to work out if it was possible to get from 13-15-14 to 13-14-15. In January 1890, a few weeks after the first puzzle went on sale, a dentist in Rochester, New York, placed an ad in the local paper offering a $100 prize and a set of false teeth to anyone who could prove it either way. He believed it was impossible – but needed some help with the maths.

Bafflement with the Fifteen puzzle spread from the world's living rooms to the halls of academe, and once the professionals got involved, the puzzle went from insanity-inducing unsolvable to satisfactorily unsolvable. In April 1890 Hermann Schubert, one of the outstanding mathematicians of his day, published in a German newspaper the earliest proof that 13-15-14 was an unsolvable position. Shortly thereafter, the recently founded *American Journal of Mathematics* also published a proof, confirming that half of the total of all starting positions in the Fifteen puzzle will produce a final solution of 13-14-15, and half will end up 13-15-14. The Fifteen puzzle remains the only international craze in which the puzzle does not always have a solution. No wonder it drove people mad.

Like the tangram, the Fifteen puzzle has not totally disappeared. It was the forerunner of the sliding-block puzzles that are still found in toyshops, Christmas crackers and corporate marketing gift packs. In 1974 a Hungarian man was devising ways to improve the puzzle when he was struck with the idea of reinventing it in three dimensions. The man, Ernö Rubik, came up with his prototype, the Rubik's Cube, which went on to become the most successful puzzle in history.

In his 2002 book *The Puzzle Instinct*, the semiotician Marcel Danesi wrote that an intuitive ability to solve puzzles is part of the human condition. When we are presented with a puzzle, he explains, our instincts drive us to find a solution until we are satisfied. From the riddle-asking sphinx of Greek mythology to the detective mystery, puzzles are a common feature across time and cultures. Danesi argues that they are a form of existential therapy, showing us that challenging questions can have precise solutions. Henry Ernest Dudeney, Britain's greatest puzzle-compiler, described solving puzzles as basic human nature. 'The fact is that our lives are largely spent in solving puzzles; for what is a puzzle but a perplexing question? And from our childhood upwards we are perpetually asking questions or trying to answer them.'

Puzzles are also a wonderfully concise way of conveying the 'wow' factor of maths. Often they require lateral thinking, or rely on counter-intuitive truths. The sense of achievement gained from solving a puzzle is an addictive pleasure; the sense of failure from not solving one almost unbearably frustrating. Publishers realized pretty quickly that fun maths had a market. *Amusing and Entertaining Problems that Can Be Had with Numbers* (*very useful for inquisitive people of all kinds who use arithmetic*) by Claude Gaspard Bachet came out in France in 1612. It included sections on magic squares, card tricks, questions in non-decimal bases, and think-of-a-number problems. Bachet was a serious scholar who translated Diophantus's *Arithmetica*. But his popular maths book was arguably more influential than his academic work. All subsequent puzzle books are indebted to it, and it has kept its relevance for centuries, republished most recently in 1959. A defining feature of maths, even recreational maths, is that it never goes out of date.

In the mid nineteenth century American newspapers started to print chess problems. One of the first, and most precocious, devisers of these problems was Sam Loyd. A New Yorker, Loyd was just 14 when he had his first conundrum printed in a local paper. By 17 he was the most successful and widely celebrated deviser of chess problems in the US.

He moved on from chess to mathematically based puzzles, and by the end of the century was the world's first professional puzzle-compiler and impresario. He published widely in the American media, and once claimed that his columns attracted 100,000 letters a day. However, we should take his figure with a pinch of salt. Loyd cultivated the kind of playful attitude towards the truth that one expects of a professional riddler. For a start, he claimed he had invented the Fifteen puzzle, which was taken to be true for more than a century until in 2006 historians Jerry Slocum and Dic Sonneveld properly traced its origin to Noyes Chapman. Loyd also revived interest in the tangram with *The 8th Book of Tan Part I*, a version of an ancient text about the supposed 4000-year history of the puzzle. The book was a spoof, even though it was initially taken seriously by academics.

Loyd had a unique brilliance at turning mathematical problems into entertaining, distinctively illustrated puzzles. His most genial creation was invented for the *Brooklyn Daily Eagle* in 1896. The 'Get off the Earth' puzzle was so popular that it was later adopted as a publicity gimmick by several brands, including *The Young Ladies Home Journal*, the Great Atlantic & Pacific Tea Company and the

Republican platform for the 1896 presidential election. (Although its message was not a manifesto pledge.) The puzzle is an image of Chinese warriors positioned around the Earth, which is on a disc made out of card that can be spun around its centre. When the arrow is pointing NE there are 13 warriors, but turn the disc so the arrow points NW and there are only 12. The puzzle is confusing. There really are 13 warriors, then – in a flash – just 12. Which man has vanished and where does he go?

The trick operating in this puzzle is known as a geometrical vanish. It can also be demonstrated in the following way. The image on page 240 shows a piece of paper with ten vertical lines on it. When the piece of paper is cut along the diagonal line, the two sections can be realigned so that only nine lines are created. Where has the tenth

line gone? What has happened is that the segments have been rearranged to form nine lines that are *longer* than the original line. If the lines in the first image have the length 10 units, then the length of the lines in the second image is $11\frac{1}{9}$, since one of the original lines has been equally shared among the nine others.

What Sam Loyd did with 'Get off the Earth' is to curve the geometrical vanish so it was in a circular form, and in place of lines put Chinese warriors. There are 12 positions in his puzzle, akin to the 10 lines in the example overleaf. The position in the bottom left corner, where there are originally two warriors, is equivalent to the end lines of the vanish. When the arrow is moved from NE to NW, all the positions gain a bit more warrior, apart from the position with two warriors, which shrinks drastically,

Geometrical vanish: ten lines become nine.

giving the impression that a whole warrior has been lost. In fact, it has just been redistributed around the others. Sam Loyd claimed that ten million copies of 'Get off the Earth' were produced. He became rich and famous, revelling in his reputation as the puzzle king of America.

Meanwhile, in Great Britain, Henry Ernest Dudeney was acquiring a similar reputation. If Loyd's capitalist chutzpah and gift for self-publicity reflected the cut-and-thrust of turn-of-the-century New York, Dudeney embodied the more reserved English way of life.

From a family of Sussex sheep farmers, Dudeney started working at 13 as a clerk in the civil service in London. Bored with the job, he began to submit short stories and puzzles to various publications. Eventually he was able to devote himself to journalism full time. His wife Alice wrote bestselling romantic novels about life in rural Sussex – where, thanks to her royalties, the couple was able to live in luxury. The Dudeneys, dividing their time between the countryside and London, were part of a highbrow literary scene that included Sir Arthur Conan Doyle, the creator of Sherlock Holmes, probably the most iconic puzzle-solver in all literature.

It is thought that Dudeney and Sam Loyd first made contact in 1894, when Loyd posed a chess problem in the firm belief that no one would discover his 53-move solution. Dudeney, who was 17 years Loyd's junior, found a solution in 50 moves. The men subsequently collaborated, but fell out when Dudeney discovered Loyd was plagiarizing his work. Dudeney despised Loyd so intensely that he equated him with the Devil.

While both Loyd and Dudeney were self-taught, Dudeney had a much finer mathematical mind. Many of his puzzles touched on deep problems – frequently predating academic interest. In 1962, for example, the mathematician Mei-Ko Kwan investigated a problem about the route a postman should take in a grid of streets so as to walk on every street and do it in the shortest route possible. Dudeney had framed – and solved – the same problem in a puzzle about a mine inspector walking through underground shafts almost 50 years earlier.

Dudeney also made unintended contributions to number theory. One of his puzzles, called *Root Extraction*, plays on the fact that the cube roots of the following numbers are also equal to the sum of the digits that make up those numbers:

$1 = 1 \times 1 \times 1$ $\quad$ $1 = 1$
$512 = 8 \times 8 \times 8$ $\quad$ $8 = 5 + 1 + 2$
$4913 = 17 \times 17 \times 17$ $\quad$ $17 = 4 + 9 + 1 + 3$
$5832 = 18 \times 18 \times 18$ $\quad$ $18 = 5 + 8 + 3 + 2$
$17{,}576 = 26 \times 26 \times 26$ $\quad$ $26 = 1 + 7 + 5 + 7 + 6$
$19{,}683 = 27 \times 27 \times 27$ $\quad$ $27 = 1 + 9 + 6 + 8 + 3$

Numbers with this property – and there are only six of them – are now known as Dudeney numbers. Another particular forte of Dudeney's was the geometrical dissection, which is when a shape is cut up into pieces and reassembled into another shape, like the principle behind the tangram. Dudeney found a way of converting a square into a pentagon in six pieces. His method became a popular classic because for many years it had been thought that the minimum dissection of a square into a pentagon required seven pieces.

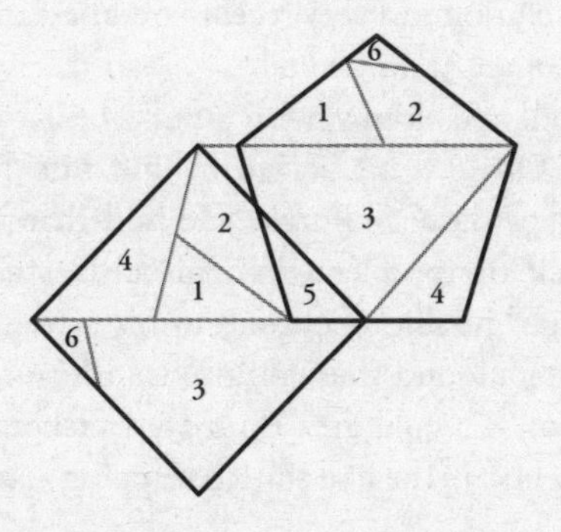

Dudeney also discovered a novel way to cut up a triangle and make it into a square in four pieces. And he realized that if the four pieces in his solution were hinged together, they could be assembled in a chain – such that folding one way gets you a triangle and folding the other way gets you the square. He called this the Haberdasher's Puzzle, since the shapes look like the leftover pieces of cloth that a haberdasher might have in his shop. The puzzle thus introduced the concept of a 'hinged dissection', and aroused such interest that Dudeney made one out of mahogany with brass hinges and presented it in 1905 at a meeting of the Royal Society in London. The Haberdasher's Puzzle was Dudeney's greatest legacy and has fascinated and delighted mathematicians for more than a century.

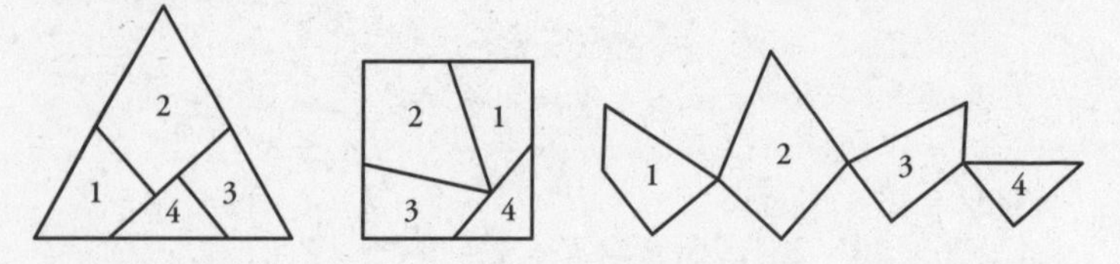

The Haberdasher's Puzzle.

One mind that was especially enthralled by the Haberdasher's Puzzle belonged to the Canadian teenager Erik Demaine. Demaine, who was such a prodigy that by the time he was 20 he was already a professor at MIT, was most interested in the 'universality' of the problem. Is it possible, he wondered, to dissect *any* straight-sided shape and then hinge the pieces together in a chain so it can be folded into *any other* straight-sided shape of equal area? He spent ten years working on the problem and in March 2008, aged 27, announced the solution to a very receptive audience of puzzle-lovers in the ballroom of an Atlanta hotel.

Demaine is tall and skinny, with a fluffy beard and fuzzy dark-blond ponytail. On to a big screen behind him he projected an image of the Haberdasher's Puzzle. He said that he had recently decided to attack the problem with his PhD students. 'I didn't believe it was true,' he said. Contrary to his expectation, however, he and his students found that you *can* transform any polygon to any other polygon of equal area through a Haberdasher's Puzzle-style hinged dissection. The hall started clapping – a rare occurrence in the upper reaches of computational geometry. But in puzzle land this was about as exciting a breakthrough as you can get – the solution to an iconic problem by one of the cleverest minds of his generation.

The Atlanta conference, called the Gathering for Gardner, was the most appreciative audience possible for Demaine's talk. The Gathering is the world's premier jamboree for mathematicians, magicians and puzzle people. It is a biannual homage to the man who revolutionized recreational mathematics in the second half of the last century. Martin Gardner, now 93 years old, wrote a monthly maths column in *Scientific American* between 1957 and 1981. This was a period of great scientific advances – space travel, information technology and genetics – yet it was Gardner's lively and lucid prose that really caught readers' imaginations. His column covered subjects from board games to magic tricks, from numerology to early computer games, and often ventured into tangential areas such as linguistics and design. 'I thought [Gardner] had a playful respect for mathematics that is often lost in mathematical circles,' Demaine told me when I spoke to him after his talk. 'People tend to be too serious. My aim is to make everything I do fun.'

As a boy, Demaine was introduced to Gardner's columns through his father, a glass-blower and sculptor. The Demaines, who often publish mathematical papers together, embody Gardner's interdisciplinary spirit. Erik is a pioneer of computational origami, a field both mathematical and artistic, and some of the Demaines' origami models have even been exhibited in New York's Museum of Modern Art. Demaine considers maths and art parallel activities, which share an 'aesthetic about simplicity and beauty'.

In Atlanta Demaine didn't explain the details of his proof of the universality of Haberdasher's Puzzle-style dissections to the audience, but he did say that dissecting one polygon so it can be rearranged and hinged to form another polygon isn't always pretty – and will often be completely impracticable. Demaine is now applying his theoretical work on hinged dissections to make robots that can transform from one shape into another through folding – just like the heroes of the comic book and movie franchise *Transformers*, where robots morph into different types of machine.

The conference was the eighth Gathering for Gardner, or G4G, and its logo, designed by Scott Kim, is known as an inversion, or ambigram.

If you turn it upside-down, it reads exactly the same. Kim, a computer scientist turned puzzle-designer, invented this style of symmetrical calligraphy in the 1970s. Ambigrams do not have to be the same when rotated 180 degrees – any symmetry, or concealed writing, will do.

The writer Isaac Asimov called Kim 'the Escher of the alphabet', comparing him to the Dutch artist who played with perspective and symmetries to create self-contradictory images, most famously steps that appear to rise and rise until they reach where they began. Another similarity between Escher and Kim is that their work first reached a mass audience thanks to Martin Gardner.

Ambigrams were independently, and contemporaneously, conceived by the typographer and artist John Langdon. Mathematicians especially love this type of lettering since it is a witty take on their own search for patterns and symmetry. The author Dan Brown was introduced to ambigrams through his father Richard Brown, a maths teacher. Dan Brown commissioned Langdon to design the phrase Angels & Demons as an ambigram for his bestselling book of the same title, and named the lead character Robert Langdon in his honour. Langdon reappeared as the hero of *The Da Vinci Code* and *The Lost Symbol*. Ambigrams have also found a new niche – as body art. The quasi-gothic flourishes, often added to aid symmetry, together with the mystic energy of reading a name backwards and forwards, or upside-down and the right way up, coincides perfectly with the aesthetics of tattoos.

At the G4G it was impossible not to think that maths wards off the onset of dementia. Many of the guests were over 70 – some were in their eighties and even nineties. For more than half a century Gardner corresponded with thousands of readers, many of them

In this tattoo designed by Mark Palmer, Angel becomes Devil when upside-down.

famous mathematicians, and some became close friends. Raymond Smullyan, aged 88, is the world's foremost expert on logical paradoxes. He began his talk: 'Before I begin speaking, there is something I want to say.' Willowy and charmingly scruffy, with flowing white hair and a feathery beard, Smullyan was frequently entertaining guests on the hotel piano. He also performed magic tricks on unsuspecting passers-by, and over dinner one evening brought the house down with a stand-up comedy routine.

Aged 76, Solomon Golomb was less physically energetic than Smullyan but able to converse without talking in paradoxes. A soft-spoken grandfatherly figure, Golomb has made important discoveries in space communications, mathematics and electrical engineering. With the helping hand of Martin Gardner, he has also contributed to global pop culture. Early in his academic career Golomb came up with the idea of polyominoes, which are dominoes made out of more than two squares. A triomino is made of three, a tetromino out of four, and so on. An early Gardner column on how they fit together caused such international interest that Golomb's book, *Polyominoes*, was translated into Russian, where it became a best-seller. One fan made a game that involved falling tetrominoes. That game, Tetris, became one of the world's most enduring and best-loved computer games. Golomb, of course, has played Tetris for no longer than half an hour.

Another attendee, Ivan Moscovich, is the spitting image of an elderly Vincent Price. Impeccably dressed in a sharp dark suit, he had sparkling eyes, a pencil moustache and full head of brushed-back grey hair. For Moscovich, the attraction of puzzles is the creative thinking they require. He was born in what is now Serbia, and during the Second World War was interned at both Auschwitz and Bergen-Belsen. He believes he survived because of an innate creativity – he was continuously creating situations that ended up saving him. After the war, he turned into a workaholic puzzle-inventor. He likes to think constantly outside the box, to sidestep the inevitable. The motivation to continually come up with new ideas, he said, was an after-effect of the trauma of his own lucky escape.

Moscovich has had about 150 puzzles licensed and produced over the last half century, and compiled a book of puzzles that has been hailed as the greatest collection since the era of Loyd and Dudeney. Now 82, he clutched his latest creation: a sliding block puzzle called You and Einstein. The idea of the game is to slide blocks around a square grid to create a picture of Einstein. Moscovich's clever twist is that each block has a slanted mirror that reflects the box to its side, meaning that what you think is the block is actually the reflection of something else. Moscovich told me he was excited that You and Einstein could be a global success.

The dream of Moscovich, like everyone in his industry, is, of course, to discover a new puzzle craze. There have been only four international puzzle crazes with a mathematical slant: the tangram, the Fifteen puzzle, the Rubik's Cube and Sudoku. So far, the Cube has been the most lucrative. More than 300 million have been sold since Ernö Rubik came up with the idea in 1974. Apart from its commercial success, the gaudily coloured cube is a popular-culture evergreen. It is the nonpareil of puzzledom and, unsurprisingly, its presence was felt at the 2008 G4G. A talk on the Rubik's Cube in four dimensions drew huge rounds of applause.

The original Rubik's Cube is a $3 \times 3 \times 3$ array made up of 26 smaller cubes, or cubies. Each horizontal and vertical 'slice' can be rotated independently. Once the pattern of the cubies is jumbled, the aim of the puzzle is to twist the slices so that each side of the cube has cubies of just one colour. There are six colours, one for each side. Moscovich told me Ernö Rubik was doubly brilliant. Not only was the idea of the cube a stroke of genius, but the way he made the blocks fit together was an outstandingly clever piece of engineering. When you dismantle a Rubik's Cube there is no separate mechanical device holding it all together – each cubie contains a piece of a central, interlocking sphere.

As an object, the cube itself is sexy. It is a Platonic solid, a shape that has had iconic, mystical status since at least the ancient Greeks. The brand name was also a dream: catchy, with delicious assonance and consonance. The Rubik's Cube had an Eastern exoticism too, not from Asia this time but from Cold War Eastern Europe. It sounded a lot like Sputnik, the original showpiece of Soviet space technology.

Another ingredient in its success was the fact that while solving the cube was not easy, the challenge did not put people off. Graham Parker, a builder from Hampshire, kept at it for 26 years until he achieved his dream. 'I have missed important events to stay in and solve it and I would lay awake at night thinking about it,' he said, after an estimated 27,400 hours of cube time. 'When I clicked that last bit into place and each face was a solid colour I wept. I cannot tell you what a relief it was.' Those who solved it over a more manageable period invariably wanted to solve it again, but quicker. Reducing one's Rubik's record became a competitive sport.

Speedcubing has only really taken off, however, since around 2000. One of the reasons is thanks to a sport even more quirky than the timed solving of mechanical puzzles. Speedstacking is the practice of stacking plastic cups in set patterns as fast as you can. It is both mesmerizing and awesome – the top stackers move so fast it is as if they are painting the air with plastic. The sport was invented in California in the 1980s as a way of improving children's hand-eye coordination and general fitness. It is claimed that 20,000 schools worldwide now include it in their physical education curriculum. Speedstacking uses specialized mats that have a touch sensor linked to a stopwatch. The mats provided the speedcubing community for the first time with a standardized method to measure the time it takes to solve the cube, and are now used in all competitions.

Every week or so, somewhere around the world now hosts an official speedcubing tournament. To make sure that the starting position is sufficiently difficult in these competitions, the regulations stipulate that cubes must be scrambled by a random sequence of moves generated by a computer program. The current record of 7.08 seconds was set in 2008 by Erik Akkersdijk, a 19-year-old Dutch student. Akkersdijk also holds the record for the 2 × 2 × 2 cube (0.96secs), the 4 × 4 × 4 cube (40.05secs) and the 5 × 5 × 5 cube (1min 16.21 secs). He can also solve the Rubik's Cube with his feet – his time of 51.36secs is fourth-best in the world. However, Akkersdijk really must improve his performance at solving the cube one-handed (33rd in the world) and blindfolded (43rd). The rules for blindfolded solving are as follows: the timer starts when the cube is shown to the competitor. He must then study it, and put on a blindfold. When he thinks it is solved he tells the judge to stop the stopwatch. The current record of 48.05secs was set by Ville Seppänen of Finland in 2008. Other speedcubing disciplines include solving the Rubik's Cube on a rollercoaster, under water, with chopsticks, while idling on a unicycle, and during freefall.

The most mathematically interesting cube-solving category is how to solve it in the fewest moves possible. The contestant is given an officially scrambled cube and has 60 minutes to study the position before describing the shortest solving sequence he can come up with. In 2009 Jimmy Coll of Belgium claimed the world record: 22 moves. Yet this was just how many moves a very smart human needed to solve a jumbled-up cube after 60 minutes of thinking about it. Might he have been able to find a solution from the same configuration in a smaller number of moves if he had had 60 hours? The question that has most intrigued mathematicians about the Rubik's Cube is this: what is the smallest number of moves, *n*, such that every configuration can be solved in *n* moves or less? As a mark of reverence, *n* in this instance is nicknamed 'God's number'.

Finding God's number is extremely complex because the numbers are so large. There are about 43×10^{18} (or 43 followed by 18 zeros) cube positions. If every unique cube position were stacked on top of each other, the tower of cubes would go to the sun and back more than eight million times. It would take far too long to analyse each position one by one. Instead, mathematicians have looked at subgroups of positions. Tomas Rokicki, who has been studying the problem for around two decades, has analysed a collection of 19.5 billion related positions and found ways of solving them in 20 moves or less. He has now looked at a million or so similar collections, each containing 19.5 billion positions, and found again that 20 moves was sufficient for a solution. In 2008 he proved that every other remaining Rubik's Cube position is only two moves away from a position in one of his collections, giving an upper bound for God's number of 22.

Rokicki is convinced that God's number is 20. 'I've solved, at this point, approximately 9 per cent of all cube positions, and none of them has required 21 moves. If there are any positions that require 21 or more moves, they are exceptionally rare.' Rokicki's challenge is not so much theoretical but logistical. Running through sets of cube positions uses an incredible amount of computer memory time. 'With my current technique, I would need about 1000 modern computers for about one year to prove [that God's number] is 20,' he said.

Cube maths has been a long-term hobby for Rokicki. When I asked him whether he has thought of investigating the maths of other puzzles, like Sudoku, he joked: 'Don't try to distract me with other shiny problems. Cube math is challenging enough!'

Ernö Rubik still lives in Hungary and rarely gives interviews. I did, however, get a chance to meet one of his former students, Dániel Erdély, in Atlanta. We met in a room in the hotel devoted to 'mathematical objects'. Origami models, geometrical shapes and elaborate puzzles were laid out on tables. Erdély was there looking after his own creations: light blue objects about the size of cricket balls, ridged with intricate, swirling patterns. Erdély treated them with the affection that a dog breeder has for a litter of his puppies. He picked one up, pointed to the palm-sized planet's crystalline landscape and said: 'Spidrons.'

Erdély, like Rubik, is not a mathematician. Rubik is an architect, and Erdély is a graphic designer who studied graphic design at the Budapest College of Applied Arts, where Rubik was a professor. In 1979 Erdély attended classes given by Rubik. As homework for these classes he devised a new shape made out of a sequence of alternate, and shrinking, equilateral and isosceles triangles. He called the shape a 'spidron' because it curved like a spiral. By the time he left university, spidrons were his obsession. He played around with them endlessly, noticing that they could fit together like tiles in

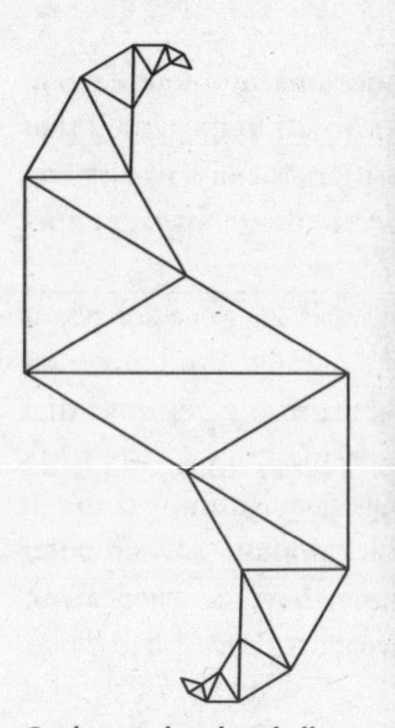

Spidron and spidron ball.

many aesthetically satisfying ways, in both two and three dimensions. About five years ago, a Hungarian friend helped write a program to generate spidrons on the computer. Their tessellating properties have subsequently captivated mathematicians, engineers and sculptors, and Erdély has made himself the shape's globetrotting chaperon. He believes it could have applications in the design, for example, of solar panels. At the G4G he had met a man who runs a company that launches rockets. The spidron, he told me, may be about to go into space.

One afternoon the delegates of the conference relocated to the home of Tom Rodgers in the Atlanta suburbs. Rodgers, a businessman in late middle age, organized the first G4G in 1993. An admirer of Gardner's since childhood, Rodgers' initial idea was to have an event where the famously shy Gardner could meet some of the many readers he had corresponded with. He decided to invite guests from three specific areas of Gardner's interest – maths, magic and puzzles. The gathering was such a success that a second one was organized in 1996. Gardner showed up to the first two but since then has been too frail to attend. Rodgers lives in a bungalow designed in Japanese style, surrounded by a forest of bamboo, pine and fruit trees that were in blossom when I visited. In the garden several guests were forming teams to build geometrical sculptures out of wood and metal. Others were attempting to solve a bespoke puzzle hunt whose clues were stuck to the house's outside walls.

Suddenly, the shriek of Princeton University maths professor John Horton Conway grabbed everyone's attention. Conway had a messy beard, a full head of silvery hair and was wearing a T-shirt with an equation on it. He is one of the most outstanding mathematicians of the last 50 years. He asked for everyone to bring him ten pine cones each so that he could count their spirals. Cone-classifying is a recent hobby of his; he has counted about 5000 of them since he started a few years ago.

Inside the house I met Colin Wright, an Australian who lives in Port Sunlight on the Wirral. With his schoolboyish ginger hair and glasses, he looks just how you might expect a mathematician to look. Wright is a juggler, which 'seemed like the obvious thing to do after I learned to ride a unicycle', he said. He also helped develop a mathematical notation for juggling, which might not sound like much, but has electrified the international juggling community. It turns out that with a language, jugglers have been able to discover tricks that had eluded them for thousands of years. 'Once you have a language to talk about a problem, it aids your thought process,' said Wright, as he took out some bean balls to demonstrate a recently invented three-ball juggle. 'Maths is not sums, calculations and formulae. It is pulling things apart to understand how things work.'

I asked him if there was something self-indulgent, pointless or even wasteful about the finest minds in mathematics spending their time working on inconsequential pastimes like juggling, counting pine cones or even puzzle-solving. 'You need to let mathematicians do what they do,' he replied. 'You genuinely never know what is going to be useful.' He quoted the example of Cambridge professor G.H. Hardy, who in 1940 famously (and proudly) declared that number theory had no practical applications; in fact, it is now the basis of many internet security programs. Mathematicians have, according to Wright, been 'unreasonably successful' in finding applications to apparently useless theorems, and often years after the theorems were first discovered.

One of the most charming aspects of the G4G is that all guests are asked to bring a gift – 'something you would want to give to Martin'.

In fact, you are asked to bring 300 of your gift, since each guest is given a goody bag at the end, containing one of everyone else's gift. The year I visited the goody bag included puzzles, magic tricks, books, CDs, gadgets and piece of plastic that can make a Coke can talk. One bag was for Martin Gardner, and I took it to him.

Gardner lives in Norman, Oklahoma. The day I arrived storms were moving across the state. After a few wrong turns off the interstate, I found his home, an Assisted Living Center next to a Texan fast-food joint. The door to his room is only a few steps from the entrance, past a common area where a couple of the elderly residents was sitting and chatting. By Gardner's door was a box of correspondence. He does not use email. He sends more letters than the rest of the home combined.

Gardner opened the door and invited me in. On the wall there was a portrait of him made out of dominoes, a large photograph of Einstein and an Escher original. Gardner was casually dressed in a green shirt and slacks. He had a soft, open complexion, with wisps of white hair, large tortoiseshell glasses and alert eyes. There was an ethereal aspect to him. He was slim and had excellent posture, possibly because he works every day standing up at his desk.

Visiting Gardner felt straight out of *The Wizard of Oz*. I was in the hurricane-strewn Midwest on a quest to meet an elderly magician. It turned out that Dorothy & co were an especially pertinent reference. I had not known this before I met him, but Gardner is a world expert on L. Frank Baum, the writer of *The Wizard of Oz*. Gardner told me that a decade previously he had even written a sequel in which Dorothy and friends go to Manhattan. It was reviewed in serious newspapers, if not very favourably. 'It is written mainly for Oz fans,' he said.

I gave him the G4G goody bag and asked how it felt to be the subject of a conference. 'I am quite honoured, and surprised,' he replied. 'I am amazed at how it has grown.' It quickly became clear that he was not entirely comfortable talking about his illustriousness among mathematicians. 'I am not a mathematician,' he said. 'I am basically a journalist. Beyond calculus I am lost. That was the secret of my column's success. It took me so long to understand what I was writing about that I knew how to write about it so most readers would understand it.' When I learned that Gardner was not a proper mathematician I initially felt a little disappointed, as though the Wizard had pulled away the curtain.

Gardner's preferred subject is magic. He described it as his principal hobby. He subscribes to magic magazines and – as much as his arthritis allows – practises tricks. He offered to show me what he said was the only sleight-of-hand card magic he had invented, called a 'wink change', in which the colour of a card is changed 'in a wink'. He took a pack of cards and lodged a black card between the deck and the palm of his hand. Instantly, the black card became a red one. Gardner became interested in maths through 'mathematical' magic tricks, and it was magicians, not mathematicians, who formed his main social circle as a young adult. He said he liked magic because it developed a sense of wonder about the world. 'You see a woman levitated and that reminds you that it is just as miraculous that she falls to the ground by gravity … you don't realize that gravity is just as mysterious as a woman levitating.' I asked him if maths gave him that same wonder. He replied, 'Absolutely, yes.'

Gardner may be best known for his writings on maths, but they represent only a portion of his output. His first book was *Fads and Fallacies*, the first popular book to debunk pseudoscience. He has written on philosophy and published a serious novel about religion. His bestselling book is *The Annotated Alice*, a timeless compendium of footnotes to *Alice in Wonderland* and *Through the Looking-Glass*. At 93, the output shows no signs of slowing down. He is due to publish a book of essays on G.K. Chesterton, and among his many other projects he is compiling a bumper book of word games.

Thanks to Gardner, recreational mathematics remains in very good shape. It is an exciting and diverse field that continues to give pleasure to people of all ages and nationalities, as well as inspiring serious research on serious problems. I had been slightly disheartened to learn that Gardner was not a mathematician, but as I left the Assisted Living Center, it struck me, after all, that it was brilliantly in the spirit of recreational maths that the man who now personifies it was only ever an enthusiastic amateur.

CHAPTER SEVEN

Secrets of Succession

In Atlanta I met a man with an unusual hobby. Neil Sloane collects numbers. Not individual numbers, that would be silly, but families of numbers in ordered lists called sequences. For example, the natural numbers are a sequence, which can be defined by saying that the *n*th term in the sequence is *n*:

1, 2, 3, 4, 5, 6, 7 …

Sloane started his collection in 1963, as a graduate student at Cornell, where he first wrote the sequences on cards. It made perfect sense for someone who liked ordered lists to make an ordered list of them. By 1973 he had reached 2400 sequences, which he published in a book entitled *A Handbook of Integer Sequences*. By the mid nineties he was up to 5500. Only with the invention of the internet, however, did the collection find its ideal medium. Sloane's list blossomed into the *On-Line Encyclopedia of Integer Sequences*, a compendium that now has more than 160,000 entries, and expands by about 10,000 a year.

On first acquaintance, Sloane appears a typical indoors type. He is slight, bald and wears thick, square glasses. Yet he is also sinewy and tough, carrying himself with a Zen-like poise – a benefit of his other passion, which is rock-climbing. Sloane likes the challenges of ascending geological formations just as much as he likes ascending numerical ones. In Sloane's opinion, the similarity between studying sequences and rock-climbing is that they both demand shrewd puzzle-solving skills. I'd say there's another parallel: sequences encourage the number equivalent of mountaineering – whenever you reach term *n* the natural inclination is to find term *n* + 1. The desire to reach the next term is like the desire to climb higher and higher peaks; although mountaineers, of course, are restricted by geography, while sequences will often carry on for ever and ever.

Like a record collector who stacks the old favourites by the colourful rarities, Sloane embraces the common as well as the bizarre for his *Encyclopedia*. His collection contains, for example, the sequence below, the 'zero sequence', which consists of only 0s. (Each sequence in the *Encylopedia* is given a reference number prefixed by the letter A. The zero sequence was the fourth sequence Sloane collected, and hence is known as A4.)

(A4) 0, 0, 0, 0, 0 …

As the simplest-possible unending sequence, it is the least dynamic in the collection, although it does have a certain nihilistic charm.

Maintaining the *On-Line Encyclopedia* is a full-time job for Sloane, which he does in addition to his real job as a mathematician at AT&T Labs in New Jersey. But he no longer needs to spend time rooting around for new sequences. With the *Encyclopedia*'s success, he is constantly receiving submissions. They come from professional mathematicians and, in larger numbers, numerically obsessed laypeople. Sloane has only one criterion for letting a sequence join the club: it must be 'well defined and interesting'. The former just means that each term in the sequence can be described, either algebraically or rhetorically. The latter is a matter of his judgement, although his tendency is to accept if he's ever unsure. Being well defined and interesting, however, does not mean there is something mathematical going on. History, folklore and quirkiness are all fair game.

Among the sequences included in the *Encyclopedia* is the ancient sequence:

(A100000) 3, 6, 4, 8, 10, 5, 5, 7

The sequence numbers are the translation into digits of marks made on one of the oldest-known mathematical objects: the Ishango bone, a 22,000-year-old artefact found in what is now the Democratic Republic of the Congo. The monkey bone was initially thought to be a tally stick, but it has since been suggested that the pattern of 3, followed by its double, then 4, followed by its double, then 10, followed by its half, indicates more sophisticated arithmetical reasoning. There's also a hateful sequence in the collection:

(A51003) 666, 1666, 2666, 3666, 4666, 5666, 6660, 6661 …

This sequence is also known as the Beastly numbers, since they are the numbers containing the string 666 in their decimal expansion.

On a lighter note, here's a nursery sequence:

(A38674) 2, 2, 4, 4, 2, 6, 6, 2, 8, 8, 16

These are the numbers from the Latin American children's song 'La Farolera': *'Dos y dos son quatro, cuatro y dos son seis. Seis y dos son ocho, y ocho dieciseis.'*

But perhaps the most classic sequence of all is the prime numbers:

(A40) 2, 3, 5, 7, 11, 13, 17, 19, 23, 29, 31, 37 …

Prime numbers are the natural numbers greater than 1 that are divisible only by themselves and 1. They are simple to describe but the sequence exhibits some rather spectacular, and sometimes mysterious, qualities. First, as Euclid proved, there is an infinite number of them. Think of a number, any number, and you will always be able to find a prime number higher than that number. Second, every natural number above 1 can be written as a unique product of primes. In other words, every number is equal to a unique set of prime numbers multiplied by each other. For example, 221 is 13 × 17. The next number, 222, is 2 × 3 × 37. The one after that, 223, is prime, so produced only by 223 × 1, and 224 is 2 × 2 × 2 × 2 × 2 × 7. We could carry on for ever and each number could be winnowed down to a product of primes in only one possible way. For example, a billion is 2 × 2 × 2 × 2 × 2 × 2 × 2 × 2 × 2 × 5 × 5 × 5 × 5 × 5 × 5 × 5 × 5 × 5. This characteristic of numbers is known as the *fundamental theorem of arithmetic*, and is why primes are considered the indivisible building blocks of the natural number system.

Primes are also building blocks when we add them together. Every even number bigger than 2 is the sum of two primes:

4 = 2 + 2
6 = 3 + 3
8 = 5 + 3
10 = 5 + 5
12 = 5 + 7
...
222 = 199 + 23
224 = 211 + 13
...

This proposition, that every even number is the sum of two primes, is known as the Goldbach Conjecture, named after the Prussian mathematician Christian Goldbach, who corresponded with Leonhard Euler about it. Euler was 'entirely certain' that the conjecture was true. In almost 300 years of looking and trying, no one has found an even number that is *not* the sum of two primes, but so far no one has actually been able to prove that the conjecture is true. It is one of the oldest and most famous unsolved problems in mathematics. In 2000, so confident were they that a proof was still beyond the limits of mathematical knowledge that the publishers of the mathematical detective story *Uncle Petros and Goldbach's Conjecture* offered a $1,000,000 prize for anyone who could solve it. No one did.

The Goldbach Conjecture is not the only unresolved issue regarding the primes. Another focus of study is how they seem to be scattered unpredictably along the number line, with no obvious pattern to the sequence. In fact, the search for the harmonies that underpin the distribution of the primes is one of the richest areas of enquiry in number theory, and it has led to many deep results and suppositions.

For all their pre-eminence, however, the primes do not have exclusive claim among the sequences to holding special secrets of mathematical order (or disorder). All sequences contribute in some way to a greater appreciation of how numbers behave. Sloane's *On-Line Encyclopedia of Integer Sequences* can also be considered a compendium of patterns, a Domesday Book of mathematical DNA, a directory of the underlying numerical order of the world. It might have sprung from Neil Sloane's personal obsession, but the project has become a truly important scientific resource.

Sloane compares the *Encyclopedia* to a maths equivalent of the FBI fingerprint database. 'When you go to a crime scene and you take a fingerprint, you then check it against the file of fingerprints to identify the suspect,' he said. 'It's the same thing with the *Encyclopedia*. Mathematicians will come up with a sequence of numbers that occurs naturally in their work, and then they look it up in the database – and it's lovely for them if they find it there already.' The database's usefulness is not restricted to pure mathematics. Engineers, chemists, physicists and astronomers have also looked up, and found, sequences in the *Encyclopedia*, making unexpected connections and gaining mathematical insights into their own fields. For anyone who is working in an area that spews out unfathomable number sequences that they hope to make some sense of, the database is a goldmine.

Through the *Encyclopedia*, Sloane sees a lot of new mathematical ideas, and he also spends time inventing his own. In 1973 he came up with the concept of the 'persistence' of a number. This is the number of steps that it takes to get to a single digit by multiplying all the digits of the preceding number to obtain a second number, then multiplying all the digits of that number to get a third number, and so on until you get down to a single digit. For example:

$$88 \rightarrow 8 \times 8 = 64 \rightarrow 6 \times 4 = 24 \rightarrow 2 \times 4 = 8$$

So, according to Sloane's system, 88 has persistence 3, since it takes three steps to get to a single digit. It would seem likely that the bigger a number is, the bigger its persistence. For example, 679 has persistence 5:

$$679 \rightarrow 378 \rightarrow 168 \rightarrow 48 \rightarrow 32 \rightarrow 6$$

Likewise, if we worked it out here, we would find that 277777788888899 has persistence 11. Yet here's the thing: Sloane has never discovered a number that has a persistence greater than 11, even after checking every number all the way up to 10^{233}, which is 1 followed by 233 zeros. In other words, whatever 233-digit number you choose, if you follow the steps of multiplying all the digits together according to the rules for persistence, you will get to a single-digit number in 11 steps or fewer.

This is splendidly counter-intuitive. It would seem to follow that if you have a number with 200 or so digits consisting of lots of high digits, say 8s and 9s, then the product of these individual digits would be sufficiently large that it would take well over 11 steps to reduce to a single digit. Large numbers, however, collapse under their own weight. This is because if a zero ever appears in the number, the product of all the digits is zero. If there are no zeros in the number to start with, a zero will *always* appear by the eleventh step, unless the number has already been reduced to a single digit by then. In persistence Sloane found a wonderfully efficient giant-killer.

Not stopping there, Sloane has compiled the sequence in which the *n*th term is the smallest number with persistence *n*. (We are considering only numbers with at least two digits.) The first such term is 10, since:

10 → 0 and 10 is the smallest two-digit number that reduces in one step.

The second term is 25, since:

25 → 10 → 0 and 25 is the smallest number that reduces in two steps.

The third term is 39, since:

39 → 27 → 14 → 4 and 39 is the smallest number that reduces in three steps.

The full list is:

(A3001) 10, 25, 39, 77, 679, 6788, 68889, 2677889, 26888999, 3778888999, 277777788888899

I find this list of numbers strangely fascinating. There is a distinct order to them, yet they also are a bit of an asymmetric jumble. Persistence is sort of like a sausage machine that produces only 11 very curiously shaped sausages.

Sloane's good friend Princeton professor John Horton Conway also likes to amuse himself by coming up with offbeat mathematical concepts. In 2007 he invented the concept of a powertrain. For any number written *abcd*..., its powertrain is $a^b c^d$... In the case of numbers where there is an odd number of digits, the last digit has no exponent, so *abcde* goes to $a^b c^d e$. Take 3462. It reduces to $3^4 6^2 = 81 \times 36 = 2916$. Reapply the powertrain until only a single digit is left:

$$3462 \rightarrow 2916 \rightarrow 2^9 1^6 = 512 \times 1 = 512 \rightarrow 5^1 2 = 10 \rightarrow 1^0 = 1$$

Conway wanted to know if there were any indestructible digits – numbers that *did not* reduce to a single digit under the powertrain. He could find only one:

$$2592 \rightarrow 2^5 9^2 = 32 \times 81 = 2592$$

Not one to sit idly by, Neil Sloane took up the chase and uncovered a second:*

24547284284866560000000000

Sloane is now confident that there are no other indestructible digits.

Consider that for a moment: Conway's powertrain is such a lethal machine that it annihilates every number in the universe apart from 2592 and 24547284284866560000000000 – two seemingly unrelated, fixed points in the never-ending expanse of numbers. 'The result is spectacular,' said Sloane. Big numbers die fairly quickly under the powertrain calculation for the same reason that they do under persistence – a zero appears and the whole thing reduces to naught. I asked Sloane if the robustness of the two numbers to survive the powertrain might have any application in the real world. He didn't think so. 'It is just amusing. Nothing wrong with that. You have to have fun.'

And Sloane does have fun. He has studied so many sequences that he's developed his own number aesthetics. One of his favourite

* Using the convention that $0^0 = 1$, since if $0^0 = 0$ the number would collapse immediately.

sequences was devised by the Colombian mathematician Bernardo Recamán Santos, called the Recamán sequence:

(A5132) 0, 1, 3, 6, 2, 7, 13, 20, 12, 21, 11, 22, 10, 23, 9, 24, 8, 25, 43, 62, 42, 63, 41, 18, 42, 17, 43, 16, 44, 15, 45 …

Look at the numbers and try to see a pattern. Follow them carefully. They jump around neurotically. It's all messed up: one up here, one down there, one over there.

In fact, though, the numbers are generated using the following simple rule: 'subtract if you can, otherwise add'. To get the *n*th term, we take the *previous* term and either add or subtract *n* from it. The rule is that subtraction must be used *unless* that results in either a negative number or in a number that is already in the sequence. Here's how the first eight terms are calculated.

Start with 0

The first term is the zeroth term *plus* 1	= 1	*We must add, since subtracting 1 from 0 leaves us with –1, which is not allowed*
The second term is the first term *plus* 2	= 3	*Again, we must add, since subtracting 2 from 1 leaves us with –1, which is not allowed*
The third term is the second term *plus* 3	= 6	*We must add, since subtracting 3 from 3 leaves us with 0, which is already in the sequence*
The fourth term is the third term *minus* 4	= 2	*We must subtract because 6–4 is positive and not in the sequence*
The fifth term is the fourth term *plus* 5	= 7	*We must add, since subtracting 5 from 2 gives –3, which is not allowed*
The sixth term is the fifth term *plus* 6	= 13	*We must add, since subtracting 6 from 7 gives 1, which is already in the sequence*

The seventh term is the sixth term *plus* 7	= 20	*We must add, since subtracting 7 from 13 gives 6, which is already in the sequence*
The eighth term is the seventh term *minus* 8	= 12	*We must subtract, since 20 minus 8 is positive, and not in the sequence*

And so on.

This rather plodding process takes the integers and calculates answers that look totally haphazard. But a way to see the pattern that emerges is to plot the sequence as a graph, as shown below. The horizontal axis is the position of the terms, so the *n*th term is at *n*, and the vertical axis is the value of the terms. The graph of the first thousand terms of the Recamán sequence is probably unlike any other graph you have seen. It is like the spray of a garden sprinkler, or a child trying to join up dots. (The thick lines in the graph are clumps of dots, since the scale is so big.) 'It is interesting to see how much order you can bring into chaos,' Sloane remarked. 'The Recamán sequence is right on the borderline between chaos and beautiful maths and that's why it is so fascinating.'

The clash between order and disorder in the Recamán sequence can also be appreciated musically. The *Encyclopedia* has a function

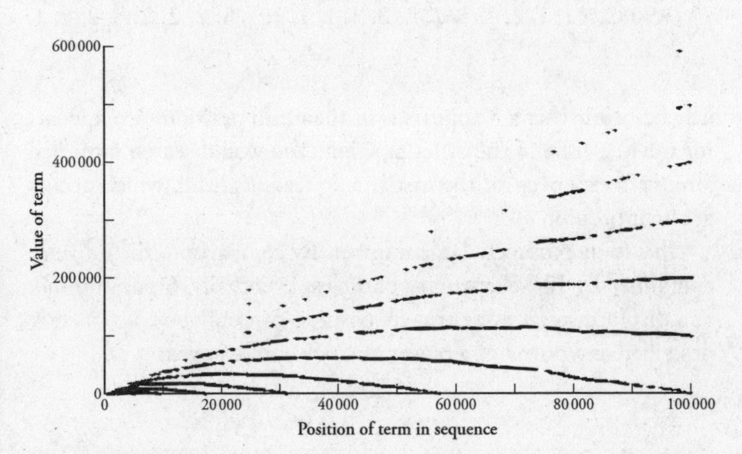

The Recamán sequence.

that allows you to listen to any sequence as musical notes. Imagine a piano keyboard with 88 keys, which comprise a spread of just under eight octaves. The number 1 makes the piano play its lowest note, the number 2 makes it play the second-lowest note, and so on all the way up to 88, which commands the highest note. When the notes run out, you start at the bottom again, so 89 is back to the first key. The natural numbers 1, 2, 3, 4, 5 … sound like a rising scale set on an endless loop. The music created by the Recamán sequence, however, is chilling. It sounds like the soundtrack of a horror movie. It is dissonant, but it does not sound random. You can hear noticeable patterns, as if there is a human hand mysteriously present behind the cacophony.

The question that interests mathematicians about Recamán is whether the sequence contains every number. After 10^{25} terms of the sequence the smallest missing number is 852,655. Sloane suspects that every number will eventually appear, including 852,655, but this remains unproved. It's not hard to understand why Sloane finds Recamán so compelling.

Another favourite of Sloane's is Gijswijt's sequence,* because, unlike many sequences that grow gloriously fast, Gijswijt's increases at a mind-bogglingly dawdling pace. It's a wonderful metaphor for never giving up:

(A90822) 1, 1, 2, 1, 1, 2, 2, 2, 3, 1, 1, 2, 1, 1, 2, 2, 2, 3, 2, 1, 1, 2 …

The first time that a 3 appears is in the ninth position. A 4 appears for the first time in the 221st position. You would search until hell almost freezes over for the first time 5 rears its head, which occurs at about position $10^{100000000000000000000000}$.

This is an extremely large number. By comparison, the universe contains only 10^{80} elementary particles. Eventually, 6 pops up too, at a distance so far away that its position can only be conveniently described as a power of a power of a power of a power:

$$2^{2^{3^{4^{5}}}}$$

* The definition of the sequence appears as an appendix, p. 421.

The other numbers will also eventually appear, although – it must be stressed – with no sense of urgency. 'The land is dying, even the oceans are dying,' said Sloane with poetic flourish, 'but one can take refuge in the abstract beauty of sequences like Dion Gijswijt's A090822.'

As well as paying serious attention to prime numbers, the Greeks were even more enthralled by what they called *perfect* numbers. Consider the number 6: the numbers that divide it – its *factors* – are 1, 2 and 3. If you add 1, 2 and 3, *voilà*, you get 6 again. A perfect number is any number, like 6, that is equal to the sum of its factors. (Strictly speaking, 6 is also a factor of 6, but in discussions of perfection it only makes sense to include the factors of a number *less* than the given number.) After six, the next perfect number is 28 because the numbers that divide it are 1, 2, 4, 7 and 14, the sum of which is 28. Not only the Greeks, but Jews and Christians too attached cosmological significance to such numerical perfection. The ninth-century Benedictine theologian Rabanus Maurus wrote, 'Six is not perfect because God has created the world in 6 days; rather, God has perfected the world in 6 days because the number was perfect.'

The practice of adding the factors of a number leads to the most whimsical concepts in maths. Two numbers are *amicable* if the sum of the factors of the first number equals the second number, and if the sum of the factors of the second number equals the first. For example, the factors of 220 are 1, 2, 4, 5, 10, 11, 20, 22, 44, 55 and 110. Added they equal 284. The factors of 284 are 1, 2, 4, 71 and 142. Together they make 220. Sweet! The Pythagoreans saw 220 and 284 as symbols of friendship. During the Middle Ages talismans with these numbers were made, to promote love. One Arab wrote that he tried to test the erotic effect of eating something labelled with the number 284, while a partner was eating something labelled 220. It was only in 1636 that Pierre de Fermat discovered the second set of amicable numbers: 17,296 and 18,416. Because of the advent of computer processing, more than 11 million amicable pairs are now known. The largest pair has more than 24,000 digits each, which makes them tricky to write on a slice of baklava.

In 1918 the French mathematician Paul Poulet coined the term *sociable* for a new type of numerical friendship. The five numbers listed below are sociable because if you add up the factors of the first one, you get the second. If you add up the factors of the second, you get the third. If you add up the factors of the third, you get the fourth, the factors of the fourth give you the fifth, and the factors of the fifth get you back to where you started: they add up to the first:

12,496
14,288
15,472
14,536
14,264

Poulet discovered only two chains of sociable numbers – the five numbers above and a less exclusive gang of 28 numbers beginning with 14,316. The next set of sociable numbers was discovered by Henri Cohen, but not until 1969. He found nine sociable chains of just four numbers each, of which the chain with the lowest values is 1,264,460, 1,547,860, 1,727,636 and 1,305,184. Currently, 175 chains of sociable numbers are known, and almost all are chains of four numbers. None are chains of three (particularly poetic, since we all know that three's a crowd, and a group of four is much more sociable). The longest chain remains Poulet's 28, which is curious, as 28 is also a perfect number.

It was the Greeks who worked out an unexpected link between perfect numbers and prime numbers, which led to many further numerical adventures. Consider the sequence of doubles starting at 1:

(A79) 1, 2, 4, 8, 16 …

In *The Elements*, Euclid showed that whenever the sum of doubles is a prime number, then you can create a perfect number by multiplying the sum by the highest double that you added. This sounds like a mouthful, so let's start adding doubles to see what he means:

$1 + 2 = 3$. 3 is prime, so, we multiply 3 with the highest double, which is 2. $3 \times 2 = 6$, and 6 is a perfect number.

$1 + 2 + 4 = 7$. Again, 7 is prime. So we multiply 7 by 4 to get another perfect number: 28.

$1 + 2 + 4 + 8 = 15$. This is not prime. No perfect numbers here.

$1 + 2 + 4 + 8 + 16 = 31$. This is prime, and $31 \times 16 = 496$, which is perfect.

$1 + 2 + 4 + 8 + 16 + 32 = 63$. This is not prime.

$1 + 2 + 4 + 8 + 16 + 32 + 64 = 127$. This is prime and $127 \times 64 = 8128$, which is perfect.

$1 + 2 + 4 + 8 + 16 + 32 + 64 + 128 = 255$. This is not prime.

Euclid's proof was, of course, done through geometry. He did not write it out in terms of numbers, instead using line segments. If he'd had the luxury of modern algebraic notation, he would have noticed that he could express the sum of doubles $1 + 2 + 4 + \ldots$ as the sum of powers of two, $2^0 + 2^1 + 2^2 + \ldots$ (Any number to the power 0 is always 1, by convention, and any number to the power 1 is itself.) It then becomes clear that any sum of doubles is equal to the next-largest double minus 1. For example:

$1 + 2 = 3 = 4 - 1$
or
$2^0 + 2^1 = 2^2 - 1$

$1 + 2 + 4 = 7 = 8 - 1$
or
$2^0 + 2^1 + 2^2 = 2^3 - 1$

This can be generalized according to the formula: $2^0 + 2^1 + 2^2 + \ldots + 2^{n-1} = 2^n - 1$, in other words that the sum of the first n terms of the doubling sequence starting at 1 is equal to $2^n - 1$.

So, using Euclid's original declaration that 'whenever the sum of doubles is a prime number, the product of the sum multiplied by the highest double is a perfect number', and adding modern algebraic notation, we can arrive at the much more concise statement:

Whenever $2^n - 1$ *is prime, then* $(2^n - 1) \times 2^{n-1}$ *is a perfect number.*

For civilizations that prized perfect numbers, Euclid's proof was terrific news. If perfect numbers could be generated whenever $2^n - 1$ was prime, all that needed to be done in order to find new perfect numbers was to find new primes that were written $2^n - 1$. The hunt for perfect numbers was reduced to the hunt for a certain type of prime.

Mathematical interest in prime numbers written $2^n - 1$ might have originated because of their link to perfect numbers, but by the seventeenth century the primes had became objects of fascination in their own right. In the same way that some mathematicians were obsessed with finding pi to more and more decimal places, others were preoccupied with finding higher and higher primes. The activities are similar but opposite: whereas finding digits in pi is about trying to see smaller and smaller objects, pursuing primes is about reaching towards the sky. They are missions that have been undertaken as much for the romance of the journey, as for the possible uses of the numbers discovered along the way.

In the quest for primes, the '$2^n - 1$' generating method took on a life of its own. It wasn't going to produce primes for every value of n, but for the low numbers the success rate was pretty good. As we saw above, when $n = 2, 3, 5$ and 7 then $2^n - 1$ is prime.

The mathematician most fixated on using $2^n - 1$ to generate primes was the French friar Marin Mersenne. In 1644 he made the sweeping claim that he knew all the values of n up to 257 such that $2^n - 1$ is prime. He claimed they were:

(A109461) 2, 3, 5, 7, 13, 17, 19, 31, 67, 127, 257

Mersenne was an able mathematician, but his list was largely based on guesswork. The number $2^{257} - 1$ is 78 digits long, far too big for the human mind to check whether it is prime or not. Mersenne realized that his numbers were stabs in the dark. He said of the list: 'all time would not suffice to determine whether they are prime'.

Time did suffice, though, as it often does in the case of maths. In 1876, two and a half centuries after Mersenne wrote his list, the French number theorist Edouard Lucas devised a method that was able to check whether numbers written $2^n - 1$ are prime, and he found that Mersenne was wrong about 67 and that he had left out 61, 89 and 107.

Amazingly, however, Mersenne had been right about 127. Lucas used his method to prove that $2^{127} - 1$, or 170,141,183,460,469,231, 731,687,303,715,884,105,727, was prime. This was the highest-known prime number until the computer age. Lucas, however, was unable to determine if $2^{257} - 1$ was prime or not; the number was simply too large to work on with pencil and paper.

Despite its patches of error, Mersenne's list immortalized him; and now a prime that can be written in the form $2^n - 1$ is known as a *Mersenne prime*.

The proof of whether $2^{257} - 1$ is prime would take until 1952 to be proven, using the Lucas method, but with a big assist. A team of scientists gathered one day that year at the Institute for Numerical Analysis in Los Angeles to watch a 24ft scroll of tape be inserted into an early digital computer called the SWAC. Simply putting in the tape took several minutes. The operator then input the number to be tested: 257. In a fraction of a second came the result. Computer said no: $2^{257} - 1$ is not prime.

On the same evening in 1952 on which it was discovered that $2^{257} - 1$ is not prime, new potential Mersenne numbers were inserted into the machine. The SWAC rejected the first 42 as not prime. Then, at 10 p.m., came a result. Computer said yes! It announced that $2^{521} - 1$ is prime. The number was the highest Mersenne prime identified in 75 years, making the corresponding perfect number, $2^{520} (2^{521} - 1)$, only the thirteenth discovered in almost twice as many centuries. But the number $2^{521} - 1$ had only two hours to enjoy its status as top of the pile. Shortly before midnight the SWAC confirmed that $2^{607} - 1$ was also prime. Over the next few months, SWAC worked to the limit of its capacities, finding three more primes. Between 1957 and 1996 another 17 Mersenne primes were discovered.

Since 1952, the largest-known prime number has always been a Mersenne prime (apart from a three-year interlude between 1989 and 1992 when the largest prime was $(391581 \times 2^{216193}) - 1$, which is a related type of prime). Among all the primes that exist, and we know there is an infinite number of them, Mersenne primes dominate the table of highest discovered primes because they give prime-hunters a target to aim for. The best technique for finding high primes is to look for Mersenne primes, in other words, to put the number $2^n - 1$ into a computer for higher and higher values of n and use the Lucas-Lehmer test, an improved version of Edouard Lucas's method mentioned earlier, to see if it is prime.

Mersenne primes also have an aesthetic loveliness. For example, in binary notation any number 2^n is written as 1 followed by n zeros. For example, $2^2 = 4$, which in binary notation is written 100, and $2^5 = 32$, which is written 100000. Since all Mersenne primes are 1 less than 2^n, all binary expansions of Mersenne primes are strings of digits that contain only 1s.

The most influential prime-hunter of modern times was inspired in his mission by the markings on an envelope. When George Woltman was a boy, in the 1960s, his father showed him a postmark with the expression $2^{11213} - 1$, then the most recently calculated prime number. 'I was amazed that a number so large could be proven prime,' he remembered.

Woltman was later responsible for writing some software that has made an enormous contribution to the quest for primes. All projects that involved massive number-crunching used to be carried out on 'supercomputers', access to which was limited. Since the 1990s, however, many big tasks have been salami-sliced by dividing up the work among thousands of smaller machines connected to each other by the internet. In 1996 Woltman wrote a piece of software that users can download for free and that, once installed, allocates a small part of the uninvestigated number line where your machine can look for primes. The software uses the processor only when your computer is otherwise idle. While you are fast asleep, your machine is busy churning through numbers on the frontier of science. The Great Internet Mersenne Prime Search, or GIMPS, currently links about 75,000 computers. Some of these are in academic institutions, some are in businesses and some are personal laptops. GIMPS was one of the first 'distributed computing' projects and has been one of the most successful. (The largest similar project, Seti@home, is deciphering cosmic noise for signs of extraterrestrial life. It claims three million users but, so far, has discovered nothing.) Only a few months after GIMPS went online a 29-year-old French programmer netted the 35th Mersenne prime: $2^{1398269} - 1$. Since then, GIMPS has revealed another 11 Mersenne primes, which is an average of about one a year. We are living in a golden age of high prime numbers.

The current record for largest prime is held by the 45th Mersenne prime: $2^{43112609} - 1$, which is a number almost 13 million digits long, found in 2008 by a computer connected to GIMPS at the University of California, Los Angeles. The 46th and 47th Mersennes found were actually *smaller* than the 45th. This happened because computers are working at different speeds on different sections of the number line at the same time, so it is possible that primes in higher sections will be discovered before primes in lower ones.

GIMPS's message of mass voluntary cooperation for scientific advancement has made it an icon of the liberal web. Woltman has unintentionally turned the search for primes into a quasi-political pursuit. As a mark of the symbolic importance of the project, the Electronic Frontier Foundation, a digital-rights campaign group, has since 1999 offered money for each prime whose digits reach the next order of magnitude. The 45th Mersenne prime was the first to hit ten million digits and the prize money won was $100,000. The EFF is offering $150,000 for the first prime with 100 million digits, and $250,000 for the first prime with a billion digits. If you plot the largest-known primes discovered since 1952 on a graph with a logarithmic scale against the time of discovery, they fall on what is almost a straight line. As well as showing how the growth of processing power has advanced remarkably consistently over time, the line also allows us to estimate when the first billion-digit prime will be discovered. I'd put money on it being found by 2025. Writing out this number in type where each digit is a millimetre would stretch further than from Paris to Los Angeles.

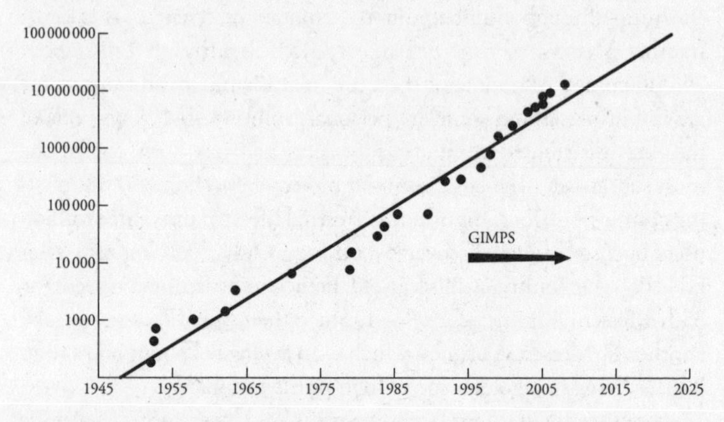

Digits in highest-known prime by year of discovery.

With an infinite number of primes (whether there is an infinite number of Mersenne primes, however, is not yet known), the search for higher and higher primes is a never-ending task. Whatever prime number we reach, no matter how large, there will always be a prime number even larger taunting us for our lack of ambition.

Endlessness is probably the most profound and challenging idea of basic maths. The mind finds it difficult to cope with the idea of something going on for ever. What, for example, would happen if we start counting 1, 2, 3, 4, 5 … and never stop? I remember asking this seemingly simple question as a child, and receiving no straightforward answer. The default response from parents and schoolteachers was that we get to 'infinity' but this answer essentially just restates the question. Infinity is simply defined as being the number that we get to when we start counting and never stop.

Nevertheless, we are told from a relatively early age to treat infinity like a number, a weird number, but a number all the same.

We are shown the symbol for infinity, the endless loop ∞ (called a 'lemniscate'), and taught its peculiar arithmetic. Add any finite number to infinity and we get infinity. Subtract any finite number from infinity and we get infinity. Multiply or divide infinity by a finite number, as long as it isn't zero, and the result is also infinity. The ease with which we are told that infinity is a number disguises more than 2000 years of struggling to come to terms with its mysteries.

The first person to showcase the trouble with infinity was the Greek philosopher Zeno of Elea, who lived in the fifth century BC. In one of his famous paradoxes, he described a theoretical race between Achilles and a tortoise. Achilles is faster than the tortoise, so the tortoise is given a head start. The famous warrior starts at a point A and his reptile challenger is ahead of him at a point B. Once the race starts, Achilles zooms forward and soon reaches point B, but by the time he gets there, the tortoise has already advanced to point C. Achilles then ploughs on to point C. But once again, when he reaches this point, the tortoise has shuffled forward to point D. Achilles must reach D, of course, but when he does, the tortoise is already at E. Zeno argued that the game of catch-up must carry on for ever and, therefore, that swift Achilles is never able to overtake his slower four-footed rival. The athlete is much faster than the tortoise, but he cannot beat him in a race.

Like this one, all of Zeno's paradoxes draw apparently absurd conclusions by dissecting continuous motion into discrete events. Before Achilles can reach the tortoise, he must complete an infinite number of these discrete dashes. The paradox stems from the assumption that it is impossible to complete an infinite number of dashes in a finite amount of time.

The Greeks, though, didn't have the depth of mathematical understanding of infinity to see that this assumption is a fallacy. It *is* possible to complete an infinite number of dashes in a finite amount of time. The crucial requirement is that the dashes are getting shorter and taking less time, and that both distance and time are approaching zero. Although this is a necessary condition, it's not sufficient; the dashes also need to be shrinking at a fast enough rate.

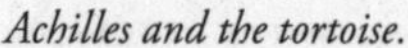

Achilles and the tortoise.

This is what is happening with Achilles and the tortoise. For example, say that Achilles is running at twice the speed of the tortoise and that B is 1m ahead of A. When Achilles reaches B, the turtle has moved $\frac{1}{2}$m to C. When Achilles reaches C, the turtle has moved another $\frac{1}{4}$m to D. And so on. The total distance in metres that Achilles is running before he reaches the tortoise is:

$$1 + \frac{1}{2} + \frac{1}{4} + \frac{1}{8} + \frac{1}{16} + \ldots$$

If it takes Achilles one second to complete each of these intervals then it will take him for ever to complete the distance. But this is not the case. Assuming constant speed, it will take him a second to go a metre, it will take half a second to go half a metre, a quarter of a second to go quarter of a metre, and so on. So, the time in seconds it takes him to reach the tortoise is described by the same addition:

$$1 + \frac{1}{2} + \frac{1}{4} + \frac{1}{8} + \frac{1}{16} + \ldots$$

When both time and distance are described by the halving sequence they simultaneously converge at a fixed, finite value. In the above case, at 2 seconds and 2 metres. So, it turns out that Achilles can overtake the tortoise after all.

Not all of Zeno's paradoxes, however, are solved by the maths of infinite series. In the 'dichotomy paradox' a runner is going from A to B. In order to get to B, however, the runner needs to pass through the halfway point between A and B, which we'll call C. But to get to C he first needs to get to the halfway point between A and C. It follows that there can be no 'first point' that the runner passes, since there will always be a point that he must pass before he reaches it, such as the halfway point. If there is no first point that the runner passes, Zeno argued, the runner cannot ever leave A.

According to lore, to refute this paradox Diogenes the Cynic silently stood up and walked from A to B, thereby demonstrating that such motion was possible. But Zeno's dichotomy paradox cannot be dismissed so easily. In two and a half thousand years of scholarly head-scratching, no one has been able to solve the riddle totally. Part of the confusion is that a continuous line is not perfectly represented by a sequence of an infinite number of points, or an infinite number of small intervals. Likewise, the unbroken passage of time is not perfectly represented by an infinite number of discrete moments. The concepts of continuity and discreteness are not entirely reconcilable.

The decimal system throws up an excellent example of a Zeno-inspired paradox. What is the largest number less than 1? It is not 0.9, since 0.99 is larger and still less than 1. It is not 0.99 since 0.999 is larger still and also less than 1. The only possible candidate is the recurring decimal 0.9999... where the '...' means that the nines go on for ever. Yet this is where we come to the paradox. It cannot be 0.9999... since the number 0.9999... is identical to 1!

Think of it this way. If 0.9999... is a different number from 1, then there must be space between them on the number line. So it must be possible to squeeze a number in the gap that is larger than 0.9999... and smaller than 1. Yet what number could this be? You cannot get closer to 1 than 0.9999.... So, if 0.9999... and 1 cannot be different, they must be the same. Counter-intuitive though it is, 0.9999... = 1.

So what is the largest number less than one? The only satisfactory conclusion to the paradox is that the largest number less than 1 *doesn't exist.* (Likewise, there is no largest number less than 2, or less than 3, or indeed less than any number at all.)

The paradox of Achilles' race against the tortoise was resolved by writing the durations of his dashes as a sum with an infinite amount of terms, which is also known as an infinite series. Whenever the terms of a sequence are added together it is called a *series*. There are both finite and infinite series. For example, if you add up the sequence of the first five natural numbers, you get the finite series:

$$1 + 2 + 3 + 4 + 5 = 15$$

Obviously we can work out this sum in our heads, but when a series has many more terms, the challenge is to find a shortcut. One famous example was worked out by the German mathematician Carl Friedrich Gauss when he was a young boy. As the story goes, a schoolteacher is said to have asked him to calculate the sum of the series of the first hundred natural numbers:

$$1 + 2 + 3 + \dots + 98 + 99 + 100$$

To the teacher's disbelief, Gauss replied almost instantly: '5050.' The prodigy had worked out the following formula. If you pair off numbers judiciously, by taking the first with the last, the second with the second-last, and so on, then the series can be rewritten as:

$(1 + 100) + (2 + 99) + (3 + 98) + \ldots + (50 + 51)$
which is:
$101 + 101 + 101 + 101 + \ldots + 101$

There are fifty terms, each with a value 101, so the sum is $50 \times 101 = 5050$. We can generalize this to get the result that for any number n, the sum of the first n numbers is $n + 1$ added $\frac{n}{2}$ times in a row, which is $\frac{n(n+1)}{2}$. In the above case n is 100, so the sum is $\frac{100(100+1)}{2} = 5050$.

When you add up the terms in a finite series you always get a finite number, that's obvious. However, when you add up the terms of an infinite series there are two possible scenarios. The *limit*, which is the number that the sum approaches as more and more terms are added, is either a finite number or it is infinite. If the limit is finite, the series is called *convergent*. If not, the series is called *divergent*.

For example, we have already seen that the series

$$1 + \frac{1}{2} + \frac{1}{4} + \frac{1}{8} + \frac{1}{16} + \ldots$$

is convergent, and converges on 2. We have also seen that there are many infinite series that converge on pi.

On the other hand, the series

$$1 + 2 + 3 + 4 + 5 + \ldots$$

is divergent, heading off towards infinity.

The Greeks may have been wary of infinity, but by the seventeenth century mathematicians were happy to take it on. An understanding of infinite series was required for Isaac Newton to invent calculus, which was one of the most significant developments in mathematics.

When I studied maths one of my favourite exercises was being presented with an infinite series and being asked to work out whether it converged or diverged. I always found it incredible that the difference between convergence and divergence was so brutal – the difference between a finite number and infinity is infinity – and yet the elements that decided which path the series took often seemed so insignificant.

Take a look at the *harmonic series*:

$$1 + \frac{1}{2} + \frac{1}{3} + \frac{1}{4} + \frac{1}{5} + \ldots$$

The nominator of every term is one, and the denominators are the natural numbers. The harmonic series looks like it should converge. Each term in the series gets smaller and smaller, so you would think that the sum of all the terms would be bounded by a fixed amount. But, bizarrely, the harmonic series is divergent, a decelerating but unstoppable snail. After 100 terms of the series, the total has only just passed 5. After 15,092,688,622,113,788,323,693,563,264,538, 101,449,859,497 terms, the total exceeds 100 for the first time. Yet this stubborn snail will continue its bid for freedom, past any distance you care to mark out. The series will eventually reach a million, then a billion, going further and further towards infinity. (The proof is included as an appendix, p. 422.)

The harmonic series appears when we consider the maths of stacking Jenga blocks. Say you have two blocks and want to position them one on top of the other so that the top one has the largest possible overhang, but doesn't topple over. The way to do this is to place the top block exactly halfway across the one underneath, as demonstrated opposite (A). In this way, the centre of gravity of the top block falls on the edge of the bottom brick.

If we had three blocks, what would be their positions so the combined overhang was as large as possible without toppling? The solution is for the top one to be placed halfway along the middle one, and for the middle one to be a quarter of the way along the bottom one, as in the diagram opposite (B).

Continuing for more and more blocks, the general pattern is that, in order to guarantee the maximum combined overhang, the top one must be halfway along the second one, which must be a quarter along the third one, which must be a sixth along the fourth one, which must be an eighth along the fifth one, and so on. This gives us a leaning tower of bricks that looks like C on the page opposite.

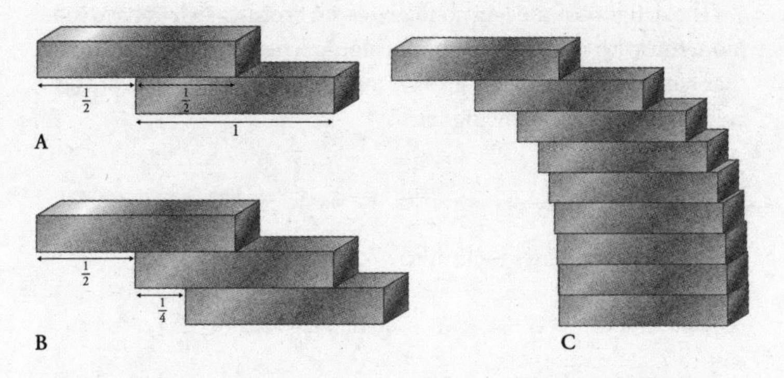

How to stack Jenga blocks with maximum overhang so that they don't topple.

The total overhang of this tower, which is the sum of all the individual overhangs, is the following series:

$$\frac{1}{2} + \frac{1}{4} + \frac{1}{6} + \frac{1}{8} + \ldots$$

Which can be rewritten as:

$$\frac{1}{2}(1 + \frac{1}{2} + \frac{1}{3} + \frac{1}{4} + \ldots$$

Which is half of the harmonic series, if we carry on for an infinite number of terms.

Now, since we know that the harmonic series increases to infinity, we also know that the harmonic series divided by two increases to infinity, because infinity divided by two is infinity. Rephrased in the context of stacking Jenga blocks, this means that it is theoretically possible to create a freestanding overhang of any length we want. If the harmonic series divided by two will eventually exceed any number we want, provided we include enough terms, then the overhang of the leaning tower of blocks will eventually exceed any length we want, provided we stack enough blocks. Although theoretically possible, however, the practicalities of constructing a tower with a large overhang are daunting. In order to achieve an overhang of 50 blocks, we would need a tower of 15×10^{42} blocks – which would be much higher than the distance from here to the edge of the observable universe.

The delights of the harmonic series are profuse, so let's have some more fun with it. Consider the harmonic series *excluding* every term that has a 9 in it, which is also an infinite series. In other words, we are extracting the following terms:

$$\frac{1}{9}, \frac{1}{19}, \frac{1}{29}, \frac{1}{39}, \frac{1}{49}, \frac{1}{59}, \frac{1}{69}, \frac{1}{79}, \frac{1}{89}, \frac{1}{90}, \frac{1}{91}, \frac{1}{92} \ldots$$

So, the depleted series looks like:

$$1 + \frac{1}{2} + \frac{1}{3} + \frac{1}{4} + \frac{1}{5} + \ldots + \frac{1}{8} + \frac{1}{10} + \ldots + \frac{1}{18} + \frac{1}{20} + \ldots$$

Recall that the harmonic series adds up to infinity, so one might think that the harmonic series with no 9s also adds up to a pretty high number. Wrong. It adds up to just under 23.

By filtering out 9 we have tamed infinity: we have slaughtered the beast of eternity and all that is left is a shrivelled carcass of about 23.

This result appears remarkable, but by looking a little closer we can understand it completely. Eliminating a 9 gets rid of only one of the first 10 terms of the harmonic series. But it gets rid of 19 of the first 100 terms, and 271 of the first thousand. By the time the numbers are very large, say with 100 digits, the vast majority of numbers contain a 9. As it turns out, thinning the harmonic series by taking out the terms with a 9 removes almost all of it.

Customizing the harmonic series gets more intriguing than this, though. The decision to extract 9s was arbitrary. If I had extracted all the terms containing 8 from the harmonic series, the remaining terms would also converge to a finite number. As it would if I extracted only the terms with a 7, or indeed with any single digit. In fact, we do not even need to limit ourselves to single digits. Remove all terms including *any* number, and the thinned-out harmonic series is convergent. This works with, say, 9 or 42 or 666 or 314159, and the same reasoning applies.

I'll use the example of 666. Between 1 and 1000 the number 666 occurs once. Between 1 and 10,000 it occurs 20 times, and between 1 and 100,000 it occurs 300 times. In other words, the percentage occurrence of 666 is 0.1 per cent in the first 1000 numbers, 0.2 percent in the first 10,000 and 0.3 per cent in the first 100,000. As you consider larger and larger numbers, the string of digits 666 is proportionately more and more common. It will eventually be the case that almost all numbers have a 666 in them. So, almost all terms in the harmonic series will eventually have a 666 in them. Exclude them from the harmonic series and the depleted series converges.

In 2008 Thomas Schmelzer and Robert Baillie calculated that the harmonic series without any term containing the number 314159 adds up to a little over 2.3 million. It is a large number, but a long, long way from infinity.

A corollary of this result is that the harmonic series with *only* the terms including 314159 must add up to infinity. In other words, the series:

$$\frac{1}{314159} + \frac{1}{1314159} + \frac{1}{2314159} + \frac{1}{3314159} + \frac{1}{4314159} + \dots$$

adds up to infinity. Even though it starts with a tiny number and the terms get only tinier, the sum of the terms will eventually surpass any number you want. The reason, again, is because once numbers get very large, almost every number has a 314159 in it. Almost all unit fractions contain a 314159.

Let's take a look at one last infinite series, one that brings us back to the mysteries of the primes. The prime harmonic series is the series of unit fractions where the denominators are the prime numbers:

$$\frac{1}{2} + \frac{1}{3} + \frac{1}{5} + \frac{1}{7} + \frac{1}{11} + \frac{1}{13} + \frac{1}{17} \dots$$

The primes get scarcer as numbers get higher, so one might expect that this series doesn't have the momentum to add up to infinity. Yet, unbelievably, it does. Counter-intuitive and spectacular, the result makes us realize the power and importance of the primes. They can be seen not only as the building blocks of natural numbers, but also as the building blocks of infinity.

CHAPTER EIGHT

Gold Finger

Sitting with me in his lounge at home, Eddy Levin handed me a sheet of white paper and asked me to write out my name in capital letters. Levin, who is 75 years old and has a donnish face with grey stubble and a long forehead, used to be a dentist. He lives in East Finchley, north London, on a street that is the epitome of prosperous and conservative suburban Britain. Expensive cars sat in the driveways of between-the-wars brick houses with freshly trimmed hedges and bright green lawns. I took the paper and wrote: ALEX BELLOS.

Levin then picked up a stainless-steel instrument that looked like a small claw, with three prongs. With a steady hand he held it up to the paper and started to analyse my script. He lined up the instrument to the E in my first name with the concentration of a rabbi preparing a circumcision.

'Pretty good,' he said.

Levin's claw is his own invention. The three prongs are positioned in such a way that the tips of the prongs stay on the same line and in the same ratio to each other when the claw opens out. He designed the instrument so that the distance between the middle prong and the prong above it is always 1.618 times the distance between the middle prong and the prong below it. Because this number is better known as the golden mean he calls his tool the Golden Mean Gauge. (Other synonyms for 1.618 include the golden ratio, the divine proportion and φ, or phi.) Levin put the gauge on my letter E so that the tip of one claw was on the top horizontal bar of the E, the middle tip was on the middle bar of the E and the bottom tip was on the bottom bar. I had assumed that when I wrote a capital E I positioned the middle bar equidistant between the top and the bottom, but Levin's gauge showed that I was subconsciously placing the bar slightly above halfway – in such a way that it divided the height of the letter into two sections with lengths of ratio 1 to 1.618.

Although I had scribbled my name with gay abandon, I had adhered to the golden mean with uncanny precision.

Levin smiled and moved on to my S. He readjusted the gauge so that the side points touched the topmost and bottommost tips of the letter and, to my further amazement, the middle one coincided exactly with the S line as it curved.

'Spot on,' Levin said calmly. 'Everybody's handwriting is in the golden proportion.'

The golden mean is the number that describes the precise ratio when a line is cut into two sections in such a way that the proportion of the entire line to the larger section is equal to the proportion of the larger section to the smaller section. In other words, when the ratio of A + B to A is equal to the ratio of A to B:

A B

A line divided into two by the golden ratio is known as a golden section, and phi, the ratio between larger and smaller sections, can be calculated as $\frac{(1+\sqrt{5})}{2}$. This is an irrational number, whose decimal expansion begins:

1.61803 39887 49894 84820...

The Greeks were fascinated by phi. They discovered it in the five-pointed star, or pentagram, which was a revered symbol of the Pythagorean Brotherhood. Euclid called it the 'extreme and mean ratio' and he provided a method to construct it with compass and straightedge. Since at least the Renaissance, the number has intrigued artists as well as mathematicians. The major work on the golden ratio was Luca Pacioli's *The Divine Proportion* in 1509, which listed the appearance of the number in many geometric constructions, and was illustrated by Leonardo da Vinci. Pacioli concluded that the ratio was a message from God, a source of secret knowledge about the inner beauty of things.

Mathematical interest in phi comes from how it is related to the most famous sequence in maths: the Fibonacci sequence, which is

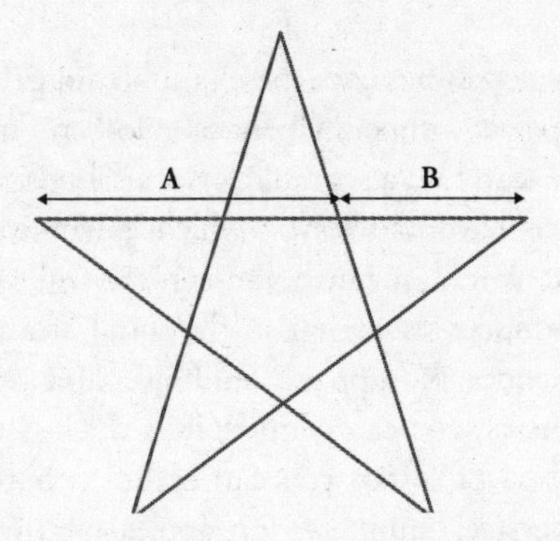

The pentagram, a mystical symbol since ancient times, contains the golden ratio.

the sequence that starts with 0, 1 and each subsequent term is the sum of the two previous terms:

0, 1, 1, 2, 3, 5, 8, 13, 21, 34, 55, 89, 144, 233, 377 …

Here is how the numbers are found:

$0 + 1 = 1$
$1 + 1 = 2$
$1 + 2 = 3$
$2 + 3 = 5$
$3 + 5 = 8$
$5 + 8 = 13$
…

Before I show how phi and Fibonacci are connected, let's investigate the numbers in the sequence. The natural world has a predilection for Fibonacci numbers. If you look in the garden, you will discover that for most flowers the number of petals is a Fibonacci number:

3 petals	lily and iris
5 petals	pink and buttercup
8 petals	delphinium
13 petals	marigold and ragwort
21 petals	aster
55 petals/89 petals	daisy

The flowers may not always have these numbers of petals, but the average number of petals will be a Fibonacci number. For example, there are usually three leaves on a stem of clover, a Fibonacci number. Only seldom do clovers have four leaves, which is why we consider them special. Four-leaf clovers are rare because 4 is not a Fibonacci number.

Fibonacci numbers also occur in the spiral arrangements on the surfaces of pine cones, pineapples, cauliflower and sunflowers. As the picture below shows, you can count spirals clockwise and anticlockwise. The numbers of spirals you can count in both directions are consecutive Fibonacci numbers. Pineapples usually have 5 and 8 spirals, or 8 and 13 spirals. Spruce cones tend to have 8 and 13 spirals. Sunflowers can have 21 and 34, or 34 and 55 spirals – although examples as high as 144 and 233 have been found. The more seeds there are, the higher up the sequence the spirals will go.

The Fibonacci sequence is so called because the terms appear in Fibonacci's *Liber Abaci*, in a problem about rabbits. The sequence only gained the name, however, more than 600 years after the book was published when, in 1877, the number theorist Edouard Lucas was studying it, and he decided to pay tribute to Fibonacci by naming the sequence after him.

A sunflower with 34 anticlockwise and 21 clockwise spirals.

The *Liber Abaci* set up the sequence like this: say that you have a pair of rabbits, and after one month the pair gives birth to another pair. If every adult pair of rabbits gives birth to a pair of baby rabbits every month, and it takes one month for the baby rabbits to become adults, how many rabbits are produced from the first pair in a year?

The answer is found by counting rabbits month by month. In the first month, there is just one pair. In the second there are two, as the original pair have given birth to a pair. In the third month there are three, since the original pair have again bred, but the first pair are only just adults. In the fourth month the two adult pairs breed, adding two to the population of three. The Fibonacci sequence is the month-on-month total of pairs:

	Total pairs
First month: 1 adult pair	1
Second month: 1 adult pair and 1 baby pair	2
Third month: 2 adult pairs and 1 baby pair	3
Fourth month: 3 adult pairs and 2 baby pairs	5
Fifth month: 5 adult pairs and 3 baby pairs	8
Sixth month: 8 adult pairs and 5 baby pairs	13
...	...

An important feature of the Fibonacci sequence is that it is *recurrent*, which means that each new term is generated by the values of previous terms. This helps explain why the Fibonacci numbers are so prevalent in natural systems. Many life forms grow by a process of recurrence.

There are many examples in nature of Fibonacci numbers, and one of my favourites concerns the reproductive patterns of bees. A male bee, or drone, has just one parent: his mother. Female bees, however, have two parents: a mother and a father. So, a drone has two grandparents, three great-grandparents, five great-great-grandparents, and so on. Plotting a chart of the drone's ancestry (as in the diagram overleaf), we find that the number of relatives he has per generation is always a Fibonacci number.

In addition to its association with fruit, promiscuous rodents and flying insects, the Fibonacci sequence has many absorbing mathematical properties. Listing the first 20 numbers will help us

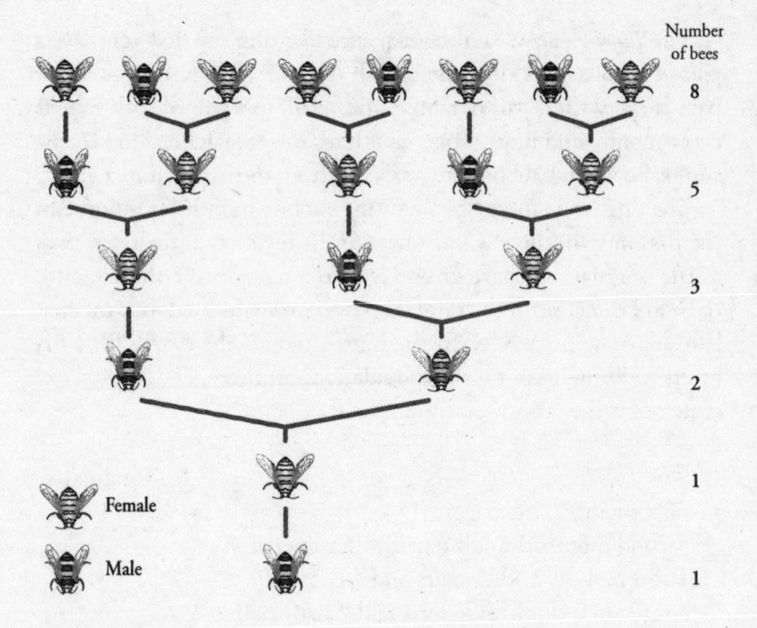

A chart tracing the ancestral history of one male bee (shown at the bottom).

see the patterns. Each Fibonacci number is traditionally written using an F with a subscript to denote the position of that number in the sequence:

(F_0	0)		
F_1	1	F_{11}	89
F_2	1	F_{12}	144
F_3	2	F_{13}	233
F_4	3	F_{14}	377
F_5	5	F_{15}	610
F_6	8	F_{16}	987
F_7	13	F_{17}	1597
F_8	21	F_{18}	2584
F_9	34	F_{19}	4181
F_{10}	55	F_{20}	6765

Upon closer examination, we see that the sequence regenerates itself in many surprising ways. Look at F_3, F_6, F_9, … , in other words, every third F-number. They are all divisible by 2. Compare this with F_4, F_8, F_{12}, … , or every fourth F-number – they are all divisible by 3. Every fifth F-number is divisible by 5; every sixth F-number, divisible by 8; and every seventh number by 13. The divisors are precisely the F-numbers in sequence.

Another amazing example comes from $\frac{1}{F_{11}}$, or $\frac{1}{89}$. This number is equal to the sum of:

.0
.01
.001
.0002
.00003
.000005
.0000008
.00000013
.000000021
.0000000034
.00000000055
.000000000089
.0000000000144

So, the Fibonacci sequence pops its head up again.

Here's another interesting mathematical property of the sequence. Take any three consecutive F-numbers. The first one multiplied by the third one is always one different from the second one squared:

For F_4, F_5, F_6:
$F_4 \times F_6 = F_5 \times F_5 - 1$ … *since 24 = 25 – 1*

For F_5, F_6, F_7:
$F_5 \times F_7 = F_6 \times F_6 + 1$ … *since 65 = 64 + 1*

For F_{18}, F_{19}, F_{20}
$F_{18} \times F_{20} = F_{19} \times F_{19} - 1$ … *since 17,480,760 = 17,480,761 – 1*

This property is the basis of a centuries-old magic trick, in which it is possible to cut up a square of 64 unit squares into four pieces and reassemble them to make a rectangle of 65 pieces. Here's how it's done: draw a square of 64 unit squares. It has a side length of 8. In the sequence, the two F-numbers preceding 8 are 5 and 3. Divide the square up using the lengths of 5 and 3, as in the first image below. The pieces can be reassembled to make a rectangle with sides the length of 5 and 13, which has an area of 65.

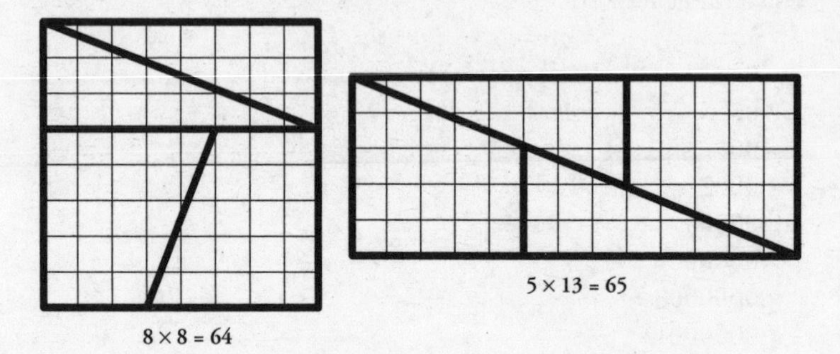

$8 \times 8 = 64$ $5 \times 13 = 65$

The trick is explained by the fact that the shapes are not a perfect fit. Though it is not that obvious to the naked eye, there is a long thin gap along the middle diagonal with an area of one unit.

It follows that a square of 169 unit squares (13 × 13) can be rearranged to 'make' a rectangle of 168 squares (8 × 21). In this case the segments overlap slightly along the middle diagonal.

In the early seventeenth century, the German astronomer Johannes Kepler wrote that: 'As 5 is to 8, so 8 is to 13, approximately, and as 8 to 13, so 13 is to 21, approximately.' In other words, he noticed that the ratios of consecutive F-numbers were similar. A century later the Scottish mathematician Robert Simson saw something even more incredible. If you take the ratios of consecutive F-numbers and put them in the sequence:

$$\frac{F_2}{F_1}, \frac{F_3}{F_2}, \frac{F_4}{F_3}, \frac{F_5}{F_4}, \frac{F_6}{F_5}, \frac{F_7}{F_6}, \frac{F_8}{F_7}, \frac{F_9}{F_8}, \frac{F_{10}}{F_9} \ldots$$

which is:

$$\frac{1}{1}, \frac{2}{1}, \frac{3}{2}, \frac{5}{3}, \frac{8}{5}, \frac{13}{8}, \frac{21}{13}, \frac{34}{21}, \frac{55}{34} \ldots$$

or (to three decimal places):

1, 2, 1.5, 1.667, 1.6, 1.625, 1.615, 1.619, 1.618 …

then the values of these terms get closer and closer to phi, the golden ratio.

In other words, the golden ratio is approximated by the ratio of consecutive Fibonacci numbers, with the approximation increasing in accuracy further down the sequence.

Now let's continue with this line of thought and consider a Fibonacci-like sequence, starting with two random numbers, and then adding consecutive terms to continue the sequence. So, just say we start with 4 and 10, the following term will be 14 and the one after that 24. Our example gives us:

4, 10, 14, 24, 38, 62, 100, 162, 262, 424 …

Look at the ratios of consecutive terms:

$$\frac{10}{4}, \frac{14}{10}, \frac{24}{14}, \frac{38}{24}, \frac{62}{38}, \frac{100}{62}, \frac{162}{100}, \frac{262}{162}, \frac{424}{262} \ldots$$

or:

2.5, 1.4, 1.714, 1.583, 1.632, 1.612, 1.620, 1.617, 1.618 …

The Fibonacci recurrence algorithm of adding two consecutive terms in a sequence to make the next one is so powerful that *whatever* two numbers you start with, the ratio of consecutive terms always converges to phi. I find this a totally enthralling mathematical phenomenon.

The ubiquity of Fibonacci numbers in nature means that phi is also ever present in the world. Which brings us back to the retired dentist, Eddy Levin. Early in his career he spent a lot of time making false teeth, which he found a very frustrating job because no matter how he arranged the teeth he could not make a person's smile look right. 'I sweated blood and tears,' he said. 'Whatever I did the teeth looked artificial.' But at around that time Levin started attending a maths and spirituality class, where he learned about phi. Levin was made aware of Pacioli's *The Divine Proportion* and was inspired. What if phi, which Pacioli claimed revealed true beauty, also held the secret of divine dentures? 'It was a Eureka moment,' he said. It was 2 a.m. and he rushed to his study. 'I spent the rest of the night measuring teeth.'

Levin scoured photographs and discovered that in the most attractive sets of teeth, the big top front tooth (the central incisor) was wider than the one next to it (the lateral incisor) by a factor of phi. The lateral incisor was also wider than the adjacent tooth (the canine) by a factor of phi. And the canine was wider than one next to it (the first premolar) by a factor of phi. Levin wasn't measuring the size of actual teeth, but the size of teeth in pictures when taken head-on. Still, he felt like he had made an historic discovery: the beauty of a perfect smile was prescribed by phi.

'I was very excited,' remembered Levin. At work, he mentioned his findings to colleagues, but they dismissed him as an oddball. He continued to develop his ideas nonetheless, and in 1978 he published an article expounding them in the *Journal of Prosthetic Dentistry*. 'From then, people got interested in it,' he said. 'Now there is not a lecture that is given on [dental] aesthetics that doesn't include a section on the golden proportion.' Levin was using phi so much in his work that in the early 1980s he asked an engineer to design him an instrument that could tell him if two teeth were in the golden proportion. The result was the three-pronged Golden Mean Gauge. He still sells them to dentists around the world.

I couldn't tell if Levin's own teeth were in the golden proportion, although there was certainly a fair amount of gold in them. Levin told me his gauge became more than a work tool, and he started to measure objects other than teeth. He found phi in the patterns of flowers, in the spread of branches along stems, and in leaves along branches. He took it with him on holiday and found phi in the proportions of buildings. He also found phi in the rest of the human body, in the length of knuckles to fingers and in the relative positions of the nose, teeth and chin. Additionally, he noticed that most people use phi in their handwriting, just as he had shown in mine.

The more Levin looked for phi, the more he found it. 'I found so many coincidences, I started to wonder what it was all about.' He opened his laptop and showed me a slideshow of images, each with the three points of the gauge showing exactly where the ratio was to be found. I saw pictures of butterfly wings, peacock feathers and animal colourings, the ECG reading of a healthy human heart, paintings by Mondrian and a car.

When a rectangle is constructed so that the ratio between its sides is phi, you get what is known as a 'golden rectangle'. This rectangle has the convenient property that if we were to cut it vertically so that one side is a square, then the other side is also a golden rectangle. The mother gives birth to a baby daughter.

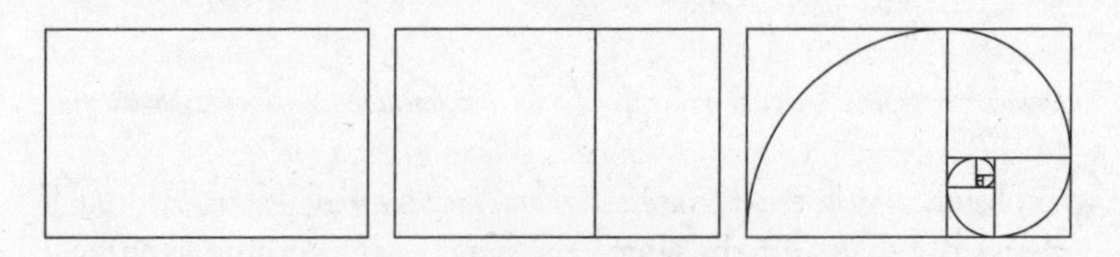

Golden rectangle and logarithmic spiral.

We can continue this process to create granddaughters, great-granddaughters, ad infinitum. Now, let's draw a quarter-circle in the largest square by using a compass, placing the point at the bottom right corner and moving the pencil from one adjacent corner to the other. Repeat in the second-largest square with the compass point at the bottom left corner, with the pencil continuing the curve for another quarter-circle, and then carry on with the smaller squares. The curve is an approximation of a *logarithmic spiral.*

A true logarithmic spiral will pass through the same corners of the same squares, yet it will wind itself smoothly, unlike the curve in the diagram, which will have small jumps in curvature where the quarter-circle sections meet. In a logarithmic spiral, a straight line from the centre of the spiral – the 'pole' – will cut the spiral curve at the same angle at all points, which is why Descartes called the logarithmic spiral an 'equiangular spiral'.

The logarithmic spiral is one of the most bewitching curves in maths. In the seventeenth century Jakob Bernoulli was the first mathematician to investigate its properties thoroughly. He called it the *spira mirabilis*, the wonderful spiral. He asked to have one engraved on his tombstone, but the sculptor engraved an Archimedean spiral by mistake.

The fundamental property of the logarithmic spiral is that it never changes shape the more it grows. Bernoulli expressed this on his

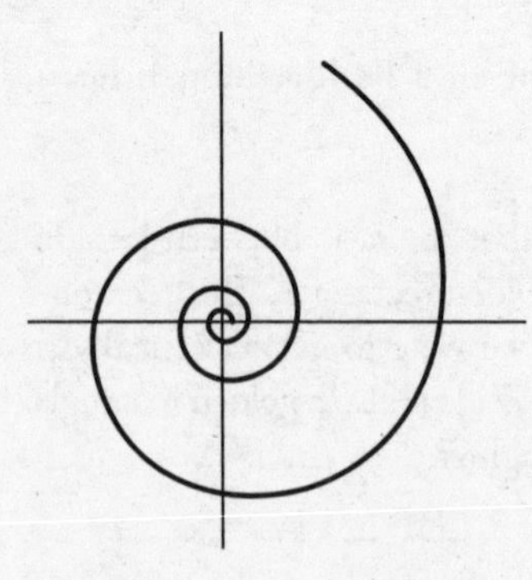

Logarithmic spiral – wonderful.

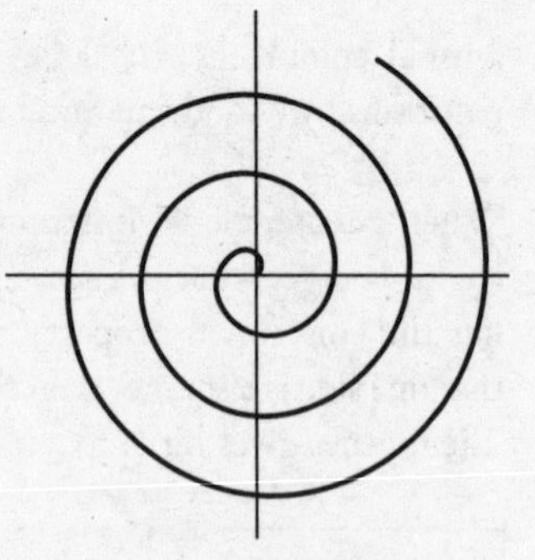

Archimedean spiral – not so wonderful.

tombstone with the epitaph: *Eadem mutata resurgo*, or 'Although changed, I shall arise the same'. The spiral rotates an infinite number of times before reaching its pole. If you took a microscope and looked at the centre of a logarithmic spiral, you would see the same shape that you would see if the logarithmic spiral above was continued until it was as big as a galaxy and you were looking at it from a different solar system. In fact, many galaxies are in the shape of logarithmic spirals. Just like a fractal, a logarithmic spiral is self-similar, that is, any smaller piece of a larger spiral has identical shape to the larger piece.

The most stunning example of a logarithmic spiral in nature is the nautilus shell. As the shell grows, each successive chamber is larger, but has the same shape as the chamber before. The only spiral that can accommodate chambers of different sizes with the same relative dimensions is Bernoulli's *spira mirabilis*.

As Descartes noted, a straight line from the pole of a logarithmic spiral always cuts the curve at the same angle, and this feature explains why the spiral is used by Peregrine falcons when they attack their prey.

Nautilus shell.

Falcons descend on their prey in a logarithmic spiral.

Peregrines do not swoop in a straight line, but rather bear down on prey by spiralling around it. In 2000 Vance Tucker of Duke University figured out why this is so. Falcons have eyes at the sides of their head, which means that if they want to look in front of themselves, they need to turn their head 40 degrees. Vance tested falcons in a wind tunnel and showed that with their head at such an angle, the wind drag on a falcon is 50 per cent greater than it would be if they were looking straight ahead. The path that lets the bird keep its head in the most aerodynamic position possible, while also enabling it to constantly look at the prey at the same angle, is a logarithmic spiral.

Plants, as well as birds of prey, move to the music of phi. When a plant grows, it needs to position its leaves around the stem in such a way as to maximize the amount of sunlight that falls on each leaf. That's why plant leaves aren't directly above each other; if that were the case, the bottom ones would get no sunlight at all.

As the stem goes higher, each new leaf appears at a fixed angle around the stem from the previous leaf. The stem sprouts a leaf at a predetermined rotation, as in the diagram on the following page.

What is the fixed angle that maximizes sunlight for the leaves, the angle that will spread out the leaves around the stem so that they overlap as little as possible? It is not 180 degrees, or a half turn, because the third leaf would be directly above the first. The angle is not 90 degrees, or a quarter turn, because if this were the case, the fifth leaf would be directly over the first – and also the first three

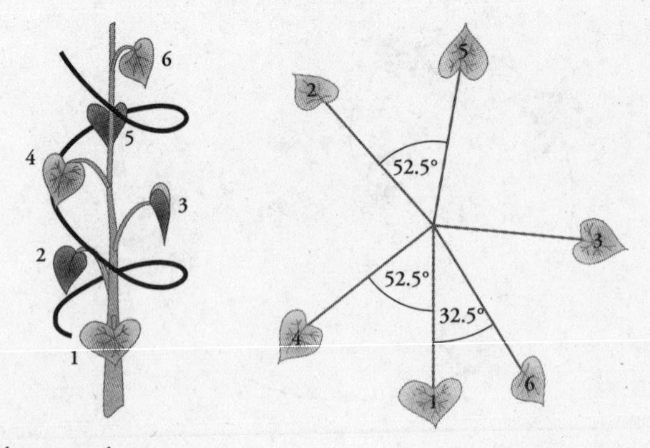

How leaves spiral up a stem.

leaves would be using only one side of the stem, which would be a waste of the sunlight available on the other side. The angle that provides the best arrangement is 137.5 degrees, and the diagram above shows where the leaves would be positioned if successive leaves are always separated by this angle. The first three leaves are positioned well apart from each other. The next two, leaves four and five, are separated by more than 50 degrees from their nearest leaves, which still gives them a good amount of room. The sixth leaf is at 32.5 degrees from the first. This is closer to a leaf than any previous one, which it has to be since there are more leaves, yet the distance is still a pretty wide berth.

The angle of 137.5 degrees is known as the golden angle. It is the angle we get when we divide the full rotation of a circle according to the golden ratio. In other words when we divide 360 degrees into two angles such that the ratio of the larger angle to the smaller angle is phi, or 1.618. The two angles are 222.5 degrees and 137.5 degrees, to one decimal place. The smaller one is known as the golden angle.

The mathematical reason why the golden angle produces the best leaf arrangement around a stem is linked to the concept of irrational numbers, which are those numbers that cannot be expressed as fractions. If an angle is an irrational number, no matter how many times you turn it around a circle you will never get back to where you started. It may sound Orwellian, but some irrational numbers are more irrational than others. And no number is more

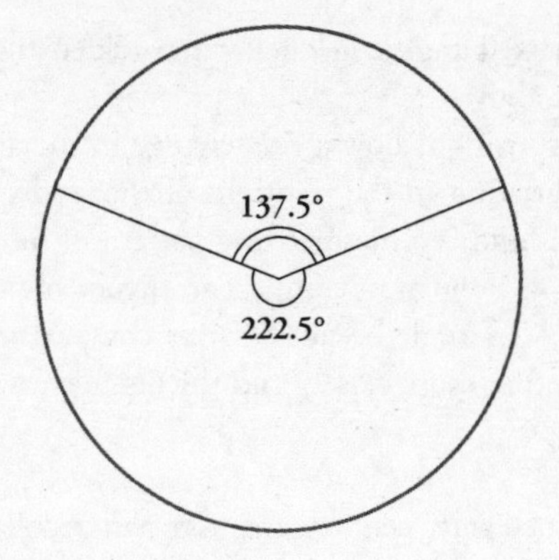

The golden angle.

irrational than the golden ratio. (There's a brief explanation why in the appendix on p. 423.)

The golden angle explains why you generally find on a plant stem that the number of leaves and number of turns before a leaf sprouts more or less directly above the first one is a Fibonacci number. For example, roses have 5 leaves every 2 turns, asters have 8 leaves for every 3 turns and almond trees have 13 leaves every 5 turns. Fibonacci numbers occur because they provide the nearest whole-number ratios for the golden angle. If a plant sprouts 8 leaves for every 3 turns, each leaf occurs every $\frac{3}{8}$ turn, or every 135 degrees, a very good approximation for the golden angle.

The unique properties of the golden angle are most strikingly seen in seed arrangements. Imagine that a flower head produces seeds from the centre point at a fixed angle of rotation. When new seeds emerge, they push the older seeds further out from the centre. These three diagrams show the patterns of seeds that emerge with

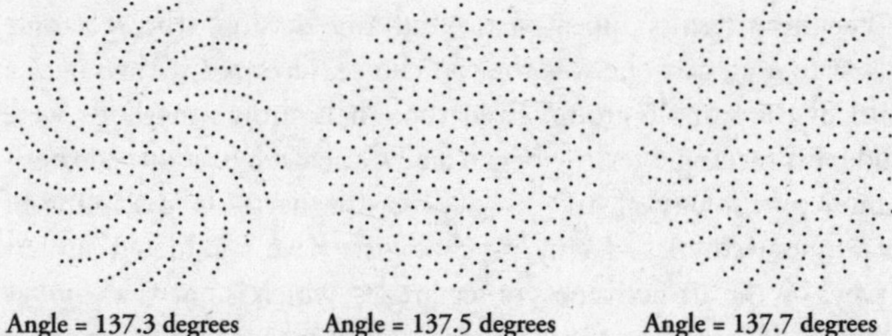

Angle = 137.3 degrees
Just under golden angle

Angle = 137.5 degrees
The golden angle

Angle = 137.7 degrees
Just over golden angle

three different fixed angles: just below the golden angle, the golden angle, and just above.

What is surprising is how a tiny change in the angle can cause such a huge variation in the positions of the seeds. At the golden angle, the seed head is a mesmerizing pattern of interlocking logarithmic spirals. It is the most compact arrangement possible. Nature chooses the golden angle because of this compactness – the seeds are bound together more closely and the organism will be stronger because of it.

In the late nineteenth century the German Adolf Zeising most forcefully put forth the view that the golden proportion is beauty incarnate, describing the ratio as a universal law 'which permeates, as a paramount spiritual ideal, all structures, forms and proportions, whether cosmic or individual, organic or inorganic, acoustic or optical; which finds its fullest realization, however, in the human form'. Zeising was the first person to claim that the front of the Parthenon is in the shape of a golden rectangle. In fact, there is no documentary evidence that those in charge of the architectural project, who included the sculptor Phidias, used the golden ratio. Nor, if you look closely, is the golden rectangle a precise fit. The edges of the pedestal fall outside. Yet it was Phidias's connection to the Parthenon that, in around 1909, inspired the American mathematician Mark Barr to name the golden ratio phi.

Despite the eccentric tone of Zeising's work, he was taken seriously by Gustav Fechner, one of the founders of experimental psychology. In order to discover if there was any empirical evidence that humans thought the golden rectangle more beautiful than any other sort of rectangle, Fechner devised a test in which subjects were shown a number of different rectangles and asked which they preferred.

Fechner's results appeared to vindicate Zeising. The rectangle closest to a golden one was the top choice, favoured by just over a third of the sample group. Even though Fechner's methods were crude, his rectangle-testing began a new scientific field – the experimental psychology of art – as well as the narrower discipline of 'rectangle aesthetics'. Many psychologists have conducted similar surveys on the attractiveness of rectangles, which is not as absurd as it sounds. If there were a 'sexiest' rectangle, this shape would be of use to the designers of commercial products. Indeed, credit cards, cigarette packets and books often approach the proportions of a golden rectangle. Unfortunately for phi-philes, the most recent and detailed piece of research, by a team led by Chris McManus of University College London, suggests that Fechner was wrong. The 2008 paper stated that 'more than a century of experimental work has suggested that the golden section actually plays little normative role in subjects' preferences for rectangles'. Yet the authors did not conclude that analysing rectangle preference is a waste of time. Far from it. They claimed that while no one rectangle is universally preferred by humans, there are important individual differences in the aesthetic appreciation of rectangles that merit further investigation.

Lesser scientists than Fechner were also inspired by Zeising's theories. Frank A. Lonc from New York measured the height of 65 women compared to the height of their belly buttons and discovered that the ratio was 1.618. He had excuses when it wasn't: 'Subjects whose measurements did not fall within this ratio testified to hip injuries or other deforming accidents in childhood.' The French architect Le Corbusier created 'Modulor man' in order to use proper proportions in architecture and design. In his man, the ratio of the height of the man to the height of his belly button is 1829/1130, or 1.619, and the ratio of the distance between his belly button and his upheld right hand and his belly button and his head is 1130/698, again 1.619. Le Corbusier's intention for the Modulor was to use these proportions in architecture and design.

Gary Meisner is a 53-year-old business consultant from Tennessee. He calls himself the Phi Guy and on his website sells merchandise including phi T-shirts and mugs. His bestselling product, however, is the PhiMatrix, a piece of software that creates a grid on your computer screen to check images for the golden ratio. Most purchasers use it as a design tool to make cutlery, furniture and homes. Some customers use it for financial speculation by superimposing the grid on graphs of indices and using phi to predict future trends. 'A guy in the Caribbean was using my matrix to trade in oil, a guy in China was using it to trade in currencies,' he said. Meisner was drawn to the golden mean because he is spiritual and says it helped him understand the universe, but even the Phi Guy thinks that his fellow travellers can go too far. He is, for example, unconvinced by the traders. 'When you look back on the market it is pretty easy to find relationships that conform to phi,' he said. 'The challenge is that looking backwards is completely different to looking out the front window.' Meisner's website has made him the go-to guy for every flavour of phi aficionado. He told me that a month ago he received an email from an unemployed man who believed that the only way to get a job interview was to design his résumé in the proportions of the golden ratio. Meisner felt the man was deluded and took pity on him. He gave him some phi design tips, but suggested that it would be more fruitful investing in more traditional job-hunting methods such as business networking. 'I got a letter from him this morning,' Meisner blurted. 'He said he has a job interview. He is giving credit to the résumé's new design!'

Back in London I told Eddy Levin the story of the golden résumé as an example of excessive eccentricity. Levin, however, didn't think it was funny. In fact, he agreed that a phi-proportioned résumé was better than a regular one. 'It would look more beautiful, and so the reader would be more attracted to it.'

After 30 years of studying the golden ratio, Levin is convinced that wherever there is beauty, there will be phi. 'Any art which looks good, the dominant proportions are the golden proportion,' he said. He knows this is an unpopular viewpoint, as it prescribes a formula for beauty, but he guarantees he will be able to find phi in any piece of art.

My instinctive reaction to Levin's phi obsession was one of scepticism. For a start, I was unconvinced that his gauge was accurate enough to measure 1.618 sufficiently precisely. It was not surprising to find a ratio of 'approximately phi' in a painting or a building, especially if you could select which parts to choose. Also, since the ratio of consecutive Fibonacci numbers makes a good approximation to 1.618, whenever there is a grid of 5×3 or 8×5 or 13×8 and so on, you will see a golden rectangle. Of course the ratio will be a common one.

Yet there was something compelling about Levin's examples. I felt the thrill of wonder with each new image he showed me. Phi really was everywhere. Yes, the golden ratio has always attracted cranks, but this in itself did not mean that all the theories were crankish. Some very respectable academics have claimed that phi creates beauty, particularly in the structure of musical compositions. The argument that human beings might be drawn to a proportion that best expresses natural growth and regeneration does not seem too far-fetched.

It was a sunny summer's day and Levin and I relocated to his garden. We sat on two lawn chairs and sipped tea. Levin told me that the limerick was a successful form of poetry because the syllables in its lines (8, 8, 5, 5, 8) are Fibonacci numbers. Then I had an idea. I asked Levin if he knew what an iPod was. He didn't. I had one in my pocket and I took it out. It was a beautiful object, I said, and according to his reasoning, it should contain the golden ratio.

Levin took my shiny white iPod and held it in his palm. Yes, he replied, it was beautiful, and it should. Not wanting to get my hopes up, he warned me that factory-produced objects often do not follow the golden ratio perfectly. 'The shape shifts slightly for the convenience of manufacture,' he said.

Levin opened his callipers and started measuring between all the significant points.

'Ooh, yes,' he grinned.

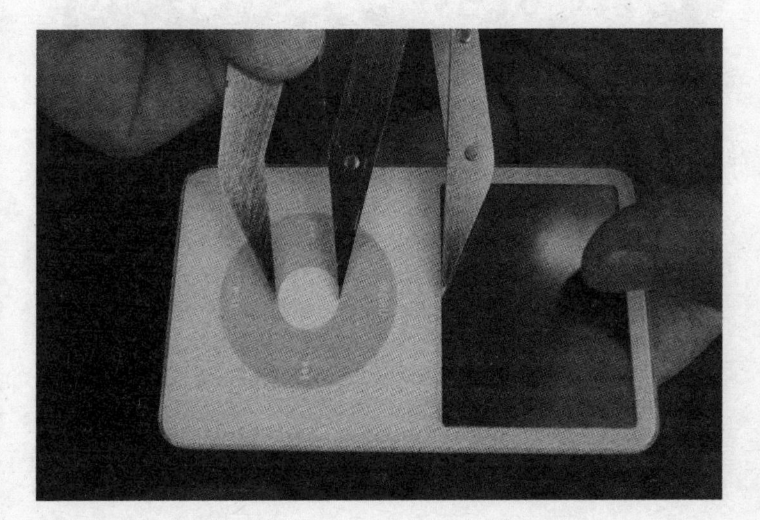

CHAPTER NINE

Chance is a Fine Thing

It used to be said that you go to Las Vegas to get married and to Reno to get a divorce. Now you can visit both cities for their slot machines. With 1900 slot machines, Reno's Peppermill Casino isn't even the largest casino in town. Walking through its main hall, the roulette and blackjack tables were shadowy and subdued in comparison to the flashing, spinning, beeping regiments of slots. Technological evolution has deprived most one-armed bandits of their lever limbs and mechanical insides. Players now bet by pressing lit buttons or touch-screen displays. Occasionally I heard the rousing sound of clattering change, but this came from pre-recorded samples since coins have been replaced by electronic credits.

Slots are the casino industry's cutting edge; its front line and its bottom line. The machines make $25 billion a year in the United States (*after* they've paid out all their prize money), which is about two and a half times the total value of movie tickets sold in the country annually. In Nevada, the global centre of casino culture, slots now make up almost 70 per cent of gambling revenue – and the number nudges higher every year.

Probability is the study of chance. When we flip a coin, or play the slots, we do not know how the coin will land, or where the spinning reels will stop. Probability gives us a language to describe how likely it is that the coin will come up heads, or that we will hit the jackpot. With a mathematical approach, unpredictability becomes very predictable. While we take this idea for granted in our daily lives – it is implicit, for example, whenever we read the weather forecast – the realization that maths can tell us about the future was a very profound and comparatively recent idea in the history of human thought.

I had come to Reno to meet the mathematician who sets the odds for more than half of the world's slot machines. His job has historical pedigree – probability theory was first conceived in the sixteenth century by the gambler Girolamo Cardano, our Italian friend we met earlier when discussing cubic equations. Rarely, however, has a mathematical breakthrough arisen from such self-loathing. 'So surely as I was inordinately addicted to the chess board and the dicing table, I know that I must rather be considered deserving of the severest censure,' he wrote. His habit yielded a short treatise called *The Book on Games of Chance*, the first scientific analysis of probability. It was so ahead of its time, however, that it was not published until a century after his death.

Cardano's insight was that if a random event has several equally likely outcomes, the chance of any individual outcome occurring is equal to the proportion of that outcome to all possible outcomes. This means that if there is a one-in-six chance of something happening, then the chance of it happening is one sixth. So, when you roll a die, the chance of getting a six is $\frac{1}{6}$. The chance of throwing an even number is $\frac{3}{6}$. which is the same as $\frac{1}{2}$. Probability can be defined as the likelihood of something happening expressed as a fraction. Impossibility has a probability of 0; certainty, a probability of 1; and the rest is in between.

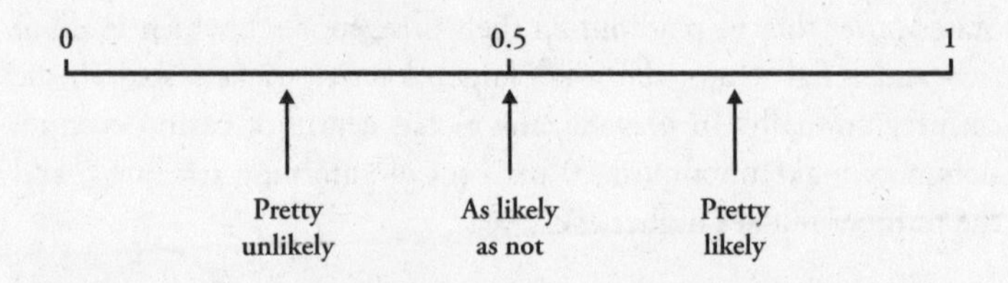

This seems straightforward, but it isn't. The Greeks, the Romans and the ancient Indians were all obsessive gamblers. None of them, though, attempted to understand how randomness is governed by mathematical laws. In Rome, for example, coins were flipped as a way of settling disputes. If the side with the head of Julius Caesar came up, it meant that he agreed with the decision. Randomness was not seen as random, but as an expression of divine will. Throughout history, humans have been remarkably imaginative in finding ways to interpret random events. Rhapsodomancy, for example, was the practice of seeking guidance through chance selection of a passage in a literary work. Similarly, according to the Bible, picking the short straw was an impartial way of selecting only in so far as God let it be that way: 'The lot is cast into the lap; but the whole disposing thereof is of the Lord' (Proverbs 16: 33).

Superstition presented a powerful block against a scientific approach to probability, but after millennia of dice-throwing, mysticism was overcome by perhaps a stronger human urge – the desire for financial profit. Girolamo Cardano was the first man to take Fortune hostage. It could be argued, in fact, that the invention of probability was the root cause of the decline, over the last few centuries, of superstition and religion. If unpredictable events obey mathematical laws, there is no need to have them explained by deities. The secularization of the world is usually associated with thinkers such as Charles Darwin and Friedrich Nietzsche, yet quite possibly the man who set the ball rolling was Girolamo Cardano.

Games of chance have most commonly involved dice. A model popular in antiquity was the *astragalus* – an anklebone from a sheep or a goat – that had four distinctly flat faces. Indians liked dice in the shape of rods and Toblerones, and they marked the different faces with pips, the most likely explanation for which is that dice predate any formal system of numerical notation, and this tradition has survived. The fairest dice have identical sides, and if one imposes the further condition that each side must also be a regular polygon, there are only five shapes that fit the bill, the Platonic solids. All the Platonic solids have been used as dice. Ur, possibly the world's oldest known game, which dates from at least as far back as the third century BC, used a tetrahedron, which, however, is the worst of the five choices because the tetrahedron hardly rolls and has only four sides. Octahedrons (eight sides) were used in ancient Egypt, and dodecahedrons (twelve) and icosahedrons (twenty) are still found in fortune-tellers' handbags.

By far the most popular dice shape has been the cube. It is the easiest to make, its span of digits is neither too big nor too small, it rolls nicely but not too easily, and it's obvious on which number it lands. Cubical dice with pips are a cross-cultural symbol of luck and chance, as comfortable bashed around in Chinese mah-jong salons as they are dangling from the rear-view mirrors of British cars.

As I mentioned earlier, roll one die and the chance of a six is $\frac{1}{6}$. Roll another die and the chance of a six is also $\frac{1}{6}$. What is the chance of rolling a pair of dice and getting a pair of sixes? The most basic rule of probability is that the chances of two independent events happening is the same thing as the chance of one event happening *multiplied* by the chance of the second event happening. When you throw a pair of dice the outcome of the first die is independent of the outcome of the second, and vice versa. So, the chances of throwing two sixes is $\frac{1}{6} \times \frac{1}{6}$, or $\frac{1}{36}$. You can see this visually by counting all the possible combinations of two dice: there are 36 equally likely outcomes and only one of them is a six and a six.

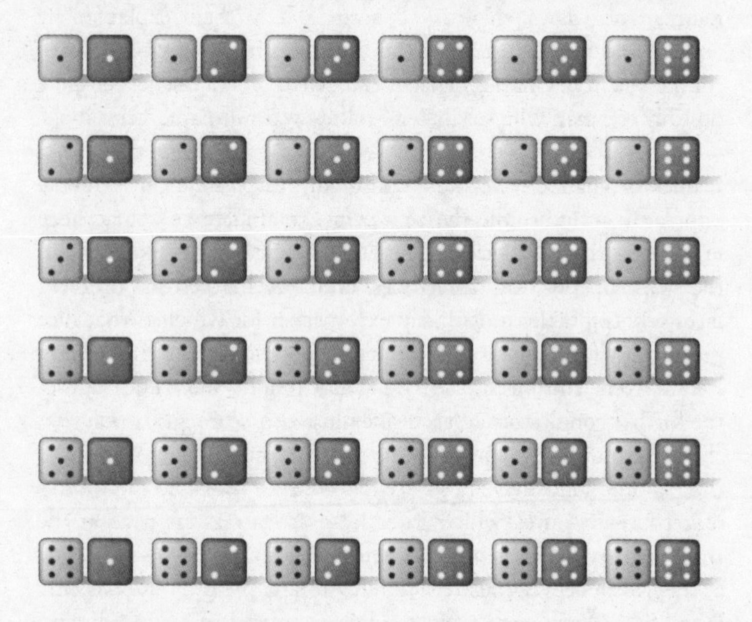

Conversely, of the 36 possible outcomes, 35 of them are not double six. So, the probability of *not* throwing a six and a six is $\frac{35}{36}$. Rather than counting up 35 examples, you can equally start with the full set and then subtract the instances of double sixes. In this example, $1 - \frac{1}{36} = \frac{35}{36}$. The probability of something not happening, therefore, is 1 minus the probability of that thing happening.

The dicing table was an early equivalent of the slot machine, where gamblers placed wagers on the outcome of dice throws. One classic gamble was to roll four dice and bet on the chance of at least one six appearing. This was a nice little earner for anyone willing to put money on it, and we already have enough maths knowledge to see why:

Step 1: The probability of rolling a six in four rolls of dice is the same as 1 minus the probability of not getting a six on any of the four dice.

Step 2: The probability of not getting a six on one die is $\frac{5}{6}$, so if there are four dice, the probability is $\frac{5}{6} \times \frac{5}{6} \times \frac{5}{6} \times \frac{5}{6} = \frac{625}{1296}$, which is 0.482.

Step 3: So, the probability of rolling a six is 1 – 0.482 = 0.518.

A probability of 0.518 means that if you threw four dice a thousand times, you could expect to get at least one six about 518 times, and get no sixes about 482 times. If you gambled on the chance of at least one six, you would win on average more than you would lose, so you would end up profiting.

The seventeenth-century writer Chevalier de Méré was a regular at the dicing table, as he was at the most fashionable salons in Paris. The chevalier was as interested in the mathematics of dicing as he was in winning money. He had a couple of questions about gambling, though, that he was unable to answer himself, so in 1654 he approached the distinguished mathematician Blaise Pascal. His chance enquiry was the random event that set in motion the proper study of randomness.

Blaise Pascal was only 31 when he received de Méré's queries, but he had been known in intellectual circles for almost two decades. Pascal had shown such gifts as a young child that at 13 his father had let him attend the scientific salon organized by Marin Mersenne, the friar and prime-number enthusiast, which brought together many famous mathematicians, including René Descartes and Pierre de Fermat. While still a teenager, Pascal proved important theorems in geometry and invented an early mechanical calculation machine, which was called the Pascaline.

The first question de Méré asked of Pascal concerned double sixes. As we saw above, there is a $\frac{1}{36}$ chance of getting double six when you throw two dice. The overall chance of getting a double six increases the more times you throw the pair of dice. The chevalier wanted to know how many times he needed to throw the dice to make gambling on the occurrence of a double six a good bet.

His second question was more complex. Say Jean and Jacques are playing a dice game consisting of several rounds in which both roll a die to see who gets the highest number. The outright winner is whoever rolls the highest number three times. Each has a stake of 32 francs, so the pot is 64 francs. If the game has to be terminated after three rounds, when Jean has rolled the highest number twice and Jacques once, how should the pot be divided up?

Pondering the answers, and feeling the need to discuss them with a fellow genius, Pascal wrote to his old friend from the Mersenne salon, Pierre de Fermat. Fermat lived far from Paris, in Toulouse, an appropriately named city for someone analysing a problem about gambling. He was 22 years older than Pascal and worked as a judge at the local criminal court, dabbling in maths only as an intellectual recreation. Nevertheless, his amateur ruminations had made him one of the most respected mathematicians of the first half of the seventeenth century.

The short correspondence between Pascal and Fermat about chance – which they called *hasard* – was a landmark in the history of science. Between them the men solved both of the literary bon vivant's problems, and in so doing, set the foundations of modern probability theory.

Now for the answers to Chevalier de Méré's questions. How many times do you need to throw a pair of dice so that it is more likely than not that a double six will appear? In one throw of two dice the chance of a double six is $\frac{1}{36}$, or 0.028. The chance of a double six appearing in two throws of two dice is 1 minus the probability of no double sixes appearing in two throws, or $1 - (\frac{35}{36} \times \frac{35}{36})$. This works out to be $\frac{71}{1296}$, or 0.055. (Note: the chance of a double six in two throws is *not* $\frac{1}{36} \times \frac{1}{36}$. This is the chance of a double six in both throws. The probability we are concerned with is the chance of *at least one* double six, which includes the outcomes of either a double six in the first throw, in the second throw, or in both throws. The gambler needs only one double six to win, not a double six in both.)

The chances of a double six in three throws of two dice are 1 minus the probability of no doubles, which this time is $1 - (\frac{35}{36} \times \frac{35}{36} \times \frac{35}{36}) = \frac{3781}{46,656}$, or 0.081. As we can see, the more times one throws the dice, the higher the probability of throwing a double six: 0.028 with one throw, 0.055 with two, and 0.081 with three. Therefore, the original question can be rephrased as 'After how many throws does this fraction exceed 0.5?', as a probability of more than a half means that the event is more likely than not. Pascal calculated correctly that the answer is 25 throws. If the chevalier gambled on the chance of a double six in 24 throws, he could expect to lose money, but after 25 throws, the odds shift in his favour and he could expect to win.

De Méré's second question, about divvying up the pot, is often called the *problem of points* and had been posed before Fermat and Pascal tackled it, but never correctly resolved. Let's restate the question in terms of heads and tails. Jean wins each round if the coin lands on heads, and Jacques wins if it lands on tails. The first person to win three rounds takes the pot of 64. With the score at two heads for Jean and one tails for Jacques, the game needs to come to an abrupt stop. If this is the case, what's the fairest way to divide the pot? One answer is that Jean should take the lot, since he is ahead, but this doesn't take account of the fact that Jacques still has a chance of winning. Another answer is that Jean should take twice as much as Jacques, but again this is not fair because the 2–1 score reflects past events. It's not an indication of what might happen in the future. Jean is not better at guessing coins than Jacques. Each time they throw, there is a 50:50 chance of the coin landing heads or tails. The best, and fairest, analysis is to consider what might happen in the future. If the coin is tossed another two times, the possible outcomes are:

heads, heads
heads, tails
tails, heads
tails, tails

After these two throws, the game has been won. In the first three instances Jean wins, and in the fourth Jacques does. The fairest way to divide the pot is for $\frac{3}{4}$ to go to Jean and $\frac{1}{4}$ to go to Jacques, so the cash is divided 48 francs to 16. This seems fairly straightforward now, but in the seventeenth century the idea that random events that haven't yet taken place can be treated mathematically was a momentous conceptual breakthrough. The concept underpins our scientific understanding of much of the modern world, from physics to finance and from medicine to market research.

A few months after he first wrote to Fermat about the gambler's queries, Pascal had a religious experience so intense that he scribbled a report of his trance on a piece of paper that he carried with him in a special pouch sewn into the lining of his jacket for the rest of his life. Perhaps the cause was the near-death accident in which his coach hung perilously off a bridge after the horses plunged over the parapet, or perhaps it was a moral reaction to the decadence of the dicing tables of pre-revolutionary France – in any case, it revitalized his commitment to Jansenism, a strict Catholic cult, and he abandoned maths for theology and philosophy.

Nonetheless, Pascal could not help but think mathematically. His most famous contribution to philosophy – an argument about whether or not one should believe in God – was a continuation of the new approach to analysing chance that he had first discussed with Fermat.

In simple terms, *expected value* is what you can expect to get out of a bet. For example, what could Chevalier de Méré expect to win by betting £10 on getting a six when rolling four dice? Imagine that he wins £10 if there is a six and loses everything if there is no six. We know that the chance of winning this bet is 0.518. So, just over half the time he wins £10, and just under half he loses £10. The expected value is calculated by multiplying the probability of each outcome with the value of each outcome, and then adding them up. In this case he can expect to win:

(chances of winning £10) × £10 + (chances of losing £10) × –£10
or
(0.518 × £10) + (0.482 × –£10) = £5.18 – £4.82 = 36p

(In this equation, money won is a positive number, and money lost is a negative number.) Of course, in no single bet will de Méré win

36p – either he wins £10 or he loses £10. The value of 36p is theoretical, but, on average, if he keeps on betting, his winnings will approximate 36p per bet.

Pascal was one of the first thinkers to exploit the idea of expected value. His mind, though, was occupied by much higher thoughts than the financial benefits of the dicing table. He wanted to know whether it was worth placing a wager on the existence of God.

Imagine, Pascal wrote, gambling on God's existence. According to Pascal, the expected value of such a wager can be calculated by the following equation:

(chance of God existing) × (what you win if He exists) + (chance of God not existing) × (what you win if He doesn't exist)

So, say the chances of God existing are 50:50; that is, the probability of God's existence is $\frac{1}{2}$. If you believe in God, what can you expect to get out of this bet? The formula becomes:

$$(\tfrac{1}{2} \times \text{eternal happiness}) + (\tfrac{1}{2} \times \text{nothing}) = \text{eternal happiness}$$

In other words, betting on God's existence is a very good bet because the reward is so fantastic. The arithmetic works out because half of nothing is nothing, but half of something infinite is *also* infinite. Likewise, if the chance of God existing is only a hundredth, the formula is:

$$(\tfrac{1}{100} \times \text{eternal happiness}) + (\tfrac{99}{100} \times \text{nothing}) = \text{eternal happiness}$$

Again, the rewards of believing that God exists are equally phenomenal, since one hundredth of something infinite is still infinite. It follows that, however minuscule the chance of God existing, provided that chance is not zero, if you believe in God, the gamble of believing will bring an infinite return. We have come down a complicated route and reached a very obvious conclusion. Of course Christians will gamble that God exists.

Pascal was more concerned with what happens if one doesn't believe in God. In such cases, is it a good gamble to bet on whether

God exists? If we assume the chance of God not existing is 50:50, the equation now becomes:

$$(\frac{1}{2} \times \text{eternal damnation}) + (\frac{1}{2} \times \text{nothing}) = \text{eternal damnation}$$

The expected outcome becomes an eternity in Hell, which looks like a terrible bet. Again, if the chance of God existing is only a hundredth, the equation is similarly bleak for non-believers. If there is any chance at all of God existing, for the non-believer the expected value of the gamble is always infinitely bad.

The above argument is known as Pascal's Wager. It can be summarized as follows: if there is the slightest probability that God exists, it is overwhelmingly worthwhile to believe in Him. This is because if God doesn't exist a non-believer has nothing to lose, but if He does exist a non-believer has *everything* to lose. It's a no-brainer. Be a Christian, go on, you might as well.

Upon closer examination, though, Pascal's argument, of course, doesn't work. For a start, he is only considering the option of believing in a Christian God. What about the gods of any other religion, or even of made-up religions? Imagine that, in the afterlife, a cat made of green cheese will determine whether we go to Heaven or Hell. Although this isn't very likely, it's still a possibility. By Pascal's argument, it is worthwhile to believe that this cat made from green cheese exists, which is, of course, absurd.

There are other problems with Pascal's Wager that are more instructive to the mathematics of probability. When we say that there is a 1-in-6 chance of a die landing on a six, we do this because we know that there is a six marked on the die. For us to be able to understand in mathematical terms the statement that there is a 1 in anything chance of God existing, there must be a possible world where God does in fact exist. In other words, the premise of the argument presupposes that, somewhere, God exists. Not only would a non-believer refuse to accept this premise, but it shows that Pascal's thinking is self-servingly circular.

Despite Pascal's devout intentions, his legacy is less sacred than it is profane. Expected value is the core concept of the hugely profitable gambling industry. Some historians also attribute to Pascal the invention of the roulette wheel. Whether this is true or not, the wheel was certainly of French origin, and by the end of the eighteenth century roulette was a popular attraction in Paris. The rules are as follows: a ball spins around an outer rim before losing momentum and falling towards an inner wheel, which is also rotating. The inner wheel has 38 pockets, marked by the numbers 1 to 36 (alternately red and black) and the special spots 0 and 00 (green). The ball reaches the wheel and bounces around before coming to rest in an individual pocket. Players can make many bets on the outcome. The simplest is to bet on the pocket where the ball will land. If you get it correct, the house pays you back 35 to 1. A £10 bet, therefore, wins you £350 (and the return of your £10 bet).

Roulette is a very efficient money-making machine because every bet in roulette has a negative expected value. In other words, for every gamble you make you can expect to lose money. Sometimes you win and sometimes you lose, but in the long run you will end up with less money than you started with. So, the important question is, how much can you expect to lose? When you bet on a single number, the probability of winning is $\frac{1}{38}$, as there are 38 potential outcomes. For each single number bet of £10, therefore, a player can expect to win:

(chance of landing on a number)(what you win) + (chance of not landing on a number)(what you win)
or
$(\frac{1}{38} \times £350) + (\frac{37}{38} \times -£10) = -52.6\text{p}$

In other words, you lose 52.6p for every £10 wagered. The other bets in roulette – betting on two or more numbers, on sections, colours, or columns, all have odds that result in an expected value of –52.6p, apart from the 'five number' bet on getting either 0, 00, 1, 2 or 3, which has even worse odds, with an expected loss of 78.9p.

Despite its bad odds, roulette was – and continues to be – a much-loved recreation. For many people, 52.6p is a fair payment for the thrill of potentially winning £350. In the nineteenth century casinos proliferated and, in order to make them more competitive, roulette wheels were built without the 00, making the chance of a single-number bet $\frac{1}{37}$ and reducing the expected loss to 27p per £10 bet. The change meant that you lost your money about half as quickly. European casinos tend to have wheels with just the 0, while America prefers the original style, with 0 and 00.

All casino games involve negative-expectation bets; in other words, in these games gamblers should expect to lose money. If they were arranged in any other way, casinos would go bust. Mistakes, however, have been made. An Illinois riverboat casino once introduced a promotion that changed the amount paid out on one type of hand in blackjack without realizing that the change moved the expected value of the bet from negative to positive. Instead of expecting to lose, gamblers could expect to win 20 cents per $10 bet. The casino reportedly lost $200,000 in a day.

The best deal to be found in a casino is at the craps table. The game originated from a French variant of an English dice-rolling game. Players throw two dice and the outcome depends on which numbers land and how they add up. In craps, your chances of winning are 244 out of 495 possible outcomes, or 49.2929 per cent, giving an expected loss of just 14.1p per £10 bet.

Craps is also worth mentioning because of the possibility of making a curious side bet in which you can bet with the house; that is, against the player throwing the dice. The side bettor wins when the main bettor loses, and the side bettor loses when the main bettor wins. Since the main bettor loses, on average, 14.1p per £10 bet, the side bettor stands to win, on average, 14.1p per £10 bet. But there is an extra rule preventing this neat outcome in craps side bets. If the main player rolls a double six on his first roll (which means that he loses), the side bettor does not win either, but only receives his money back. This seems like a very insignificant change. There's only a 1-in-36 chance of throwing a double six. Yet $\frac{1}{36}$ less of a chance of winning decreases the expected value by 27.8p per £10 bet, which shifts the expected value of the bet into negative territory. Instead of winning 14.1p per £10 as the house does, the side bettor will win 14.1p minus 27.8p per bet, which is –13.7p, or a loss of 13.7p. The side bet is indeed a better deal, but only marginally, by 0.4p per £10 wagered.

Another way of looking at an expected loss is to consider it in terms of *payback percentage*. If you bet £10 at craps, you can expect to receive about £9.86 back. In other words, craps has a payback percentage of 98.6 per cent. European roulette has a payback percentage of 97.3 per cent; and US roulette, 94.7 per cent. While this might seem a bad deal to gamblers, it is better value than the slots.

In 1893 the *San Francisco Chronicle* notified its readers that the city was home to one and a half thousand 'Nickel-in-the-Slot Machines That Make Enormous Profits … They are of mushroom growth, having appeared in the space of only a few months.' The machines came in many styles, but it was only at the turn of the century, when Charles Fey, a German immigrant, came up with the idea of three spinning reels, that the modern-day slot machine was born. The reels of his Liberty Bell machine were marked with a horseshoe, a star, a heart, a diamond, a spade, and an image of Philadelphia's cracked Liberty Bell. Different combinations of symbols gave different payouts, with the jackpot set to three bells. The slot machine added an element of suspense its competitors did not have because, when spun, the wheels came to rest one by one. Other companies copied, the machines spread beyond San Francisco, and by the 1930s three-reel slots were part of the fabric of American society. One early machine paid out fruit-flavoured chewing gum as a way to get round gaming laws. This introduced the classic melon and cherry symbols and is why slots are known in the UK as fruit machines.

The Liberty Bell had a payback average of 75 per cent, but these days slots are more generous than they used to be. 'The rule of thumb is, if it's a dollar denomination [machine], most people would put [the payback percentage] at 95 per cent,' said Anthony Baerlocher, the director of game design at International Game Technology (IGT), a slot-machine company that accounts for 60 per cent of the world's million or so active machines, referring to slots where the bets are made in dollars. 'If it's a nickel it's more like 90 per cent, 92 per cent for a quarter, and if they do pennies it might go down to 88 per cent.' Computer technology allows machines to accept bets of multiple denominations, so the same machine can have different payback percentages according to the size of the bet. I asked him if there was a cut-off percentage below which players will stop using the machine because they are losing too much. 'My personal belief is that once we start getting down around 85 per cent it's extremely difficult to design a game that's fun to play. You have to get really

The size of a cash register, Charles Fey's Liberty Bell was an immediate success when it was first manufactured at the very end of the nineteenth century.

lucky. There's just not enough money to give back to the player to make it exciting. We can do a pretty good job at 87.5 per cent, 88 per cent. And when we start doing 95, 97 per cent games they can get pretty exciting.'

Baerlocher and I met at IGT's head office in a Reno business park, a 20-minute drive from the Peppermill Casino. He walked me through the production line, where tens of thousands of slot machines are built every year, and past a storage hall where hundreds were neatly stacked in rows. Baerlocher was clean-shaven and preppy, with short dark hair and a dimple in his chin. Originally from Carson City, half an hour's drive away, he joined IGT after completing a maths degree at Notre Dame University, Indiana. For someone who loved inventing games as a child, and discovered a talent for probability at college, the job was a perfect fit.

When I wrote earlier that the core concept of gambling was the notion of expected value, I was giving only half the story. The other half is what mathematicians call the *law of large numbers*. If you bet only a few times on roulette or on the slots, you might come out on top. The more you play roulette, however, the more likely it is that you will lose overall. Payback percentages are only true in the long run.

The law of large numbers says that if a coin is flipped three times, it might not come up heads at all, but flip it three billion times and you can be pretty sure that it will come up heads almost exactly 50 per cent of the time. During the Second World War, the mathematician John Kerrich was visiting Denmark when he was arrested and interned by the Germans. With time on his hands, he decided to test the law of large numbers and flipped a coin 10,000 times in his prison cell. The result: 5067 heads, or 50.67 per cent of the total. Around 1900, the statistician Karl Pearson did the same thing 24,000 times. With significantly more trials, you would expect the percentage to be closer to 50 per cent – and it was. He threw 12,012 heads, or 50.05 per cent.

The results mentioned above seem to confirm what we take for granted – that in a coin flip the outcome of heads is equally likely as the outcome of tails. Recently, however, a team at Stanford University led by the statistician Persi Diaconis investigated whether heads really are as likely to show up as tails. The team built a coin-tossing machine and took slow-motion photography of coins as they spun through the air. After pages of analysis, including estimates that a nickel will land on its edge in about 1 in 6000 throws, Diaconis's results appeared to show the fascinating and surprising result that a coin will, in fact, land on the same face from which it was launched about 51 per cent of the time. So, if a coin is launched heads up, it will land on heads slightly more often than it will land on tails. Diaconis concluded, though, that what his research really proved was how difficult it is to study random phenomena and that 'for tossed coins, the classical assumptions of independence with probability $\frac{1}{2}$ are pretty solid'.

Casinos are all about large numbers. As Baerlocher explained, 'Instead of just having one machine [casinos] want to have thousands because they know if they get the volume, even though one machine maybe what we call "upside-down" or losing, the group as a collective has a very strong probability of being positive for them.' IGT's slots are designed so that the payback percentage is met, within an error of 0.5 per cent, after ten million games. At the Peppermill, where I was staying during my visit to Reno, each machine racks up about 2000 games a day. With almost 2000 machines, this makes a daily casino rate of four million games a day. After two and a half days, the Peppermill can be almost certain that it will be hitting its payback percentage within half a per cent. If the average bet is a dollar, and the percentage is set to 95 per cent, this works out to be $500,000 in profit, give or take $50,000, every 60 hours. It is little wonder, then, that slots are increasingly favoured by casinos.

The rules of roulette and craps haven't changed since the games were invented centuries ago. By contrast, part of the fun of Baerlocher's job is that he gets to devise new sets of probabilities for each new slot machine that IGT introduces to the market. First, he decides what symbols to use on the reel. Traditionally, they are cherries and bars, but they are now as likely to be cartoon characters, Renaissance painters, or animals. Then he works out how often these symbols are on the reel, what combinations result in payouts, and how much the machine pays out per winning combination.

Baerlocher drew me up a simple game, Game A opposite, which has three reels and 82 positions per reel made up of cherries, bars, red sevens, a jackpot and blanks. If you read the table, you can see that there is a $\frac{9}{82}$ or 10.976 per cent chance of cherry coming up on the first reel, and when this happens $1 wins a $4 payout. The probability of a winning combination multiplied by the payout is called the *expected contribution*. The expected contribution of cherry-anything-anything is $4 \times 10.967 = 43.902$ per cent. In other words, for every $1 put into the machine, 43.902 cents will be paid out on cherry-anything-anything. When he is designing games, Baerlocher needs to make sure the sum of expected contributions for all payouts equals the desired payback percentage of the whole machine.

The flexibility of slot design is that you can vary the symbols, the winning combinations and the payouts to make very different

Game A – Low volatility

Symbol	Reel 1	Reel 2	Reel 3
Blank	23	27	25
Cherry (CH)	9	0	0
1 Bar (1B)	19	27	25
2 Bar (2B)	12	15	16
3 Bar (3B)	12	7	10
Red Seven (R7)	5	4	4
Jackpot (JP)	2	2	1
Total	**82**	**82**	**82**

Payable scorecard

Combination			$ payout on $1 bet	Probability (%)	Contribution (%)
CH	Any	Any	4	10.976	43.902
1B	1B	1B	10	2.326	23.260
2B	2B	2B	25	0.522	13.058
3B	3B	3B	50	0.152	7.617
R7	R7	R7	100	0.015	1.451
JP	JP	JP	1000	0.001	0.725
Total hit frequency				13.992 %	
Total return to player				90.015 %	

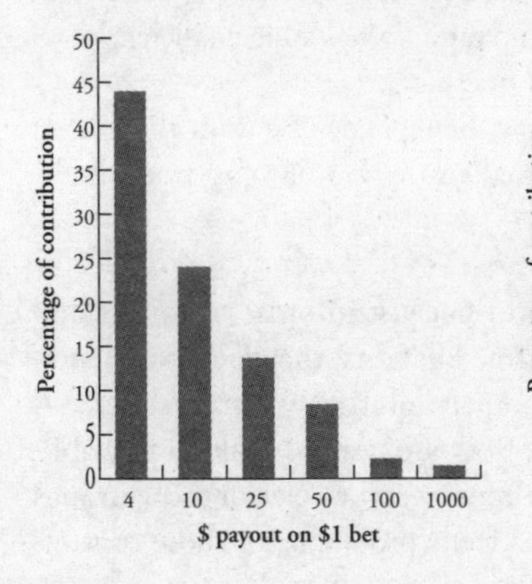

Game B – High volatility

Symbol	Reel 1	Reel 2	Reel 3
Blank	20	22	23
Cherry (CH)	6	0	0
1 Bar (1B)	18	25	19
2 Bar (2B)	13	15	14
3 Bar (3B)	12	9	13
Red Seven (R7)	9	7	10
Jackpot (JP)	4	4	3
Total	**82**	**82**	**82**

Payable scorecard

Combination			$ payout on $1 bet	Probability (%)	Contribution (%)
CH	Any	Any	4	7.317	29.268
1B	1B	1B	10	1.551	15.507
2B	2B	2B	25	0.495	12.378
3B	3B	3B	50	0.255	12.932
R7	R7	R7	100	0.114	11.426
JP	JP	JP	1000	0.009	8.706
Total hit frequency				9.740 %	
Total return to player				90.017 %	

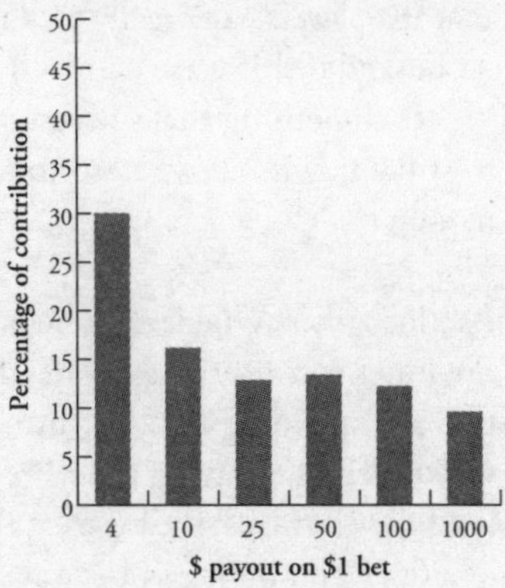

games. Game A is a 'cherry dribbler' – a machine that pays out frequently, but in small amounts. Almost half of the total payout money is accounted for in payouts of just $4. By contrast, in Game B only a third of the payout money goes on $4 payouts, leaving much more money to be won in the larger jackpots. Game A is what is called a low-volatility game, while Game B is high-volatility – you hit a winning combination less often, but the chances of a big win are greater. The higher the volatility, the more short-term risk there is for the slot operator.

Some gamblers prefer low-volatility slots, while others prefer high. The game designer's chief role is to make sure the machine pays out just enough for the gambler to want to continue playing – because the more someone plays, on average, the more he or she will lose. High volatility generates more excitement – especially in a casino, where machines hitting jackpots draw attention by erupting in spine-tingling *son et lumière*. Designing a good game, however, is not just about sophisticated graphics, colourful sounds and entertaining video narratives – it's also about getting the underlying probabilities just right. I asked Baerlocher whether by playing around with volatility it was possible to design a low-payback machine that was more attractive to gamblers than a higher-payback machine. 'My colleague and I spent over a year mapping things out and writing down some formulas and we came up with a method of hiding what the true payback percentage is,' he said. 'We're now hearing from some casinos that they are running lower-payback machines and that the players can't really pick up on it. It was a big challenge.'

I asked if this wasn't a touch unethical.

'It's something that's necessary,' he replied. 'We want the players to still enjoy it but we need to make sure that our customers make money.'

Baerlocher's pay tables are helpful not just in order to understand the inner constitutions of one-armed bandits; they are also illustrative in explaining how the insurance industry works. Insurance is very much like playing the slots. Both are systems built on probability theory in which the losses of almost everyone pay for the winnings of a few. And both can be fantastically profitable for those controlling the payback percentages.

An insurance premium is no different from a gamble. You are betting on the chance that, for example, your house will be burgled. If your house is burgled, you receive a payout, which is the reimbursement for what was stolen. If your house isn't burgled you, of course, receive nothing. The actuary at the insurance company behaves exactly like Anthony Baerlocher at IGT. He knows how much he wants to pay back to customers overall. He knows the probability for each payout event (a burglary, a fire, serious illness, etc.), so he works out how much his payouts should be per event so that the sum of expected contributions equals the total payback amount. Although compiling insurance tables is vastly more complicated than creating slot machines, the principle is the same. Since insurance companies pay out less than they receive in premiums, their payout percentage is less than 100 per cent. Buying an insurance policy is a negative-expectation bet and, as such, it is a bad gamble.

So why do people take out insurance if it's such a bad deal? The difference between insurance and gambling in casinos is that in casinos you are (hopefully) gambling with money you can afford to lose. With insurance, however, you are gambling to protect something you cannot afford to lose. While you will inevitably lose small amounts of money (the premium), this guards you against losing a catastrophic amount of money (the value of the contents of your house, for example). Insurance offers a good price for peace of mind.

It follows, however, that insuring against losing a non-catastrophic amount of money is pointless. One example is insuring against loss of a mobile phone. Mobile phones are relatively cheap (say, £100), but phone insurance is expensive (say, £7 per month). On average you will be better off if you don't take out insurance, and instead buy yourself a new phone on the occasions that you lose it. In this way, you are 'self-insuring' and keeping the insurance company's profit margins for yourself.

One reason for recent growth in the slots market is the introduction of 'progressive' machines, which have little to do with enlightened social policy and lots to do with the dream of instant wealth. Progressive slots have higher jackpots than other machines because they are joined in a network, with each machine contributing a percentage to a communal jackpot, the value of which gets progressively larger. In the Peppermill I had been struck by rows of linked machines offering prizes in the tens of thousands of dollars.

Progressive machines have high volatility, which means that in the short term casinos can lose significant sums. 'If we put out a game with a progressive jackpot, about one in every twenty [casino owners] will write us a letter telling us our game is broken. Because this thing hit two or three jackpots in the first week and the machines are $10,000 in the hole,' Baerlocher said, finding it ironic that people who are trying to profit from probability still have trouble understanding it on a basic level. 'We'll do an analysis and see the probability of it happening is, say, 200 to 1. They had [results] that should only happen half of a per cent of the time – it had to happen to someone. We tell them, ride it out, it's normal.'

IGT's most popular progressive slot, Megabucks, links together hundreds of machines across Nevada. When the company introduced it a decade ago, the minimum jackpot was $1 million. Initially, casinos didn't want the liability of having to pay out so much, so IGT underwrote the entire network by taking a percentage from all of its machines, and paid the jackpot itself. Despite paying out hundreds of millions of dollars in prize money, IGT has never suffered a loss on Megabucks. The law of large numbers is remarkably reliable: the bigger you get, the better it all works out.

The Megabucks jackpot now starts at $10 million. If it hasn't been won by the time the jackpot reaches around $20 million, casinos see queues forming at their Megabucks slots and IGT gets requests to distribute more machines. 'People think it's past that point where it normally hits so it's going to hit soon,' explained Baerlocher.

This reasoning, however, is erroneous. Every game played on a slot machine is a random event. You are just as likely to win when the jackpot is at $10, $20 or even $100 million, but it is very instinctive to feel that after a long period of holding money back, the machines are more likely to pay out. The belief that a jackpot is 'due' is known as the *gambler's fallacy*.

The gambler's fallacy is an incredibly strong human urge. Slot machines tap into it with particular virulence, which makes them, perhaps, the most addictive of all casino games. If you are playing many games in quick succession, it is only natural to think after a long run of losses: 'I'm bound to win next time.' Gamblers often talk of a machine being 'hot' or 'cold' – meaning that it is paying out lots or paying out little. Again, this is nonsense, since the odds are always the same. Still, one can see why one might attribute personality to a human-sized piece of plastic and metal often referred to as a one-armed bandit. Playing a slot machine is an intense, intimate experience – you get right up close, tap it with your fingertips, and cut out the rest of the world.

Because our brains are bad at understanding randomness, probability is the branch of maths most riddled with paradoxes and surprises. We instinctively attribute patterns to situations, even when we know there are none. It's easy to be dismissive of a slot-machine player for thinking that a machine is more likely to pay out after a losing streak, yet the psychology of the gambler's fallacy is present in non-gamblers too.

Consider the following party trick. Take two people, and explain to them that one will flip a coin 30 times and write down the order of heads and tails. The other will imagine flipping a coin 30 times, and then will write down the order of heads and tails outcomes that they have visualized. Without telling you, the two players will decide who takes which role, and then present you with their two lists. I asked my mother and stepfather to do this and was handed the following:

List 1
H T T H T H T T T H H T H H T H H H H T H T T H T H T T H H

List 2
T T H H T T T T T H H T T T H T T H T H H H H T H H T H T H

The point of this game is that it is very easy to spot which list comes from the flipping of the real coin, and which from the imagined coin. In the above case, it was clear to me that the second list was the real one, which was correct. First, I looked at the maximum runs of heads or tails. The second list had a maximum run of 5 tails. The first list had a maximum run of 4 heads. The probability of a run of 5 in 30 flips is almost two thirds, so it is much more likely than not that 30 flips gives a run of 5. The second list was already a good candidate for the real coin. Also, I knew that most people never ascribe a run of 5 in 30 flips because it seems too deliberate to be random. But to be sure I was right that the second list was the real coin I looked at how frequently both lists alternated between heads and tails. Due to the fact that each time you flip a coin the chances of heads and tails are equal, you would expect each outcome to be followed by a different outcome about half the time, and half the time to be followed by the same outcome. The second list alternates 15 times. The first list alternates 19 times – evidence of human interference. When imagining coin flips, our brains tend to alternate outcomes much more frequently than what actually occurs in a truly random sequence – after a couple of heads, our instinct is to compensate and imagine an outcome of tails, even though the chance of heads is still just as likely. Here, the gambler's fallacy appears. True randomness has no memory of what came before.

The human brain finds it incredibly difficult, if not impossible, to fake randomness. And when we are presented with randomness, we often interpret it as non-random. For example, the shuffle feature on an iPod plays songs in a random order. But when Apple launched the feature, customers complained that it favoured certain bands because often tracks from the same band were played one after another. The listeners were guilty of the gambler's fallacy. If the iPod shuffle were truly random, then each new song choice is independent of the previous choice. As the coin-flipping experiment shows, counter-intuitively long streaks are the norm. If songs are chosen randomly, it is very possible, if not entirely likely, that there will be clusters of songs by the same artist. Apple CEO Steve Jobs was totally serious when he said, in response to the outcry: 'We're making [the shuffle] less random to make it feel more random.'

Why is the gambler's fallacy such a strong human urge? It's all about control. We like to feel in control of our environments. If events occur randomly, we feel that we have no control over them. Conversely, if we do have control over events, they are not random. This is why we prefer to see patterns when there are none. We are trying to salvage a feeling of control. The human need to be in control is a deep-rooted survival instinct. In the 1970s a fascinating (if brutal) experiment examined how important a sense of control was for elderly patients in a nursing home. Some patients were allowed to choose how their rooms were arranged and allowed to choose a plant to look after. The others were told how their rooms would be and had a plant chosen and tended for them. The result after 18 months was striking. The patients who had control over their rooms had a 15 per cent death rate, but for those who had no control the rate was 30 per cent. Feeling in control can keep us alive.

Randomness is not smooth. It creates areas of empty space and areas of overlap.

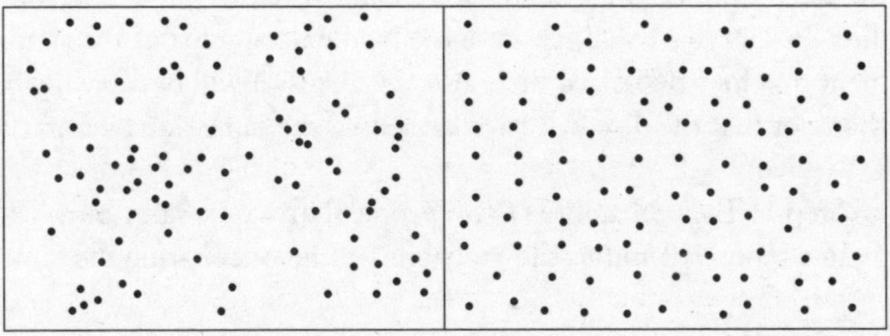

Random dots; non-random dots.

Randomness can explain why some small villages have higher than normal rates of birth defects, why certain roads have more accidents, and why in some games basketball players seem to score every free throw. It's also why in seven of the last ten World Cup finals at least two players shared birthdays:

2006 Patrick Vieira, Zinedine Zidane (France) 23 June
2002 None
1998 Emmanuel Petit (France), Ronaldo (Brazil) 22 September
1994 Franco Baresi (Italy), Claudio Taffarel (Brazil) 8 May
1990 None
1986 Sergio Batista (Argentina), Andreas Brehme (West Germany) 9 November
1982 None
1978 Rene and Willy van de Kerkhof (Holland) 16 September
Johnny Rep, Jan Jongbloed (Holland) 25 November
1974 Johnny Rep, Jan Jongbloed (Holland) 25 November
1970 Piazza (Brazil), Pierluigi Cera (Italy) 25 February

While at first this seems like an amazing series of coincidences, the list is actually mathematically unsurprising because whenever you have a randomly selected group of 23 people (such as two football teams and a referee), it is more likely than not that two people in it will share the same birthday. The phenomenon is known as the 'birthday paradox'. There is nothing self-contradictory about the result, but it does fly in the face of common sense – twenty-three seems like an absurdly small number.

Proof of the birthday paradox is similar to the proofs we used at the beginning of this chapter for rolling certain combinations of dice. In fact, we could rephrase the birthday paradox as the statement that for a 365-sided dice, after 23 throws it will be more likely than not that the dice will have landed on the same side twice.

Step 1: The probability of two people sharing the same birthday in a group is 1 minus the probability of no one sharing the same birthday.

Step 2: The probability of no one sharing the same birthday in a group of two people is $\frac{365}{365} \times \frac{364}{365}$. This is because the first person can be born on any day (365 choices out of 365) and the second can be born on any day apart from the day the first one is (364 choices out of 365). For convenience, we will ignore the extra day in a leap year.

Step 3: The probability of no one sharing the same birthday in a group of three people is $\frac{365}{365} \times \frac{364}{365} \times \frac{363}{365}$. With four people it becomes $\frac{365}{365} \times \frac{364}{365} \times \frac{363}{365} \times \frac{362}{365}$, and so on. When you multiply this out the result gets smaller and smaller. When the group contains 23 people, it finally shrinks to below 0.5 (the exact number is 0.493).

Step 4: If the probability of no one sharing the same birthday in a group is less than 0.5, the probability of at least two people sharing the same birthday is more than 0.5 (from Step 1). So it is more likely than not that in a group of 23 people two will have been born on the same day.

Football matches provide the perfect sample group to see if the facts fit the theory because there are always 23 people on the pitch. Looking at World Cup finals, however, the birthday paradox works a little too well. The probability of two people having the same birthday in a group of 23 is 0.507 or just over 50 per cent. Yet with seven out of ten positives (even excluding the van de Kerkhof twins), we have achieved a 70 per cent strike rate.

Part of this is the law of large numbers. If I analysed every match in every World Cup, I can be very confident that the result would be closer to 50.7 per cent. Yet there is another variable. Are the birthdays of footballers equally distributed throughout the year? Probably not. Research shows that footballers are more likely to be born at certain times of the year – favouring those born just after the school year cut-off point, since they will be the oldest and largest in their school years, and will therefore dominate sports. If there is a bias in the spread of birth dates, we can expect a higher chance of shared birthdays. And often there is a bias. For example, a sizeable proportion of babies are now born by caesarian section or induced. This tends to happen on weekdays (as maternity staff prefer not to work weekends), with the result that births are not spread as randomly throughout the year. If you take a section of 23 people born in the same 12-month period – say, the children in a primary-school classroom – the chance of two pupils sharing the same birthday will be significantly more than 50.7 per cent.

If a group of 23 people is not immediately accessible to test this out, just look at your immediate family. With 4 people it is 70 per cent likely that 2 will have birthdays within the same month. You only need 7 people for it to be likely that 2 of them were born in the same week, and in a group of 14 it is as likely as not that 2 people were born within a day of each other. As group size gets bigger, the probability rises surprisingly fast. In a group of 35 people, the chance of a shared birthday is 85 per cent, and with a group of 60 the chance is more than 99 per cent.

Here's a different question about birthdays with an answer as counter-intuitive as the birthday paradox: how many people do there need to be in a group for there to be a more than 50 per cent chance that someone shares *your* birthday. This is different from the birthday paradox because we are specifying a date. In the birthday paradox we are not bothered who shares a birthday with whom; we just want a shared birthday. Our new question can be rephrased as: given a fixed date, how many times do we need to roll our 365-sided dice until it lands on this date? The answer is 253 times! In other words, you would need to assemble a group of 253 people just to be more sure than not that one of them shares your birthday. This seems absurdly large – it is well over halfway between one and 365. Yet randomness is doing its clustering thing again – the group needs to be that size because the birthdays of its members are not falling in an orderly way. Among those 253 people there will be many people who double up on birthdays that are not yours, and you need to take them into account.

A lesson of the birthday paradox is that coincidences are more common than you think. In German Lotto, like the UK National Lottery, each combination of numbers has a 1-in-14-million chance of winning. Yet in 1995 and in 1986 identical combinations won: 15-25-27-30-42-48. Was this an amazing coincidence? Not especially, as it happens. Between the two occurrences of the winning combination there were 3016 lottery draws. The calculation to find how many times the draw should pick the same combination is equivalent to calculating the chances of two people sharing the same birthday in a group of 3016 people with there being 14 million possible birthdays. The probability works out to be 0.28. In other words, there was more than a 25 per cent chance that two winning combinations would be identical over that period; so the 'coincidence' was therefore not an extremely weird occurrence.

More disturbingly, a misunderstanding of coincidence has resulted in several miscarriages of justice. In one famous California case, from 1964, witnesses to a mugging reported seeing a blonde with a ponytail, a black man with a beard and a yellow getaway car. A couple fitting this description were arrested and charged. The prosecutor calculated the chance of such a couple existing by multiplying the probabilities of the occurrence of each of detail together: $\frac{1}{10}$ for a yellow car, $\frac{1}{3}$ for a blonde, and so on. The prosecutor calculated that the chance of such a couple existing was 1 in 12 million. In other words, for every 12 million people, only one couple on average would fit the exact description. The chances of the arrested couple being the guilty couple, he argued, were overwhelming. The couple was convicted.

The prosecutor, however, was doing the wrong calculation. He had worked out the chance of randomly selecting a couple that matched the witness profiles. The relevant question should have been, given there is a couple that matches the description, what is the chance that the arrested couple is the guilty couple? This probability was only about 40 per cent. More likely than not, therefore, the fact that the arrested couple fitted the description was a coincidence. In 1968 the California Supreme Court reversed the conviction.

Returning to the world of gambling, in another lottery case a New Jersey woman won her state lottery twice in four months in 1985–6. It was widely reported that the chances of this happening was 1 in 17 trillion. However, though 1 in 17 trillion was the correct chance of buying a single lottery ticket in both lotteries and scooping the jackpot on both occasions, this did not mean that the chances of someone, somewhere winning two lotteries was just as unlikely. In fact, it is pretty likely. Stephen Samuels and George McCabe of Purdue University calculated that over a seven-year period the odds of a lottery double-win in the United States are better than evens. Even over a four-month period, the odds of a double winner somewhere in the country are better than 1 in 30. Persi Diaconis and Frederick Mosteller call this the *law of very large numbers*: 'With a large enough sample, any outrageous thing is apt to happen.'

Mathematically speaking, lotteries are by far the worst type of legal bet. Even the most miserly slot machine has a payback percentage of about 85 per cent. By comparison, the UK National Lottery has a payback percentage of approximately 50 per cent. Lotteries offer no risk to the organizers since the prize money is just the takings redistributed. Or, in the case of the National Lottery, half of the takings, redistributed.

In rare instances, however, lotteries can be the best bet around. This is the case when, due to 'rollover' jackpots, the prize money has grown to become larger than the cost of buying every possible combination of numbers. In these instances, since you are covering all outcomes, you can be assured of having the winning combination. The risk here is only that there might be other people who also have the winning combination – in which case you will have to share the first prize with them. The buy-every-combination approach, however, relies on an ability to do just that – which is a significant theoretical and logistical challenge.

The UK lottery is a 6/49 lottery, which means that for each ticket the punter must select 6 numbers out of 49. There are about 14 million possible combinations. How do you list these combinations in such a way that each combination is listed exactly once, avoiding duplications? In the early 1960s, Stefan Mandel, a Romanian economist, asked himself the same question about the smaller Romanian lottery. The answer is not straightforward. Mandel cracked it, however, after spending several years on the problem and won the Romanian lottery in 1964. (In fact, in this case he did not buy every combination, since that would have cost him too much money. He used a supplementary method called 'condensing' that guarantees that at least 5 of the 6 numbers are correct. Usually getting 5 numbers means winning second prize, but he was lucky and won first prize first time.) The algorithm that Mandel had written out to decide which combinations to buy covered 8000 foolscap sheets. Shortly afterwards, he emigrated to Israel and then to Australia.

While in Melbourne, Mandel founded an international betting syndicate, raising enough money from its members to ensure that, if he wanted to buy every single combination in a lottery, then he could. He surveyed the world's lotteries for rollover jackpots that were more than three times the cost of every combination. In 1992 he identified the Virginia state lottery – a lottery with seven million combinations, each costing $1 a ticket – whose jackpot had reached almost $28 million. Mandel got to work. He printed out coupons in Australia, filled them in by computer so that they covered the seven million combinations, and then flew the coupons to the US. He won the first prize and 135,000 secondary prizes too.

The Virginia lottery was Mandel's largest jackpot, bringing his tally since leaving Romania to 13 lottery wins. The US Internal Revenue Service, the FBI and the CIA all looked into the syndicate's Virginia lottery win, but could prove no wrongdoing. There is nothing illegal about buying up every combination, even though it sounds like a scam. Mandel has now retired from betting on lotteries and lives on a tropical island in the South Pacific.

A particularly useful visualization of randomness was invented by John Venn in 1888. Venn is perhaps the least spectacular mathematician ever to become a household name. A Cambridge professor and Anglican cleric, he spent much of his later life compiling a biographical register of 136,000 of the university's pre-1900 alumni. While he did not push forward the boundaries of his subject, he did, nevertheless, develop a lovely way of explaining logical arguments with intersecting circles. Even though Leibniz and Euler had both done something very similar in previous centuries, the diagrams are named after Venn. Much less known is that Venn thought up an equally irresistible way to illustrate randomness.

Imagine a dot in the middle of a blank page. From the dot there are eight possible directions to go: north, northeast, east, southeast, south, southwest, west and northwest. Assign the numbers 0 to 7 to each of the directions. Choose a number from 0 to 7 randomly. If the number comes up, trace a line in that direction. Do this repeatedly to create a path. Venn carried this out with the most unpredictable sequence of numbers he knew: the decimal expansion of pi (excluding 8s and 9s, and starting with 1415). The result, he wrote, was 'a very fair graphical indication of randomness'.

Venn's sketch is thought to be the first-ever diagram of a 'random walk'. It is often called the 'drunkard's walk' because it is more colourful to imagine that the original dot is instead a lamp-post and the path traced is the random staggering of a drunk. One of the most obvious questions to ask is how far will the drunk wander from the point of origin before collapsing? On average, the longer he has been walking, the further away he will be. It turns out that his distance increases with the square root of the time spent walking. So, if in one hour he stumbles, on average, one block from the lamp-post, it will take him four hours, on average, to go two blocks, and nine hours to go three.

As the drunkard randomly walks, there will be times when he goes in circles and doubles back on himself. What is the chance of the drunk eventually walking back into the lamp-post? Surprisingly, the answer is 100 per cent. He might stray for years in the most remote places but it is a sure thing that, given sufficient time, the drunk will eventually return to base.

Imagine a drunkard's walk in three dimensions. Call it the buzz of the befuddled bee. The bee starts at a point suspended in space and

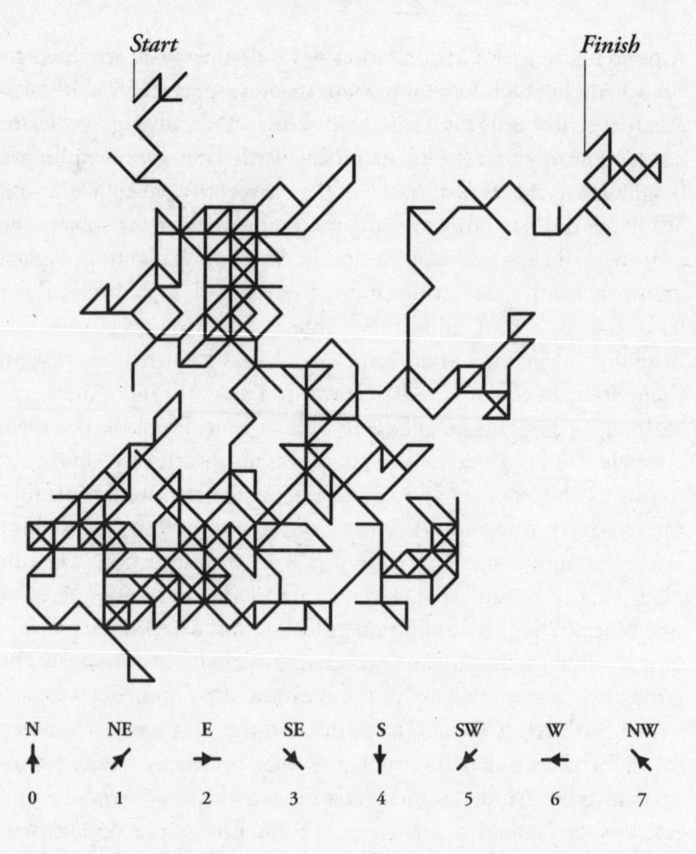

The first-ever random walk appeared in the third edition of John Venn's Logic of Chance *(1888). The rules for the direction of the walk (my addition) follow the digits 0–7 that appear in pi after the decimal point.*

then flies in a straight line in a random direction for a fixed distance. The bee stops, dozes, then buzzes off in another random direction for the same distance. And so on. What is the chance of the bee eventually buzzing back into the spot where it started? The answer is only 0.34, or about a third. It was weird to realize that in two dimensions the chance of a drunkard walking back into the lamp-post was an absolute certainty, but it seems even weirder to think that a bee buzzing for ever is very unlikely ever to return home.

In Luke Rhinehart's bestselling novel *The Dice Man*, the eponymous hero makes life decisions based on the throwing of dice.

Consider Coin Man, who makes decisions based on the flip of a coin. Let's say that if he flips heads, he moves one step up the page, and if he flips tails, he moves a step down the page. Coin Man's path is a drunkard's walk in one dimension – he can move only up and down the same line. Plotting the walk described by the second list of 30 coin flips on p. 323, you get the following graph.

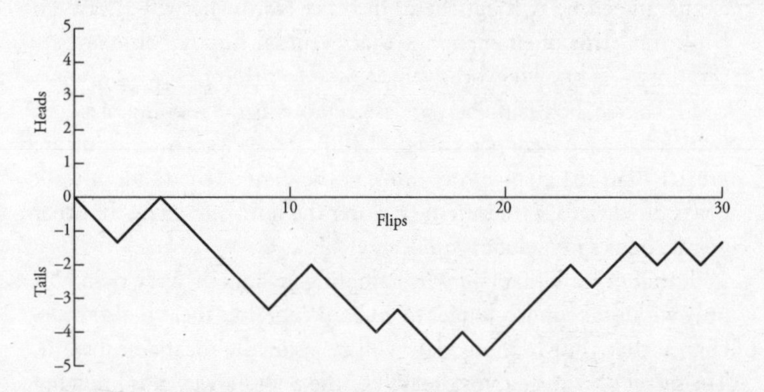

The walk is a jagged line of peaks and valleys. If you extended this for more and more flips, a trend emerges. The line swings up and down, with larger and larger swings. Coin Man roams further and further from his starting point in both directions. Below are the journeys of six coin men that I plotted from 100 flips each.

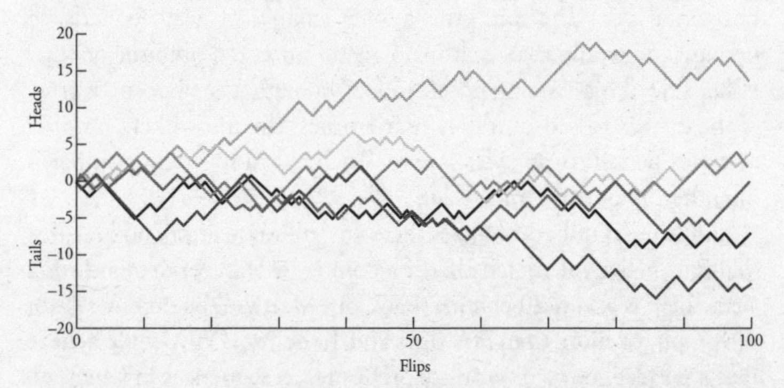

If we imagine that at a certain distance from the starting point in one direction there is a barrier, there is a 100 per cent probability that eventually Coin Man will hit the barrier. The inevitability of this collision is very instructive when we analyse gambling patterns.

Instead of letting Coin Man's random walk describe a physical journey, let it represent the value of his bank account. And let the coin flip be a gamble. Heads he wins £100, tails he loses £100. The value in his bank account will swing up and down in increasingly large waves. Let us say that the only barrier that will stop Coin Man playing is when the value of his account is £0. We know it is guaranteed he will get there. In other words, he will always go bankrupt. This phenomenon – that eventual impoverishment is a certainty – is known evocatively as *gambler's ruin*.

Of course, no casino bets are as generous as the flipping of a coin (which has a payback percentage of 100). If the chances of losing are greater than the chances of winning, the map of the random walk drifts downward, rather than tracking the horizontal axis. In other words, bankruptcy looms quicker.

Random walks explain why gambling favours the very rich. Not only will it take much longer to go bankrupt, but there is also more chance that your random walk will occasionally meander upward. The secret of winning, for the rich or the poor, however, is knowing when to stop.

Inevitably, the mathematics of random walks contains some head-popping paradoxes. In the graphs on p. 333 where Coin Man moves up or down depending on the results of a coin toss, one would expect the graph of his random walk to regularly cross the horizontal axis. The coin gives a 50:50 chance of heads or tails, so perhaps we would expect him to spend an equal amount of time either side of his starting point. In fact, though, the opposite is true. If the coin is tossed infinitely many times, the most likely number of times he will swap sides is zero. The next most likely number is one, then two, three and so on.

For finite numbers of tosses there are still some pretty odd results. William Feller calculated that if a coin is tossed every second for a year, there is a 1-in-20 chance that Coin Man will be on one side of the graph for more than 364 days and 10 hours. 'Few people believe that a perfect coin will produce preposterous sequences in which no change of [side] occurs for millions of trials in succession, and yet this is what a good coin will do rather regularly,' he wrote in *An Introduction to Probability Theory and Its Applications*. 'If a modern educator or psychologist were to describe the long-run case histories of individual coin-tossing games, he would classify the majority of coins as maladjusted.'

While the wonderful counter-intuitions of randomness are exhilarating to pure mathematicians, they are also alluring to the dishonourable. Lack of a grasp of basic probability means that you can easily be conned. If you are ever tempted, for example, by a company that claims it can predict the sex of your baby, you are about to fall victim to one of the oldest scams in the book. Imagine I set up a company, which I'll call BabyPredictor, that announces a scientific formula for predicting whether a baby will be a boy or a girl. BabyPredictor charges mothers a set fee for the prediction. Because of a formidable confidence in its formula, and the philanthropic generosity of its CEO, me, the company also offers a total refund if the prediction turns out to be wrong. Buying the prediction sounds like a good deal – since either BabyPredictor is correct, or it is wrong and you can get your money back. Unfortunately, however, BabyPredictor's secret formula is actually the tossing of a coin. Heads I say the child will be a boy, tails it will be a girl. Probability tells me that I will be correct about half the time, since the ratio of boys to girls is about 50:50. Of course, half the time I will give the money back, but so what – since the other half of the time I get to keep it.

The scam works because the mother is unaware of the big picture. She sees herself as a sample group of one, rather than as part of a larger whole. Still, baby-predicting companies are alive and well. Babies are born every minute, and so are mugs.

In a more elaborate version, this time targeting greedy men rather than pregnant women, a company we'll call StockPredictor puts up a fancy website. It sends out 32,000 emails to an investors' mailing list, announcing a new service, which, using a very sophisticated computer model, can predict whether a certain stock index will rise or fall. In half of the emails it predicts a rise the following week; and in the other half, a fall. Whatever happens to the index, 16,000 investors will have received an email with the correct prediction. So, StockPredictor then sends these 16,000 addresses another email with the following week's prediction. Again, the prediction will be correct for 8000 of them. If StockPredictor continues like this for another four weeks, eventually there will be 1000 email recipients whose six consecutive predictions have all turned out to be true. StockPredictor then informs them that in order to receive any further predictions they must pay a fee – and why wouldn't they, since the predictions have thus far been pretty good?

The stock-predicting scam can be adapted to predicting horse-races, football matches, and even the weather. If all outcomes are covered, there will be at least one person receiving a correct prediction of all matches, races or sunny days. That person might think, 'Wow, there is only a one-in-a-million chance of such a combination being true,' but if a million emails are sent out covering all possibilities, then someone, somewhere has to receive the correct one.

Scamming people is immoral and often illegal. However, trying to get one over on a casino is often seen as a just cause. For mathematicians, the challenge of beating the odds is like showing a red rag to a bull – and there is an honourable tradition of those who have succeeded.

The first method of attack is to realize that the world is not perfect. Joseph Jagger was a cotton-factory mechanic from Lancashire who knew enough about Victorian engineering to realize that roulette wheels might not spin perfectly true. His hunch was that if the wheel was not perfectly aligned, it might favour some numbers over others. In 1873, aged 43, he visited Monte Carlo to test his theory. Jagger hired six assistants, assigned each of them to one of the six tables in a casino, and instructed them to write down every number that came up over a week. Upon analysing the figures, he saw that one wheel was indeed biased – nine numbers came up more than others. The greater prevalence of these numbers was so slight that their advantage was only noticeable when considering hundreds of plays.

Jagger started betting and in one day won the equivalent of $70,000. Casino bosses, however, noticed that he gambled at only one table. To counter Jagger's onslaught, they switched the wheels around. Jagger started to lose, until he realized what the management had done. He relocated to the biased table, which he recognized because it had a distinctive scratch. He started winning again and only gave up when the casino reacted a second time by sliding the frets around the wheel after each daily session, so that new numbers would be favoured. By that time Jagger had already amassed $325,000 – making him, in today's terms, a multimillionaire. He returned home, left his factory job and invested in real estate. In Nevada, from 1949 to 1950, Jagger's method was repeated by two science grads named Al Hibbs and Roy Walford. They started off with a borrowed $200 and turned it into $42,000, allowing them to buy a 40ft yacht and sail the Caribbean for 18 months before returning to their studies. Casinos now change wheels around much more regularly than they used to.

The second way to manipulate the odds in your favour is to question what randomness is anyway. Events that are random under one set of information might well become non-random under a greater set of information. This is turning a maths problem into a physics one. A coin flip is random because we don't know how it will land, but flipped coins obey Newtonian laws of motion. If we knew exactly the speed and angle of the flip, the density of the air and any other relevant physical data, we would be able to calculate exactly the face on which it would land. In the mid 1950s a young mathematician named Ed Thorp began to ponder what set of information would be required to predict where a ball would land in roulette.

Thorp was helped in his endeavour by Claude Shannon, his colleague at the Massachusetts Institute of Technology. He couldn't have wished for a better co-conspirator. Shannon was a prolific inventor with a garage full of electronic and mechanical gadgets. He was also one of the most important mathematicians in the world, as the father of information theory, a crucial academic breakthrough that led to the development of the computer. The men bought a roulette wheel and conducted experiments in Shannon's basement. They discovered that if they knew the speed of the ball as it went around the stationary outer rim, and the speed of the inner wheel (which is spun in the opposite direction of the ball), they could make pretty good estimates regarding which segment of the wheel the ball would fall in. Since casinos allow players to place a bet after the ball has been dropped, all Thorp and Shannon needed to do was figure out how to measure the speeds and process them in the few seconds before the croupier closed the betting.

Once again, gambling pushed forward scientific advancement. In order to predict roulette plays accurately, the mathematicians built the world's first wearable computer. The machine, which could fit in a pocket, had a wire going down to the shoe, where there was a switch, and another wire going up to a pea-sized headphone. The wearer was required to touch the switch at four moments – when a point on the wheel passed a reference point, when it had made one full revolution, when the ball passed the same point, and when the ball had again made a full revolution. This was enough information for estimating the speeds of the wheel and the ball.

Thorp and Shannon divided the wheel into eight segments of five numbers each (some overlapped, though, as there are 38 pockets). The pocket-sized computer played a scale of eight notes – an octave – through the headphones and the note it stopped on determined the segment where it predicted the ball would fall. The computer could not say with total certainty which segment the ball would land in, but it didn't need to. All Thorp and Shannon desired was that the prediction would be better than the randomness of guessing. On hearing the notes, the computer-wearer would then place chips on all five numbers of the segment (which, although next to each other on the wheel, were not adjacent on the baize). The method was surprisingly accurate – for the single-number bets they estimated that they could expect to win $4.40 for every $10 wager.

When Thorp and Shannon went to Las Vegas for a test drive, the computer worked, if precariously so. The men needed to look inconspicuous, but the earpiece was prone to popping out and the wires were so fragile that they kept breaking. Still, the system worked, and they were able to convert a small pile of dimes into a few piles of dimes. Thorp was content to have beaten roulette in theory, if not in practice, because his assault on another gambling game was having much more striking success.

Blackjack, or twenty-one, is a card game in which the aim is to get a hand where the total value of the cards is as close as possible to an upper limit of 21. The dealer also deals a hand for himself. To win, you must have a hand higher than his while not exceeding 21.

Like all the classic casino games, blackjack gives a slight advantage to the house. If you play blackjack, in the long run you will lose money. In 1956 an article was published in an obscure statistics journal that claimed to have devised a playing strategy that gave the house an advantage of just 0.62 per cent. After reading the article, Thorp learned the strategy and tested it during a vacation trip to Vegas. He discovered that he lost his money much slower than the other players. He decided he would begin to think deeply about blackjack, a decision that would change his life.

Ed Thorp is now 75 but I suspect he doesn't look that different from how he looked half a century ago. Slim, with a long neck and concise features, he has a clean-cut college-boy haircut, unpretentious glasses, and a calm, upright posture. After returning from Vegas, Thorp reread the journal article. 'I saw right away, within a couple of minutes, how you could almost certainly beat this game by keeping track of the cards that were played,' he remembered. Blackjack is different from, say, roulette, since the odds change once a card has been dealt. The chance of getting a 7 in roulette is 1 in 38 every time you spin the wheel. In blackjack the probability of the first dealt card being an ace is $\frac{1}{13}$. If the first dealt card is an ace, the probability of the second card being an ace, however, is not $\frac{1}{13}$ – it is $\frac{3}{51}$, since the pack now has 51 cards and there are only 3 aces left in it. Thorp thought there must be a system that turned the odds in the favour of the player. It then became just a matter of finding it.

In a 52-card deck there are $52 \times 51 \times 50 \times 49 \times \ldots \times 3 \times 2 \times 1$ ways of the cards being ordered. This number is about 8×10^{67}, or 8 followed by 67 zeros. The number is so huge that it is very unlikely that any two randomly shuffled decks will ever have had the same order in the history of the world – even if the world's population had started playing cards at the Big Bang. Thorp reasoned that there are too many possible permutations of cards for any system of memorizing permutations to be feasible for a human brain. Instead he decided to look at how the house advantage changes depending on which cards have already been dealt. Using a very early computer, he found that by keeping track of the number-five cards of each suit – the five of hearts, spades, diamonds and clubs – a player could judge whether the deck was favourable. Under Thorp's system, blackjack morphed into a beatable game, with an expected return of up to 5 per cent depending on the cards left in the pack. Thorp had invented 'card-counting'.

He wrote up his theory and submitted it to the American Mathematical Society (AMS). 'When the abstract came through everybody thought it was ridiculous,' he remembered. 'It was gospel in the scientific world that you couldn't beat any of the major gambling games, and that had rather strong support from the research and analysis that had been done over a couple of centuries.' Proofs that demonstrate that you can beat the odds at casino games are rather like proofs that you can square the circle – surefire evidence of a crackpot. Luckily, one of the members of the AMS's submission committee was an old classmate of Thorp's, and the abstract was accepted.

In January 1961 Thorp presented his paper at the American Mathematical Society's winter meeting in Washington. It made national news, including the front page of his local paper, the *Boston Globe*. Thorp received hundreds of letters and calls, with many offers to finance gambling sprees for a share of the profit. A syndicate from New York was offering $100,000. He called the number on the New York letter and the following month a Cadillac pulled up outside his apartment. Out stepped a pint-sized senior citizen, accompanied by two spectacular blondes in mink coats.

The man was Manny Kimmel, a mathematically astute New York gangster and inveterate high-stakes gambler. Kimmel had taught himself enough about probability to know the birthday paradox – one of his favourite things to bet on was whether two people in a group shared the same birthday. Kimmel introduced himself as the owner of 64 parking lots in New York City, which was true. He introduced the girls as his nieces, which probably wasn't. I asked Thorp if he suspected Kimmel of mob ties. 'At that time I wasn't very knowledgeable about the gambling world; in fact I had no knowledge of it except the theoretical, and I also hadn't investigated the world of crime. He represented himself as a wealthy businessman and the evidence for that was overwhelming.' Kimmel invited Thorp to play blackjack at his luxurious Manhattan apartment the following week. After a few sessions, Kimmel was convinced that card-counting worked. Both men flew to Reno to try it out. They started off with $10,000 and by the end of the trip had built their pot up to $21,000.

When you are gambling in a casino, two factors come into play that determine how much money you will win or lose. *Playing strategy* is about how to win a game. *Betting strategy* is about money management – how much to bet and when. Is it worth, for example, betting your entire purse on one bet? Or is it worth dividing your money up into the smallest possible stakes? Different strategies can have a surprisingly large impact on how much money you can expect to make.

The best-known betting strategy is called the 'martingale', or doubling up, and was popular with French gamblers in the eighteenth century. The principle is to double your bet if you lose. Let's say you are betting on the toss of a coin. Heads you win $1, tails you lose $1. Just say the first flip is tails. You lose $1. For the next bet you must stake $2. Winning on the second bet wins you $2, which recoups your $1 loss from the first bet and puts you $1 in profit. Say you lose the first five flips:

Lose $1 bet so next time bet $2
Lose $2 bet so next time bet $4
Lose $4 bet so next time bet £8
Lose $8 bet so next time bet $16
Lose $16 bet

You will be 1 + 2 + 4 + 8 + 16 = $31 down, so the sixth bet must be for $32. If you win you recoup your losses, and profit. But despite risking so much money, you are only ahead by $1, your original stake.

Martingale certainly has an appeal. In a game where the odds are almost 50:50 – like, say, betting on the red at roulette, which has a probability of 47 per cent – you are very likely to win a fair percentage of plays and so have a good chance of staying ahead. But the martingale system is not fail-safe. For a start, you are only winning in small increments. And we know that in a run of 30 coin flips, a streak of five heads, or five tails, is more likely than not. If you start with a $40 bet and have a five-game losing streak, you will find yourself having to bet $1280. At the Peppermill Casino, though, you wouldn't be able to – it has a maximum bet of $1000. One reason why casinos have maximum bets is to stop systems like martingale. The exponential growth of martingale bets on a losing streak often accelerates bankruptcy, rather than insuring against it. The system's most famous champion, eighteenth-century Venetian playboy Giacomo Casanova, discovered this the hard way. 'I still played on the martingale,' he once said, 'but with such bad luck that I was soon left without a sequin.'

Still, if you stood at the Peppermill roulette table playing martingale with a $10 starting stake on red, you would have to be very unlucky not to eventually win $10. The system would break down only if you lost six times in a row, and there is only a 1-in-47 chance of that happening. Once you have won, however, it would then be advisable to cash in your winnings and leave. By continuing to gamble, the chances of an unlucky streak will eventually become more likely than not.

Let's consider a different system of betting. Imagine you are given $20,000 in a casino and told you must gamble it on red at the roulette table. What's the best strategy for doubling your money? Is it to be bold and bet the whole thing in one bet, or to be cautious and bet in the smallest possible amount, in stakes of $1? Even though it seems initially reckless, your chances of success are much better if you bet the whole amount in one go. In maths-speak, bold play is *optimal*. With a smidgen of reflection, this makes sense: the law of large numbers says that you will lose in the long run. Your best chances are to make the run as short as possible.

In fact, this is exactly what Ashley Revell, a 32-year-old from Kent, did in 2004. He sold all his possessions, including his clothes, and in a Las Vegas casino bet the total amount – $135,300 – on red. Had he lost, he would have at least become a C-list TV celebrity, as the bet was being filmed for a TV reality show. But the ball landed on red 7, and he came home with $270,600.

At blackjack, Ed Thorp was presented with a different issue. His card-counting system meant that he could tell at certain points during the game whether he had an advantage over the dealer. Thorp asked himself: what is the best betting strategy when the odds are in your favour?

Imagine there is a bet where the chance of winning is 55 per cent and the chance of losing 45 per cent. For simplicity, the game pays evens and we play it 500 times. The advantage – the *edge* – is 10 per cent. In the long run our winnings will work out on average as a $10 profit for every $100 gambled. To maximize our total profit, we obviously need to maximize the combined total of wagers. It is not immediately obvious how this is done, since maximizing wealth requires a minimizing of the risk of losing it all. This is how four betting strategies perform:

Strategy 1: Bet everything. Just like Ashley Revell, put your entire bankroll on the first bet. If you win, you have doubled your money. If you lose, you are bankrupt. If you win, put everything down again for the next bet. The only way you can avoid losing everything is if you win all 500 games. The chance of this happening, if the probability of winning each game is 0.55, is about one in 10^{130}, or 1 followed by 130 zeros. In other words, it is almost certain you will be bankrupt by the 500th game. Obviously, this is *not* a good long-term strategy.

Stategy 2: Fixed wager. Bet a fixed amount on every bet. If you win, your wealth grows by that fixed amount. If you lose, your wealth shrinks by that amount. Since you win more than you lose, your wealth will increase overall, but it does so only by increasing in jumps of the same fixed amount. As the graph overleaf shows, your money doesn't grow very fast.

Strategy 3: Martingale. This gives a faster rate than fixed wager, since losses are compensated by doubling up after a loss, but brings with it a much higher risk. With only a few losing bets, you could be bankrupt. Again, this is *not* a good long-term strategy.

Strategy 4: Proportional betting. In this case, bet a fraction of your bankroll related to the edge you have. There are several variations of proportional betting, but the system where wealth grows fastest is called the Kelly strategy. Kelly tells you to bet the fraction of your bankroll determined by $\frac{\text{edge}}{\text{odds}}$. In this case, the edge is 10 per cent and the odds are evens (or 1 to 1), making $\frac{\text{edge}}{\text{odds}}$ equal to 10 per cent. So, wager 10 per cent of the bankroll every bet. If you win, the bankroll will increase by 10 per cent,

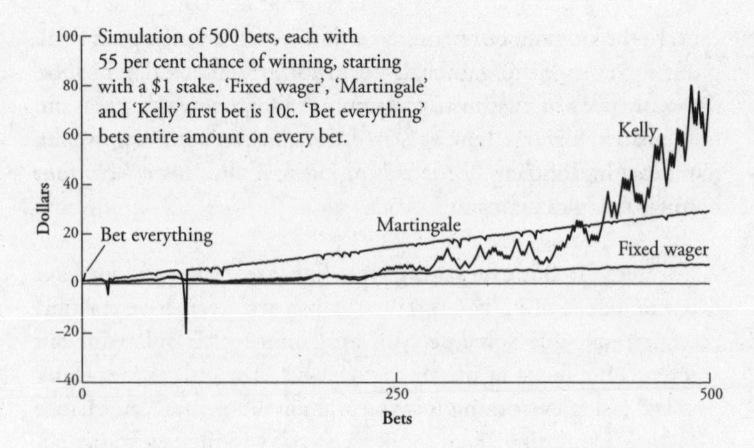

so the next wager will be 10 per cent larger than the first. If you lose, the bankroll shrinks by 10 per cent, so the second wager will be 10 per cent lower than the first.

This is a very safe strategy because if you have a losing streak, the absolute value of the wager shrinks – which means that losses are limited. It also offers large potential rewards since – like compound interest – on a winning streak one's wealth grows exponentially. It is the best of both worlds: low risk and high return. And just look how it performs: starting slowly but eventually, after around 400 bets, racing way beyond the others.

John Kelly Jr was a Texan mathematician who outlined his famous gambling strategy formula in a 1956 paper, and when Ed Thorp put it into practical use at the blackjack table, the results were striking. 'As one general said, you get there firstest with the mostest.' With small edges and judicious money management, huge returns could be achieved. I asked Thorp which method was more important to making money at blackjack – card-counting or using the Kelly criterion. 'I think the consensus after decades of examining this question,' he replied, 'is that the betting strategies may be two thirds or three quarters of what you're going to get out of it, and playing strategy is maybe a third to a quarter. So, the betting strategy is much more important.' Kelly's strategy would later help Thorp make more than $80 billion dollars on the financial markets.

Ed Thorp announced his card-counting system in his 1962 book *Beat the Dealer*. He refined the method for a second edition in 1966 that also counted the cards worth ten (the jack, queen, king and ten). Even though the ten cards shift the odds less than the five do, there are more of them, so it's easier to identify advantages. *Beat the Dealer* sold more than a million copies, inspiring – and continuing to inspire – legions of gamblers.

In order to eliminate the threat of card-counting, casinos have tried various tactics. The most common was to introduce multiple decks because with more cards, counting is made both more difficult and less profitable. The 'professor stopper', a shoe that shuffles many packs at once, is essentially named in Thorp's honour. And casinos have been forced to make it an offence to use a computer to predict at roulette.

Thorp last played blackjack in 1974. 'The family went on a trip to the World's Fair in Spokane and on the way back we stopped at Harrah's [casino] and I told my kids to give me a couple of hours because I wanted to pay for the trip – which I did.'

Beat the Dealer is not just a gambling classic. It also reverberated through the worlds of economics and finance. A generation of mathematicians inspired by Thorp's book began to create models of the financial markets and apply betting strategies to them. Two of them, Fischer Black and Myron Scholes, created the Black-Scholes formula indicating how to price financial derivatives – Wall Street's most famous (and infamous) equation. Thorp ushered in an era when the quantitative analyst, the 'quant' – the name given to the mathematicians relied on by banks to find clever ways of investing – was king. '*Beat the Dealer* was kind of the first quant book out there and it led fairly directly to quite a revolution,' said Thorp, who can claim – with some justification – to being the first-ever quant. His follow-up book, *Beat the Market*, helped transform securities markets. In the early 1970s he, together with a business partner, started the first 'market neutral' derivatives hedge fund, meaning that it avoided any market risk. Since then, Thorp has developed more and more mathematically sophisticated financial products, which has made him extremely rich (for a maths professor, anyway). Although he used to run a well-known hedge fund, he now runs a family office in which he invests only his own money.

I met Thorp in September 2008. We were sitting in his office in a high-rise tower in Newport Beach, which looks out over the Pacific Ocean. It was a delicious California day with a pristine blue sky. Thorp is scholarly without being earnest, careful and considered, but also sharp and playful. Just one week earlier, the bank Lehman Brothers had filed for bankruptcy. I asked him if he felt any sense of guilt for having helped create some of the mechanisms that had contributed to the largest financial crisis in decades. 'The problem wasn't the derivatives themselves, it was the lack of regulation of the derivatives,' he replied, perhaps predictably.

This led me to wonder, since the mathematics behind global finance is now so complicated, had the government ever sought his advice? 'Not that I know of, no!' he said with a smile. 'I have plenty here if they ever show up! But a lot of this is highly political and also very tribal.' He said that if you want to have your voice heard you need to be on the East Coast, playing golf and having lunches with bankers and politicians. 'But I'm in California with a great view ... just playing mathematical games. You're not going to run into these people, except once in a while.' But Thorp relishes his position as an outsider. He doesn't even see himself as a member of the financial world, though he has been for four decades. 'I think of myself as a scientist who has applied his knowledge to analysing financial markets.' In fact, challenging the conventional wisdom is the defining theme of his life, something he's done successfully over and over again. And he thinks that clever mathematicians will always be able to beat the odds.

I was also interested in whether having such a sophisticated understanding of probability helped him avoid the subject's many counter-intuitions. Was he ever victim, for example, of the gambler's fallacy? 'I think I'm very good at just saying no – but it took a while. I went through an expensive education when I first began to learn about stocks. I would make decisions based on less-than-rational decisions.'

I asked him if he ever played the lottery.

'Do you mean make bad bets?'

I guess, I said, that he didn't do that.

'I can't help it. You know once in a while you have to. Let's suppose that your whole net worth is your house. To insure your house is a bad bet in the expected-value sense, but it is probably prudent in the long-term survival sense.'

So, I asked, have you insured your house?

He paused for a few moments. 'Yes.'

He had stalled because he was working out exactly how rich he was. 'If you are wealthy enough you don't have to insure small items,' he explained. 'If you were a billionaire and had a million-dollar house it wouldn't matter whether you insured it or not, at least from the Kelly-criterion standpoint. You don't need to pay to protect yourself against this relatively small loss. You are better off taking the money and investing it in something better.

'Have I really insured my houses or not? Yeah, I guess I have.'

I had read an article in which it was mentioned that Thorp planned to have his body frozen when he dies. I told him it sounded like a gamble – and a very Californian one at that.

'Well, as one of my science-fiction friends said: "It's the only game in town."'

CHAPTER TEN

Situation Normal

I recently bought an electronic kitchen scale. It has a glass platform and an Easy To Read Blue Backlit Display. My purchase was not symptomatic of a desire to bake elaborate desserts. Nor was I intending my flat to become the stash house for local drug gangs. I was simply interested in weighing stuff. As soon as the scale was out of its box I went to my local bakers, Greggs, and bought a baguette. It weighed 391g. The following day I returned to Greggs and bought another baguette. This one was slightly heftier at 398g. Greggs is a chain with more than a thousand shops in the UK. It specializes in cups of tea, sausage rolls and buns plastered in icing sugar. But I had eyes only for the baguettes. On the third day the baguette weighed 399g. By now I was bored with eating a whole baguette every day, but I continued with my daily weighing routine. The fourth baguette was a whopping 403g. I thought maybe I should hang it on the wall, like some kind of prize fish. Surely, I thought, the weights would not rise for ever, and I was correct. The fifth loaf was a minnow, only 384g.

In the sixteenth and seventeenth centuries Western Europe fell in love with collecting data. Measuring tools, such as the thermometer, the barometer and the perambulator – a wheel for clocking distances along a road – were all invented during this period, and using them was an exciting novelty. The fact that Arabic numerals, which provided effective notation for the results, were finally in common use among the educated classes helped. Collecting numbers was the height of modernity, and it was no passing fad; the craze marked the beginning of modern science. The ability to describe the world in quantitative, rather than qualitative, terms totally changed our relationship with our own surroundings. Numbers gave us a language for scientific investigation and with that came a new confidence that we could have a deeper understanding of how things really are.

I was finding my daily ritual of buying and weighing bread every morning surprisingly pleasurable. I would return from Greggs with a skip in my step, eager to see just how many grams my baguette would be. The frisson of expectation was the same as the feeling when you check the football scores or the financial markets – it is genuinely exciting to discover how your team has done or how your stocks have performed. And so it was with my baguettes.

The motivation behind my daily trip to the bakers was to chart a table of how the weights were distributed, and after ten baguettes I could see that the lowest weight was 380g, the highest was 410g, and one of the weights, 403g, was repeated. The spread was quite wide, I thought. The baguettes were all from the same shop, cost the same amount, and yet the heaviest one was almost 8 per cent heavier than the lightest one.

I carried on with my experiment. Uneaten bread piled up in my kitchen. After a month or so, I made friends with Ahmed, the Somali manager of Greggs. He thanked me for enabling him to increase his daily stock of baguettes, and as a gift gave me a *pain au chocolat*.

It was fascinating to watch how the weights spread themselves along the table. Although I could not predict how much any one baguette would weigh, when taken collectively a pattern was definitely emerging. After 100 baguettes, I stopped the experiment, by which time every number between 379g and 422g had been covered at least once, with only four exceptions:

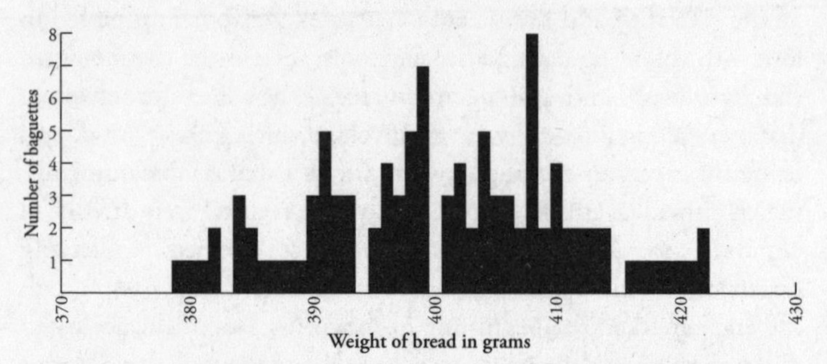

I had embarked on the bread project for mathematical reasons, yet I noticed interesting psychological side-effects. Just before weighing each loaf, I would look at it and ponder the colour, length, girth and texture – which varied quite considerably between days. I began to consider myself a connoisseur of baguettes, and would say to myself with the authority of a champion *boulanger*, 'Now, this is a heavy one' or 'Definitely an average loaf today'. I was wrong as often as I was right. Yet my poor forecasting record did not diminish my belief that I was indeed an expert in baguette-assessing. It was, I reasoned, the same self-delusion displayed by sports and financial pundits who are equally unable to predict random events, and yet build careers out of it.

Perhaps the most disconcerting emotional reaction I was having to Greggs' baguettes was what happened when the weights were either extremely heavy or extremely light. On the rare occasions when I weighed a record high or a record low I was thrilled. The weight was extra special, which made the day seem extra special, as if the exceptionalness of the baguette would somehow be transferred to other aspects of my life. Rationally, I knew that it was inevitable that some baguettes would be oversized and some undersized. Still, the occurrence of an extreme weight gave me a high. It was alarming how easily my mood could be influenced by a stick of bread. I consider myself unsuperstitious and yet I was unable to avoid seeing meaning in random patterns. It was a powerful reminder of how susceptible we all are to unfounded beliefs.

Despite the promise of certainty that numbers provided the scientists of the Enlightenment, they were often not as certain as all that. Sometimes when the same thing was measured twice, it gave two different results. This was an awkward inconvenience for scientists aiming to find clear and direct explanations for natural phenomena. Galileo Galilei, for instance, noticed that when calculating distances of stars with his telescope, his results were prone to variation; and the variation was not due to a mistake in his calculations. Rather, it was because measuring was intrinsically fuzzy. Numbers were not as precise as they had hoped.

This was exactly what I was experiencing with my baguettes. There were probably many factors that contributed to the variance in weight – the amount and consistency of the flour used, the length of time in the oven, the journey of the baguettes from Greggs' central bakery to my local store, the humidity of the air and so on. Likewise, there were many variables affecting the results from Galileo's telescope – such as atmospheric conditions, the temperature of the equipment and personal details, like how tired Galileo was when he recorded the readings.

Still, Galileo was able to see that the variation in his results obeyed certain rules. Despite variation, data for each measurement tended to cluster around a central value, and small errors from this central value were more common than large errors. He also noticed that the spread was symmetrical too – a measurement was as likely to be less than the central value as it was to be more than the central value.

Likewise, my baguette data showed weights that were clustered around a value of about 400g, give or take 20g on either side. Even though none of my hundred baguettes weighed precisely 400g, there were a lot more baguettes weighing around 400g than there were ones weighing around 380g or 420g. The spread seemed pretty symmetrical too.

The first person to recognize the pattern produced by this kind of measurement error was the German mathematician Carl Friedrich Gauss. The pattern is described by the following curve, called the bell curve:

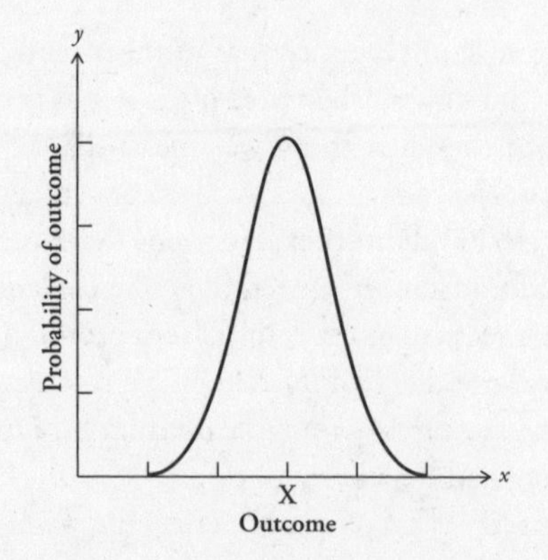

Gauss's graph needs some explaining. The horizontal axis describes a set of outcomes, for instance the weight of baguettes or the distance of stars. The vertical axis is the probability of those outcomes. A curve plotted on a graph with these parameters is known as a *distribution*. It shows us the spread of outcomes and how likely each is.

There are lots of different types of distribution, but the most basic type is described by the curve opposite. The bell curve is also known as the *normal distribution*, or the *Gaussian distribution*. Originally, it was known as the *curve of error*, although because of its distinctive shape, the term *bell curve* has become much more common. The bell curve has an average value, which I have marked X, called the *mean*. The mean is the most likely outcome. The further you go from the mean, the less likely the outcome will be.

When you take two measurements of the same thing and the process has been subject to random error you tend not to get the same result. Yet the more measurements you take, the more the distribution of outcomes begins to look like the bell curve. The outcomes cluster symmetrically around a mean value. Of course, a graph of measurements won't give you a continuous curve – it will give you (as we saw with my baguettes) a jagged landscape of fixed amounts. The bell curve is a theoretical ideal of the pattern produced by random error. The more data we have, the closer the jagged landscape of outcomes will fit the curve.

In the late nineteenth century the French mathematician Henri Poincaré knew that the distribution of an outcome that is subject to random measurement error will approximate the bell curve. Poincaré, in fact, conducted the same experiment with baguettes as I did, but for a different reason. He suspected that his local boulangerie was ripping him off by selling underweight loaves, so he decided to exercise mathematics in the interest of justice. Every day for a year he weighed his daily 1kg loaf. Poincaré knew that if the weight was less than 1kg a few times, this was not evidence of malpractice, since one would expect the weight to vary above and below the specified 1kg. And he conjectured that the graph of bread weights would resemble a normal distribution – since the errors in making the bread, such as how much flour is used and how long the loaf is baked for, are random.

After a year he looked at all the data he had collected. Sure enough, the distribution of weights approximated the bell curve. The peak of the curve, however, was at 950g. In other words, the average weight was 0.95kg, not 1kg as advertised. Poincaré's suspicions were confirmed. The eminent scientist was being diddled, by an average of 50g per loaf. According to popular legend, Poincaré alerted the Parisian authorities and the baker was given a stern warning.

After his small victory for consumer rights, Poincaré did not let it lie. He continued to measure his daily loaf, and after the second year he saw that the shape of the graph was not a proper bell curve; rather, it was skewed to the right. Since he knew that total randomness of error produces the bell curve, he deduced that some non-random event was affecting the loaves he was being sold. Poincaré concluded that the baker hadn't stopped his cheapskate, underbaking ways but instead was giving him the largest loaf at hand, thus introducing bias in the distribution. Unfortunately for the *boulanger*, his customer was the cleverest man in France. Again, Poincaré informed the police.

Poincaré's method of baker-baiting was prescient; it is now the theoretical basis of consumer protection. When shops sell products at specified weights, the product does not legally have to be that exact weight – it cannot be, since the process of manufacture will inevitably make some items a little heavier and some a little lighter. One of the jobs of trading-standards officers is to take random samples of products on sale and draw up graphs of their weight. For any product they measure, the distribution of weights must fall within a bell curve centred on the advertised mean.

Half a century before Poincaré saw the bell curve in bread, another mathematician was seeing it wherever he looked. Adolphe Quételet has good claim to being the world's most influential Belgian. (The fact that this is not a competitive field in no way diminishes his achievements.) A geometer and astronomer by training, he soon became sidetracked by a fascination with data – more specifically, with finding patterns in figures. In one of his early projects, Quételet examined French national crime statistics, which the government started publishing in 1825. Quételet noticed that the number of murders was pretty constant every year. Even the proportion of different types of murder weapon – whether it was perpetrated by a gun, a sword, a knife, a fist, and so on – stayed roughly the same.

Nowadays this observation is unremarkable – indeed, the way we run our public institutions relies on an appreciation of, for example, crime rates, exam pass rates and accident rates, which we expect to be comparable every year. Yet Quételet was the first person to notice the quite amazing regularity of social phenomena when populations are taken as a whole. In any one year it was impossible to tell who might become a murderer. Yet in any one year it was possible to predict fairly accurately how many murders would occur. Quételet was troubled by the deep questions about personal responsibility this pattern raised and, by extension, about the ethics of punishment. If society was like a machine that produced a regular number of murderers, didn't this indicate that murder was the fault of society and not the individual?

Quételet's ideas transformed the use of the word *statistics*, whose original meaning had little to do with numbers. The word was used to describe general facts about the state; as in the type of information required by statesmen. Quételet turned statistics into a much wider discipline, one that was less about statecraft and more about the mathematics of collective behaviour. He could not have done this without advances in probability theory, which provided techniques to analyse the randomness in data. In Brussels in 1853 Quételet hosted the first international conference on statistics.

Quételet's insights on collective behaviour reverberated in other sciences. If by looking at data from human populations you could detect reliable patterns, then it was only a small leap to realize that populations of, for example, atoms also behaved with predictable regularities. James Clerk Maxwell and Ludwig Boltzmann were indebted to Quételet's statistical thinking when they came up with the kinetic theory of gases, which explains that the pressure of a gas is determined by the collisions of its molecules travelling randomly at different velocities. Though the velocity of any individual molecule cannot be known, the molecules overall behave in a predictable way. The origin of the kinetic theory of gases is an interesting exception to the general rule that developments in the social sciences are the result of advances in the natural sciences. In this case, knowledge flowed in the other direction.

The most common pattern that Quételet found in all of his research was the bell curve. It was ubiquitous when studying data about human populations. Sets of data in those days were harder to come by than they are now, so Quételet scoured the world for them with the doggedness of a professional collector. For example, he came across a study published in the 1814 *Edinburgh Medical Journal* containing chest measurements of 5738 Scottish soldiers. Quételet drew up a graph of the numbers and showed that the distribution of chest sizes traced a bell curve with a mean of about 40 inches. From other sets of data he showed that the heights of men and women also plot a bell curve. To this day, the retail industry relies on Quételet's discoveries. The reason why clothes shops stock more mediums than they do smalls and larges is because the distribution of human sizes corresponds roughly to the bell curve. The most recent data on the shoe sizes of British adults, for example, throws up a very familiar shape:

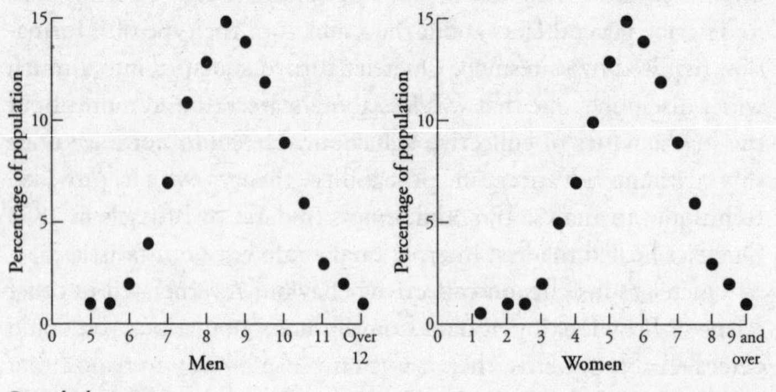

British shoe sizes.

Quételet died in 1874. A decade later, this side of the Channel, a 60-year-old man with a bald pate and fine Victorian whiskers could frequently be seen on the streets of Britain gawping at women and rummaging around in his pocket. This was Francis Galton, the eminent scientist, conducting fieldwork. He was measuring female attractiveness. In order to discreetly register his opinion on passing women he would prick a needle in his pocket into a cross-shaped piece of paper, to indicate whether she was 'attractive', 'indifferent' or 'repellent'. After completing his survey he compiled a map of the country based on looks. The highest-rated city was London and the lowest-rated was Aberdeen.

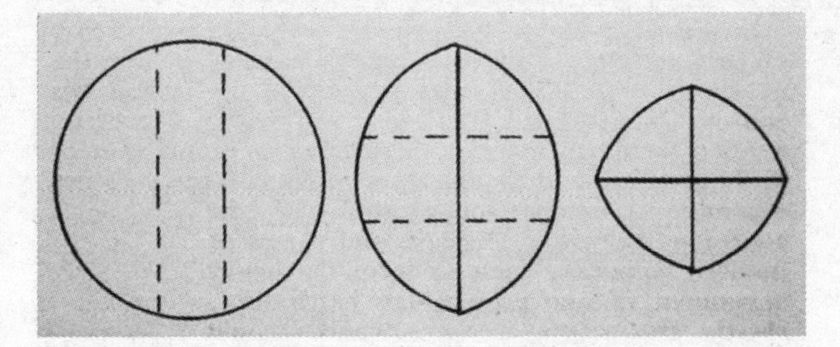

In 'Cutting a Round Cake on Scientific Principles' Galton marked intended cuts as broken straight lines, and cuts as solid lines. This method minimizes exposing the insides of the cake to become dry, which would happen if one cuts a slice in the traditional (and, he concludes, 'very faulty') way. In the second and third stages the cake is to be held together with an elastic band.

Galton was probably the only man in nineteenth-century Europe who was even more obsessed with gathering data than Quételet was. As a young scientist, Galton took the temperature of his daily pot of tea, together with such information as the volume of boiling water used and how delicious it tasted. His aim was to establish how to make the perfect cuppa. (He reached no conclusions.) In fact, an interest in the mathematics of afternoon tea was a lifelong passion. When he was an old man he sent the diagram above to the journal *Nature*, which shows his suggestion of the best way to cut a tea-cake in order to keep it as fresh as possible.

Oh, and since this is a book with the word 'number' in its title, it would be unsporting for me at this juncture not to mention Galton's 'number forms' – even if they have little to do with the subject of this chapter. Galton was fascinated that a substantial number of people – he estimated 5 per cent – automatically and involuntarily envisaged numbers as mental maps. He coined the term *number form* to describe these maps, and wrote that they have a 'precisely defined and constant position' and are such that individuals cannot think of a number 'without referring to its own particular habitat in their mental field of view'. What is especially interesting about number forms is that they generally show up very peculiar patterns. Instead of a straight line, which might be expected, they often involve rather peculiar twists and turns.

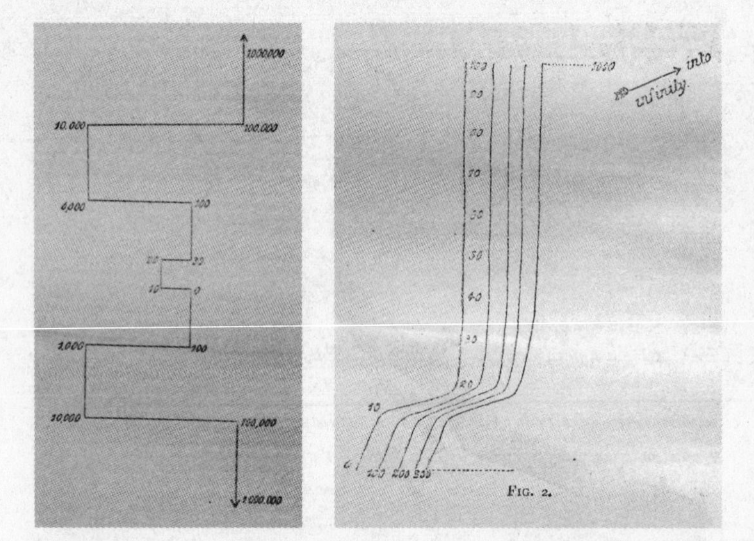

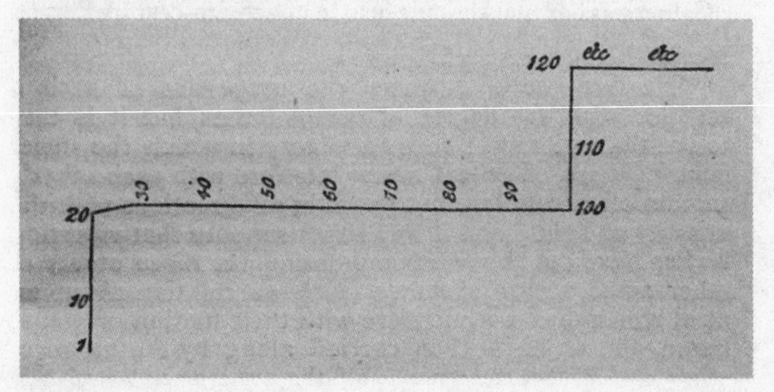

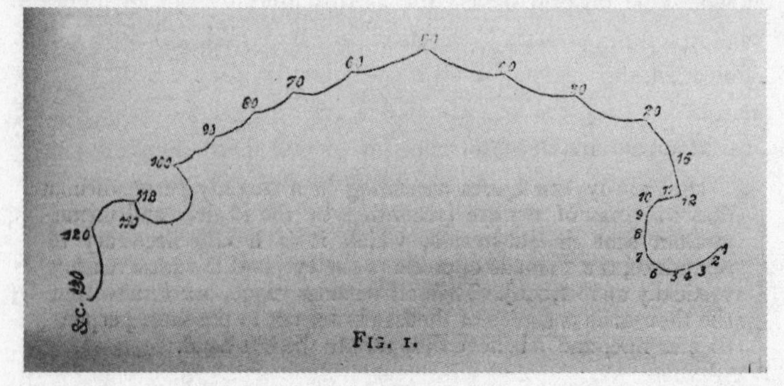

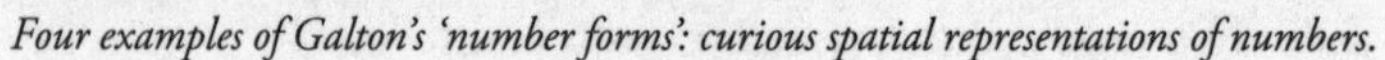
Four examples of Galton's 'number forms': curious spatial representations of numbers.

Number forms have the whiff of Victorian freakishness, perhaps evidence of repressed emotions or overindulgence in opiates. Yet a century later they are researched in academia, recognized as a type of synaesthesia, which is the neurological phenomenon that occurs when stimulation of one cognitive pathway leads to involuntary stimulation in another. In this case, numbers are given a location in space. Other types of synaesthesia include believing letters have colours, or that days of the week have personalities. Galton, in fact, underestimated the presence of number forms in humans. It is now thought that 12 per cent of us experience them in some way.

But Galton's principal passion was measuring. He built an 'anthropometric laboratory' – a drop-in centre in London, where members of the public could come to have their height, weight, strength of grip, swiftness of blow, eyesight and other physical attributes measured by him. The lab compiled details on more than 10,000 people, and Galton achieved such fame that Prime Minister William Gladstone even popped by to have his head measured. ('It was a beautifully shaped head, though low,' Galton said.) In fact, Galton was such a compulsive measurer that even when he had nothing obvious to measure he would find something to satisfy his craving. In an article in *Nature* in 1885 he wrote that while present at a tedious meeting he had begun to measure the frequency of fidgets made by his colleagues. He suggested that scientists should henceforth take advantage of boring meetings so that 'they may acquire the new art of giving numerical expression to the amount of boredom expressed by [an] audience'.

Galton's research corroborated Quételet's in that it showed that the variation in human populations was rigidly determined. He too saw the bell curve everywhere. In fact, the frequency of the appearance of the bell curve led Galton to pioneer the word 'normal' as the appropriate name for the distribution. The circumference of a human head and the size of the brain all produced bell curves, though Galton was especially interested in non-physical attributes such as intelligence. IQ tests hadn't been invented at the time, so Galton looked for other measures of intelligence. He found them in the results of the admission exams to the Royal Military Academy at Sandhurst. The exam scores, he discovered, also conformed to the bell curve. It filled him with a sense of awe. 'I know of scarcely anything so apt to impress the imagination as the wonderful form of cosmic order expressed by the [bell curve],' he wrote. 'The law would have been personified by the Greeks and deified, if they had known of it. It reigns with serenity and in complete self-effacement amidst the wildest confusion. The huger the mob, and the greater the apparent anarchy, the more perfect is its sway. It is the supreme law of unreason.'

Galton invented a beautifully simple machine that explains the mathematics behind his cherished curve, and he called it the quincunx. The word's original meaning is the ⁙ pattern of five dots on a die, and the contraption is a type of pinball machine in which each horizontal line of pins is offset by half a position from the line above. A ball is dropped into the top of the quincunx, and then it bounces between the pins until it falls out the bottom into a rack of columns. After many balls have been dropped in, the shape they make in the columns where they have naturally fallen resembles a bell curve.

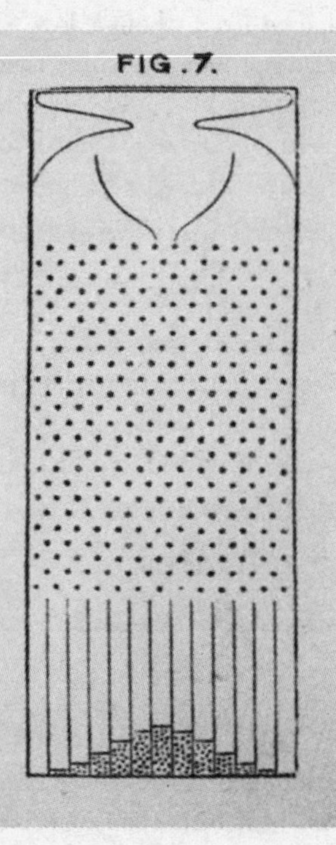

The quincunx.

Using probability, we can understand what is going on. First, imagine a quincunx with just one pin and let us say that when a ball hits the pin the outcome is random, with a 50 per cent chance that it bounces to the left and a 50 per cent chance of it bouncing to the right. In other words, it has a probability of $\frac{1}{2}$ of ending up one place to the left and a probability of $\frac{1}{2}$ of being one place to the right.

Now, let's add a second row of pins. The ball will either fall left and then left, which I will call LL, or LR or RL or RR. Since moving left and then right is equivalent to staying in the same position, the L and R together cancel each other out (as does the R and L together), so there is now $\frac{1}{4}$ of a chance the ball will end up one place to the left, $\frac{2}{4}$ chance it will be in the middle and $\frac{1}{4}$ it will be to the right.

Repeating this for a third row of pins, the equally probable options of where the ball will fall are LLL, LLR, LRL, LRR, RRR, RRL, RLR, RLL. This gives us probabilities of $\frac{1}{8}$ of landing on the far left, $\frac{3}{8}$ of landing on the near left, $\frac{3}{8}$ of landing on the near right and $\frac{1}{8}$ of landing on the far right.

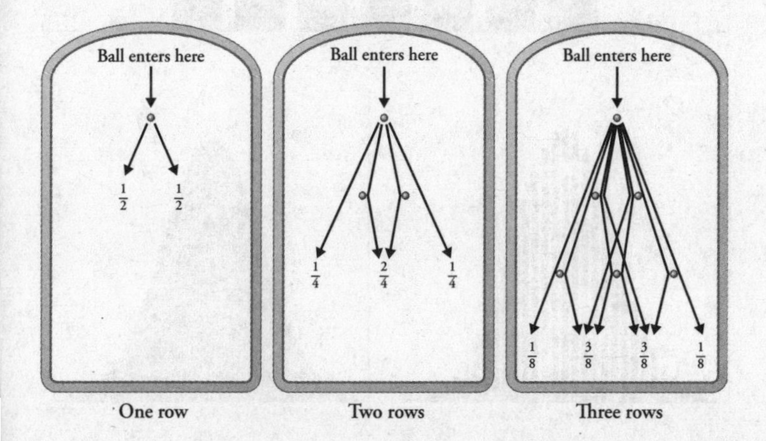

In other words, if there are two rows of pins in the quincunx and we introduce lots of balls into the machine, the law of large numbers says that the balls will fall along the bottom such as to approximate the ratio 1:2:1.

If there are three rows, they will fall in the ratio 1:3:3:1.

If there are four rows, they will fall in the ratio 1:4:6:4:1.

If I carried on working out probabilities, a ten-row quincunx will produce balls falling in the ratio 1:10:45:120:210:252:210:120:45:10:1.

Plotting these numbers gives us the first of the shapes below. The shape becomes even more familiar the more rows we include. Also below are the results for 100 and 1000 rows as bar charts. (Note that only the middle sections of these two charts are shown since the values to the left and right are too small to see.)

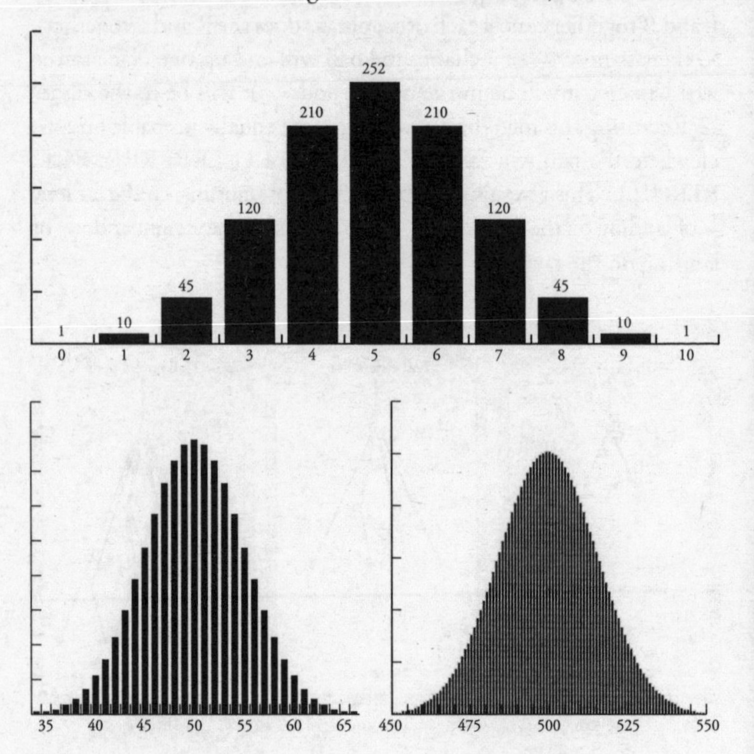

So how does this pinball game relate to what goes on in the real world? Imagine that each row of the quincunx is a random variable that will create an error in measurement. Either it will add a small amount to the correct measurement or it will subtract a small amount. In the case of Galileo and his telescope, one row of pins could represent the temperature of the equipment, another could represent whether there is a thermal front passing through, and another could represent the pollution in the air. Each variable contributes an error either one way or the other, just as in the quincunx the ball will bounce left or right. In any measurement there may be many millions of unobservable random errors – their combined errors, however, will give measurements that are distributed like a bell curve.

If the characteristics of a population are normally distributed, in other words are clustered around an average in the shape of a bell curve, and if the bell curve is produced through random error, then, Quételet argued, the variation in human characteristics can be seen as errors from a paradigm. He called this paradigm *l'homme moyen*, or 'the average man'. Populations, he said, were made up of deviations from this prototype. In Quételet's mind, being average was something to aspire to since it was a way of keeping society in check – deviations from the average, he wrote, led to 'ugliness in body as well as vice in morals'. Even though the concept of *l'homme moyen* never gained acceptance in science, its use filtered down to society at large. We often talk about morality or taste in terms of what an average representative of a population might think or feel about it: such as what is seen as acceptable 'in the eyes of the average man'.

Whereas Quételet extolled averageness, Galton looked down on it. Galton, as I mentioned before, saw that exam results were normally distributed. Most people scored about average, while a few got very high marks and a few very low.

Galton, incidentally, was himself from a very above-average family. His first cousin was Charles Darwin, and the two men corresponded regularly about their scientific ideas. About a decade after Darwin published *On the Origin of Species*, which set out the theory of natural selection, Galton started to theorize on how human evolution itself could be guided. He was interested in the heritability of cleverness and wondered how it might be possible to improve the overall intelligence of a population. He wanted to shift the bell curve to the right. To this end Galton suggested a new field of study about the 'cultivation of race', or improving the intellectual stock of a population through breeding. He had thought to call his new science *viticulture*, from the Latin *vita*, 'life', but eventually settled on *eugenics*, from the Greek *eu*, good, and *genos*, birth. (The usual meaning of 'viticulture', grape cultivation, comes from *vitis*, Latin for 'vine', and dates from around the same time.) Even though many liberal intellectuals of the late nineteenth and early twentieth centuries supported eugenics as a way to improve society, the desire to 'breed' cleverer humans was an idea that was soon distorted and discredited. In the 1930s eugenics became synonymous with murderous Nazi policies to create a superior Aryan race.

In retrospect, it is easy to see how ranking traits – such as intelligence or racial purity – can lead to discrimination and bigotry. Since the bell curve appears when human features are measured, the curve has become synonymous with attempts to classify some humans as intrinsically better than others. The highest-profile example of this was the publication in 1994 of *The Bell Curve* by Richard J. Herrnstein and Charles Murray, one of the most fiercely debated books of recent years. The book, which owes its name to the distribution of IQ scores, argues that IQ differences between racial groups are evidence of biological differences. Galton wrote that the bell curve reigned with 'serenity and in complete self-effacement'. Its legacy, though, has been anything but.

Another way to appreciate the lines of numbers produced by the quincunx is to lay them out like a pyramid. In this form, the results are better known as Pascal's triangle.

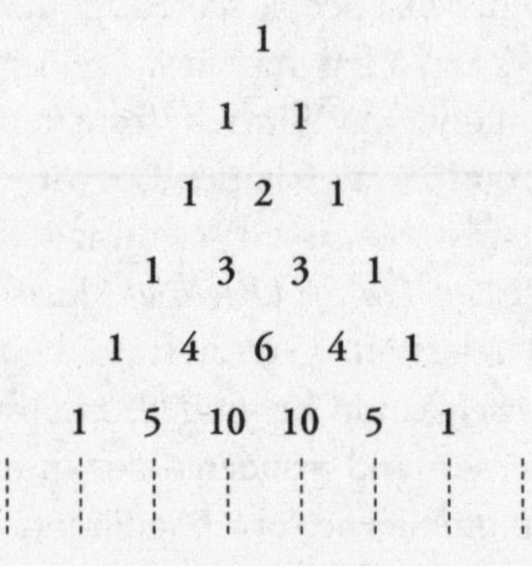

Pascal's can be constructed much more simply than by working out the distributions of randomly falling balls through a Victorian bean machine. Start with a 1 in the first row, and under it place two 1s so as to make a triangle shape. Continue with subsequent rows, always

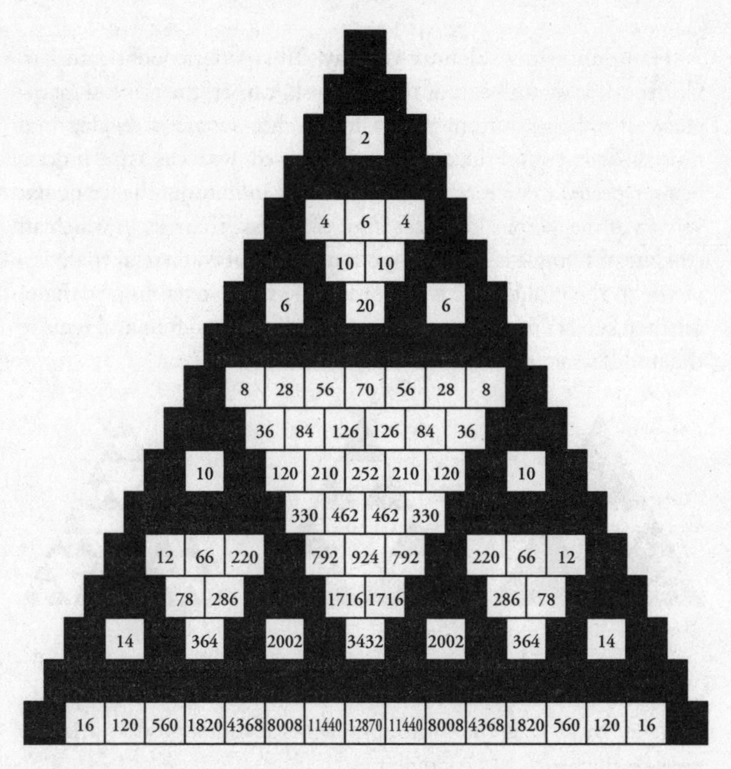

Pascal's triangle with only squares divisible by 2 in white.

placing a 1 at the beginning and end of the rows. The value of every other position is the *sum* of the two numbers above it.

The triangle is named after Blaise Pascal, even though he was a latecomer to its charms. Indian, Chinese and Persian mathematicians were all aware of the pattern centuries before he was. Unlike its prior fans, though, Pascal wrote a book about what he called *le triangle arithmétique*. He was fascinated by the mathematical richness of the patterns he discovered. 'It is a strange thing how fertile it is in properties,' he wrote, adding that in his book he had to leave out more than he could put in.

My favourite feature of Pascal's triangle is the following. Let each number have its own square, and colour all the odd number squares black. Keep all the even-number squares white. The result is the wonderful mosaic above.

Hang on minute, I hear you say. This pattern looks familiar. Correct. It is reminiscent of the Sierpinski carpet, the piece of mathematical upholstery from p. 105 in which a square is divided into nine subsquares and the central one removed, with the same process being repeated to each of the subsquares ad infinitum. The triangular version of the Sierpinski carpet is the Sierpinski triangle, in which an equilateral triangle is divided into four identical equilateral triangles, of which the middle one is removed. The three remaining triangles are then subject to the same operation – divide into four and remove the middle one. Here are the first three iterations:

If we extend the method of colouring Pascal's triangle to more and more lines, the pattern looks more and more like the Sierpinski triangle. In fact, as the limit approaches infinity, Pascal's triangle *becomes* the Sierpinski triangle.

Sierpinski is not the only old friend we find in these black-and-white tiles. Consider the size of the white triangles down the centre of Pascal's triangle. The first is made up of 1 square, the second is made up of 6 squares, the third is made of 28, and the next ones have 120 and 496 squares. Do these numbers ring any bells? Three of them – 6, 28 and 496 – are perfect numbers, from p. 265. The occurrence is a remarkable visual expression of a seemingly unrelated abstract idea.

Let's continue painting Pascal's triangle by numbers. First keep all numbers divisible by 3 as white, and make the rest black. Then repeat the process with the numbers that are divisible by 4. Repeat again with numbers divisible by 5. The results, shown opposite, are all symmetrical patterns of triangles pointing in the opposite direction to the whole.

In the nineteenth century, another familiar face was discovered in Pascal's triangle: the Fibonacci sequence. Perhaps this was inevitable,

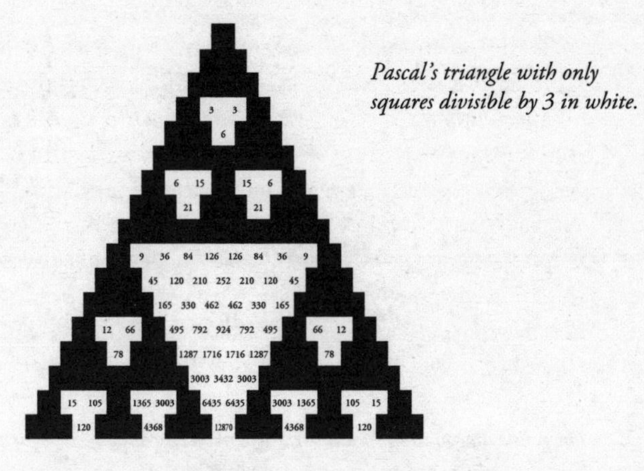

Pascal's triangle with only squares divisible by 3 in white.

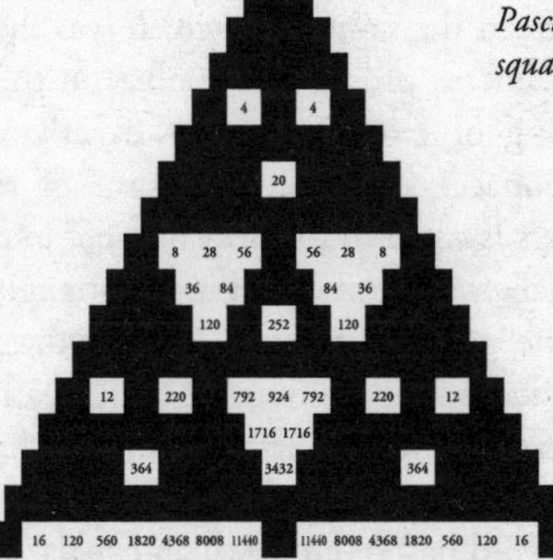

Pascal's triangle with only squares divisible by 4 in white.

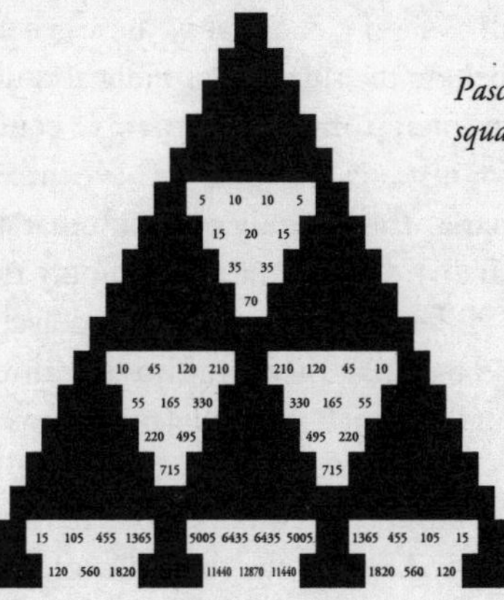

Pascal's triangle with only squares divisible by 5 in white.

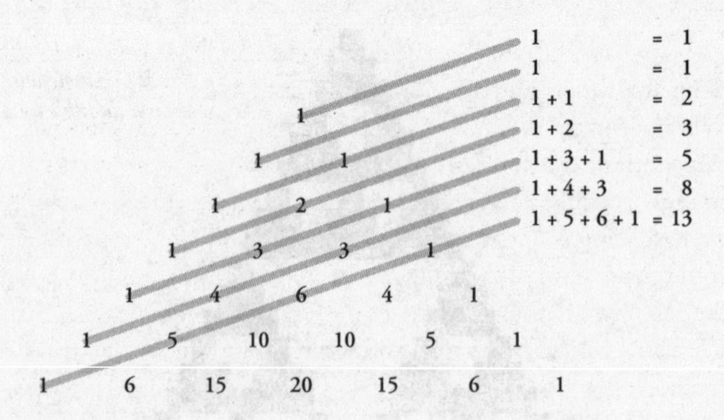

The gentle diagonals in Pascal's triangle reveal the Fibonacci sequence.

as the method of constructing of the triangle was recursive – we repeatedly performed the same rule, which was the adding of two numbers on one line to produce a number on the next line. The recursive summing of two numbers is exactly what we do to produce the Fibonacci sequence. The sum of two consecutive Fibonacci numbers is equal to the next number in the sequence.

Fibonacci numbers are embedded in the triangle as the sums of what are called the 'gentle' diagonals. A gentle diagonal is one that moves from any number to the number underneath to the left and then along one space to the left, or above and to the right and then along one space to the right. The first and second diagonals consist simply of 1. The third has 1 and 1, which equals 2. The fourth has 1 and 2, which adds up to 3. The fifth gentle diagonal gives us 1 + 3 + 1 = 5. The sixth is 1 + 4 + 3 = 8. So far we have generated 1, 1, 2, 3, 5, 8, and the next ones are the subsequent Fibonacci numbers in order.

Ancient Indian interest in Pascal's triangle concerned combinations of objects. For instance, imagine we have three fruits: a mango, a lychee and a banana. There is only one combination of three items: mango, lychee, banana. If we want to select only two fruits, we can do this in three different ways: mango and lychee, mango and banana, lychee and banana. There are also only three ways of taking the fruit individually, which is each fruit on its own. The final option is to select zero fruit, and this can happen in only one way. In other words, the number of combinations of three different fruits produces the string 1, 3, 3, 1 – the third line of Pascal's triangle.

If we had four objects, the number of combinations when taken none-at-a-time, individually, two-at-a-time, three-at-a-time and four-at-a-time is 1, 4, 6, 4, 1 – the fourth line of Pascal's triangle. We can continue this for more and more objects and we see that Pascal's triangle is really a reference table for the arrangement of things. If we had *n* items and wanted to know how many combinations we could make of *m* of them, the answer is exactly the *m*th position in the *n*th row of Pascal's triangle. (Note: by convention, the leftmost 1 of any row is taken as the zeroth position in the row.) For example, how many ways are there of grouping three fruits from a selection of seven fruits? There are 35 ways, since the third position on row seven is 35.

Now let's move on to start combining mathematical objects. Consider the term $x + y$. What is $(x + y)^2$? It is the same as $(x + y)(x + y)$. To expand this, we need to multiply each term in the first bracket by each term in the second. So, we get $xx + xy + yx + yy$, or $x^2 + 2xy + y^2$. Spot something here? If we carry on, we can see the pattern more clearly. The coefficients of the individual terms are the rows of Pascal's triangle.

$$(x + y)^2 = x^2 + 2xy + y^2$$
$$(x + y)^3 = x^3 + 3x^2y + 3xy^2 + y^3$$
$$(x + y)^4 = x^4 + 4x^3y + 6x^2y^2 + 4xy^3 + y^4$$

The mathematician Abraham de Moivre, a Huguenot refugee living in London in the early eighteenth century, was the first to understand that the coefficients of these equations will approximate a curve the more times you multiply $(x + y)$ together. He didn't call it the bell curve, or the curve of error, or the normal distribution, or the Gaussian distribution, which are the names that it later acquired. The curve made its first appearance in mathematics literature in de Moivre's 1718 book on gaming called *The Doctrine of Chances*. This was the first textbook on probability theory, and another example of how scientific knowledge flourished thanks to gambling.

I've been treating the bell curve as if it is one curve, when, in fact, it is a family of curves. They all look like a bell, but some are wider than others (see diagram overleaf).

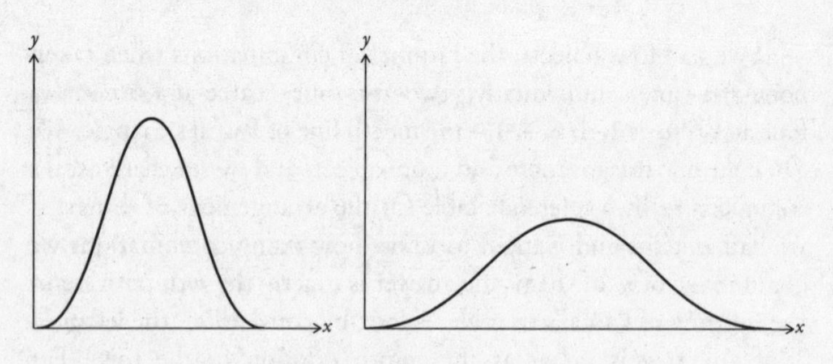

Bell curves with different deviations.

Here's an explanation for why we get different widths. If Galileo, for example, measured planetary orbits with a twenty-first-century telescope, the margin of error would be less than if he were using his sixteenth-century one. The modern instrument would produce a much thinner bell curve than the antique one. The errors would be much smaller, yet they would still be distributed normally.

The average value of a bell curve is called the mean. The width is called the *deviation*. If we know the mean and the deviation, then we know the shape of the curve. It is incredibly convenient that the normal curve can be described using only two parameters. Perhaps, though, it is too convenient. Often statisticians are overly eager to find the bell curve in their data. Bill Robinson, an economist who heads KPMG's forensic-accounting division, admits this is the case. 'We love to work with normal distributions because [the normal distribution] has mathematical properties that have been very well explored. Once we know it's a normal distribution, we can start to make all sorts of interesting statements.'

Robinson's job, in basic terms, is to deduce, by looking for patterns in huge data sets, whether someone has been cooking the books. He is carrying out the same strategy that Poincaré used when he weighed his loaves every day, except that Robinson is looking at gigabytes of financial data, and has much more sophisticated statistical tools at his disposal.

Robinson said that his department tends to work on the assumption that for any set of data the default distribution is the normal distribution. 'We like to assume that the normal curve operates because then we are in the light. Actually, sometimes it doesn't, and sometimes we probably should be looking in the dark. I think in the financial markets it is true that we have assumed a normal distribution when perhaps it doesn't work.' In recent years, in fact, there has been a backlash in both academia and finance against the historic reliance on the normal distribution.

When a distribution is less concentrated around the mean than the bell curve it is called *platykurtic*, from the Greek words *platus*, meaning 'flat', and *kurtos*, 'bulging'. Conversely, when a distribution is more concentrated around the mean it is called *leptokurtic*, from the Greek *leptos*, meaning 'thin'. William Sealy Gosset, a statistician who worked for the Guinness brewery in Dublin, drew the aide-memoire below in 1908 to remember which was which: a duck-billed platypus was platykurtic, and the kissing kangaroos were leptokurtic. He chose kangaroos because they are 'noted for "lepping", though, perhaps, with equal reason they should be hares!' Gosset's sketches are the origin of the term *tail* for describing the far-left and far-right sections of a distribution curve.

When economists talk of distributions that are *fat-tailed* or *heavy-tailed*, they are talking of curves that stay higher than normal from the axis at the extremes, as if Gosset's animals have larger than average tails. These curves describe distributions in which extreme events are more likely than if the distribution were normal. For instance, if the variation in the price of a share were fat-tailed, it would mean there was more of a chance of a dramatic drop, or hike, in price than if the variation were normally distributed. For this reason, it can sometimes be reckless to assume a bell curve over a fat-tailed curve. The economist Nassim Nicholas Taleb's position in his bestselling book *The Black Swan* is that we have tended to underestimate the size and importance of the tails in distribution curves.

Platykurtic and leptokurtic distributions.

He argues that the bell curve is a historically defective model because it cannot anticipate the occurrence of, or predict the impact of, very rare, extreme events – such as a major scientific discovery like the invention of the internet, or of a terrorist attack like 9/11. 'The ubiquity of the [normal distribution] is not a property of the world,' he writes, 'but a problem in our minds, stemming from the way we look at it.'

The desire to see the bell curve in data is perhaps most strongly felt in education. The awarding of grades from A to E in end-of-year exams is based on where a pupil's score falls on a bell curve to which the distribution of grades is expected to approximate. The curve is divided into sections, with A representing the top section, B the next section down, and so on. For the education system to run smoothly, it is important that the percentage of pupils getting grades A to E each year is comparable. If there are too many As, or too many Es, in one particular year the consequences – not enough, or too many, people on certain courses – would be a strain on resources. Exams are specifically designed in the hope that the distribution of results replicates the bell curve as much as possible – irrespective of whether or not this is an accurate reflection of real intelligence. (It might be as a whole, but is probably not in all cases.)

It has even been argued that the reverence some scientists have for the bell curve actively encourages sloppy practices. We saw from the quincunx that random errors are distributed normally. So, the more random errors we can introduce into measurement, the more likely it is that we will get a bell curve from the data – even if the phenomenon being measured is not normally distributed. When the normal distribution is found in a set of data, this could simply be because the measurements have been gathered too shambolically.

Which brings me back to my baguettes. Were their weights really normally distributed? Was the tail thin or fat? First, a recap. I weighed 100 baguettes. The distribution of their weights was shown on p. 350. The graph showed some hopeful trends – there was a mean of somewhere around 400g, and a more or less symmetrical spread between 380 and 420g. If I had been as indefatigable as Henri Poincaré, I would have continued the experiment for a year and had 365 (give or take days of bakery closure) weights to compare. With more data, the distribution would have been clearer. Still, my smaller sample was enough to get an idea of the pattern forming. I used a trick, compressing my results by redrawing the graph with a scale that grouped baguette weights in bounds of 8g rather than 1g. This created the following graph:

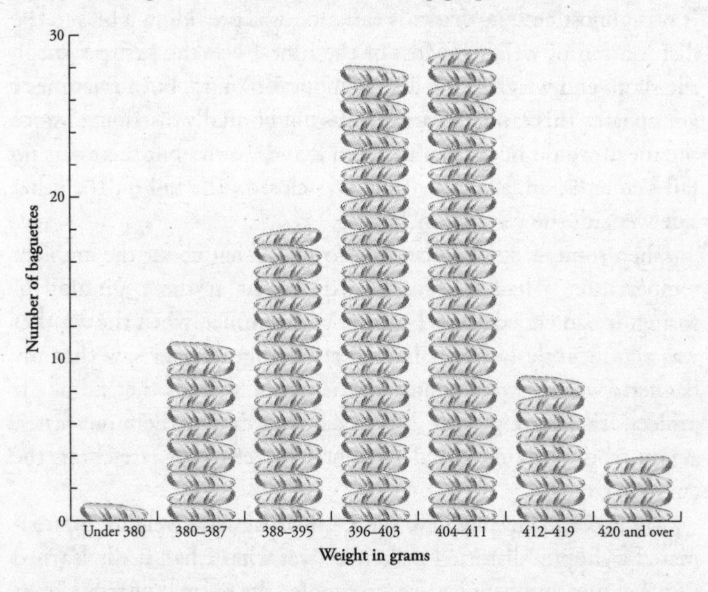

When I first drew this out I felt relief, as it really looked like my baguette experiment was producing a bell curve. My facts appeared to be fitting the theory. A triumph for applied science! But when I looked closer, the graph wasn't really like the bell curve at all. Yes, the weights were clustered around a mean, but the curve was clearly not symmetrical. The left side of the curve was not as steep as the right side. It was as if there was an invisible magnet stretching the curve a little to the left.

I could therefore conclude one of two things. Either the weights of Greggs' baguettes were not normally distributed, or they were normally distributed but some bias had crept in to my experimentation process. I had an idea of what the bias might be. I had been storing the uneaten baguettes in my kitchen, and I decided to weigh one that was a few days old. To my surprise it was only 321g

– significantly lower than the lowest weight I had measured. It dawned on me then that baguette weight was not fixed because bread gets lighter as it dries out. I bought another loaf and discovered that a baguette loses about 15g between 8 a.m. and noon.

It was now clear that my experiment was flawed. I had not taken into account the hour of the day when I took my measurements. It was almost certain that this variation was providing a bias to the distribution of weights. Most of the time I was the first person in the shop, and weighed my loaf at about 8.10 a.m., but sometimes I got up late. This random variable was not normally distributed since the mean would have been between 8 and 9 a.m., but there was no tail before 8 a.m. since the shop was closed. The tail on the other side went all the way to lunchtime.

Then something else occurred to me. What about the ambient temperature? I had started my experiment at the beginning of spring. It had ended at the beginning of summer, when the weather was significantly hotter. I looked at the figures and saw that my baguette weights were lighter on the whole towards the end of the project. The summer heat, I assumed, was drying them out faster. Again, this variation could have had the effect of stretching the curve leftwards.

My experiment may have shown that baguette weights approximated a slightly distorted bell curve, yet what I had really learned was that measurement is never so simple. The normal distribution is a theoretical ideal, and one cannot assume that all results will conform to it. I wondered about Henri Poincaré. When he measured his bread did he eliminate bias due to the Parisian weather, or the time of day of his measurements? Perhaps he had not demonstrated that he was being sold a 950g loaf instead of a 1kg loaf at all, but had instead proved that from baking to measuring, a 1kg loaf reduces in weight by 50g.

The history of the bell curve, in fact, is a wonderful parable about the curious kinship between pure and applied scientists. Poincaré once received a letter from the French physicist Gabriel Lippmann, who brilliantly summed up why the normal distribution was so widely exalted: 'Everybody believes in the [bell curve]: the experimenters because they think it can be proved by mathematics; and the mathematicians because they believe it has been established by observation.' In science, as in so many other spheres, we often choose to see what serves our interests.

Francis Galton devoted himself to science and exploration in the way that only a man in possession of a large fortune can do. His early adulthood was spent leading expeditions to barely known parts of Africa, which brought him considerable fame. A masterful dexterity with scientific instruments enabled him, on one occasion, to measure the figure of a particularly buxom Hottentot by standing at a distance and using his sextant. This incident, it seems, was indicative of a desire to keep women at arm's length. When a tribal chief later presented him with a young woman smeared in butter and red ochre in preparation for sex – Galton declined the offer, concerned she would smudge his white linen suit.

Eugenics was Galton's most infamous scientific legacy, yet it was not his most enduring innovation. He was the first person to use questionnaires as a method of psychological testing. He devised a classification system for fingerprints, still in use today, which led to their adoption as a tool in police investigations. And he thought up a way of illustrating the weather, which when it appeared in *The Times* in 1875 was the first public weather map to be published.

That same year, Galton decided to recruit some of his friends for an experiment with sweet peas. He distributed seeds among seven of them, asking them to plant the seeds and return the offspring. Galton measured the baby seeds and compared their diameters to those of their parents. He noticed a phenomenon that initially seems counter-intuitive: the large seeds tended to produce smaller offspring, and the small seeds tended to produce larger offspring. A decade later he analysed data from his anthropometric laboratory and recognized the same pattern with human heights. After measuring 205 pairs of parents and their 928 adult children, he saw that exceptionally tall parents had kids who were generally shorter than they were, while exceptionally short parents had children who were generally taller than their parents.

After reflecting upon this, we can understand why it must be the case. If very tall parents always produced even taller children, and if very short parents always produced even shorter ones, we would by now have turned into a race of giants and midgets. Yet this hasn't happened. Human populations may be getting taller as a whole – due to better nutrition and public health – but the distribution of heights within the population is still contained.

Galton called this phenomenon 'regression towards mediocrity in hereditary stature'. The concept is now more generally known as *regression to the mean*. In a mathematical context, regression to the mean is the statement that an extreme event is likely to be followed by a less extreme event. For example, when I measured a Greggs baguette and got 380g, a very low weight, it was very likely that the next baguette would weigh more than 380g. Likewise, after finding a 420g baguette, it was very likely that the following baguette would weigh less than 420g. The quincunx gives us a visual representation of the mechanics of regression. If a ball is put in at the top and then falls to the furthest position on the left, then the next ball dropped will probably land closer to the middle position – because most of the balls dropped will land in the middle positions.

Variation in human height through generations, however, follows a different pattern from variation in baguette weight through the week or variation in where a quincunx ball will land. We know from experience that families with above-average-sized parents tend to have above-average-sized kids. We also know that the shortest guy in the class probably comes from a family with adults of correspondingly diminutive stature. In other words, the height of a child is not totally random in relation to the height of his parents. On the other hand, the weight of a baguette on a Tuesday probably *is* random in relation to the weight of a baguette on a Monday. The position of one ball in a quincunx is (for all practical purposes) random in relation to any other ball dropped.

In order to understand the strength of association between parental height and child height, Galton came up with another idea. He plotted a graph with parental height along one axis and child height along the other, and then drew a straight line through the points that best fitted their spread. (Each set of parents was represented by the height midway between mother and father – which he called the 'mid-parent'). The line had a gradient of $\frac{2}{3}$. In other words for every inch taller than the average that the mid-parent was, the child would only be $\frac{2}{3}$ of an inch taller than the average. For every inch shorter than the average the mid-parent was, the child would only be $\frac{2}{3}$ of an inch shorter than the average. Galton called the gradient of the line the *coefficient of correlation*. The coefficient is a number that determines how strongly two sets of variables are related. Correlation was more fully developed by Galton's protégé Karl Pearson, who in 1911 set up the world's first university statistics department, at University College London.

Regression and correlation were major breakthroughs in scientific thought. For Isaac Newton and his peers, the universe obeyed deterministic laws of cause and effect. Everything that happened had a reason. Yet not all science is so reductive. In biology, for example, certain outcomes – such as the occurrence of lung cancer – can have multiple causes that mix together in a complicated way. Correlation provided a way to analyse the fuzzy relationships between linked sets of data. For example, not everyone who smokes will develop lung cancer, but by looking at the incidence of smoking and the incidence of lung cancer mathematicians can work out your chances of getting cancer if you do smoke. Likewise, not every child from a big class in school will perform less well than a child from a small class, yet class sizes do have an impact on exam results. Statistical analysis opened up whole new areas of research – in subjects from medicine to sociology, from psychology to economics. It allowed us to make use of information without knowing exact causes. Galton's original insights helped make statistics a respectable field: 'Some people hate the very name of statistics, but I find them full of beauty and interest,' he wrote. 'Whenever they are not brutalized, but delicately handled by the higher methods, and are warily interpreted, their power of dealing with complicated phenomena is extraordinary.'

In 2002 the Nobel Prize in Economics was not won by an economist. It was won by the psychologist Daniel Kahneman, who had spent his career (much of it together with his colleague Amos Tversky) studying the cognitive factors behind decision-making. Kahneman has said that understanding regression to the mean led to his most satisfying 'Eureka moment'. It was in the mid 1960s and Kahneman was giving a lecture to Israeli air-force flight instructors. He was telling them that praise is more effective than punishment for making cadets learn. On finishing his speech, one of the most experienced instructors stood up and told Kahneman that he was mistaken. The man said: 'On many occasions I have praised flight cadets for clean execution of some aerobatic manœuvre, and in general when they try it again, they do worse. On the other hand, I have often screamed at cadets for bad execution, and in general they do better the next time. So please don't tell us that reinforcement works and punishment does not, because the opposite is the case.' At that moment, Kahneman said, the penny dropped. The flight instructor's opinion that punishment is more effective than reward was based on a lack of understanding of regression to the mean. If a cadet does an extremely bad manœuvre, then of course he will do better next time – irrespective of whether the instructor admonishes or praises him. Likewise, if he does an extremely good one, he will probably follow that with something less good. 'Because we tend to reward others when they do well and punish them when they do badly, and because there is regression to the mean, it is part of the human condition that we are statistically punished for rewarding others and rewarded for punishing them,' Kahneman said.

Regression to the mean is not a complicated idea. All it says is that if the outcome of an event is determined at least in part by random factors, then an extreme event will probably be followed by one that is less extreme. Yet despite its simplicity, regression is not appreciated by most people. I would say, in fact, that regression is one of the least grasped but most useful mathematical concepts you need for a rational understanding of the world. A surprisingly large number of simple misconceptions about science and statistics boil down to a failure to take regression to the mean into account.

Take the example of speed cameras. If several accidents happen on the same stretch of road, this could be because there is one cause – for example, a gang of teenage pranksters have tied a wire across the road. Arrest the teenagers and the accidents will stop. Or there could be many random contributing factors – a mixture of adverse weather conditions, the shape of the road, the victory of the local football team or the decision of a local resident to walk his dog. Accidents are equivalent to an extreme event. And after an extreme event, the likelihood is of less extreme events occurring: the random factors will combine in such a way as to result in fewer accidents. Often speed cameras are installed at spots where there have been one or more serious accidents. Their purpose is to make drivers go more slowly so as to reduce the number of crashes. Yes, the number of accidents tends to be reduced after speed cameras have been introduced, but this might have very little to do with the speed camera. Because of regression to the mean, whether or not one is installed, after a run of accidents it is already likely that there will be fewer accidents at that spot. (This is not an argument against speed cameras, since they may indeed be effective. Rather it is an argument about the argument for speed cameras, which often displays a misuse of statistics.)

My favourite example of regression to the mean is the 'curse of *Sports Illustrated*', a bizarre phenomenon by which sportsmen suffer a marked drop in form immediately after appearing on the cover of America's top sports magazine. The curse is as old as the first issue. In August 1954 baseball player Eddie Mathews was on the cover after he had led his team, the Milwaukee Braves, to a nine-game winning streak. Yet as soon as the issue was on the news-stands, the team lost. A week later Mathews picked up an injury that forced him to miss seven games. The curse struck most famously in 1957, when the magazine splashed on the headline 'Why Oklahoma is Unbeatable' after the Oklahoma football team had not lost in 47 games. Yet sure enough, on the Saturday after publication, Oklahoma lost 7–0 to Notre Dame.

One explanation for the curse of *Sports Illustrated* is the psychological pressure of being on the cover. The athlete or team becomes more prominent in the public eye, held up as the one to beat. It might be true in some cases that the pressure of being a favourite is detrimental to performance. Yet most of the time the curse of *Sports Illustrated* is simply an illustration of regression to the mean. For someone to have earned their place on the cover of the magazine, they will usually be on top form. They might have had an exceptional season, or just won a championship or broken a record. Sporting performance is due to talent, but it is also reliant on many random factors, such as whether your opponents have the flu, whether you get a puncture, or whether the sun is in your eyes. A best-ever result is comparable to an extreme event, and regression to the mean says that after an extreme event the likelihood is of one less extreme.

Of course, there are exceptions. Some athletes are so much better than the competition that random factors have little sway on their performances. They can be unlucky and still win. Yet we tend to underestimate the contribution of randomness to sporting success. In the 1980s statisticians started to analyse scoring patterns of basketball players. They were stunned to find that it was completely random whether or not a particular player made or missed a shot. Of course, some players were better than others. Consider player A, who scores 50 per cent of his shots, on average; in other words, he has an equal chance of scoring or missing. Researchers discovered that the sequence of baskets and misses made by player A appeared to be totally random. In other words, instead of shooting he might as well have flipped a coin.

Consider player B, who has a 60 per cent chance of scoring and a 40 per cent chance of missing. Again, the sequence of baskets was random, as if the player was flipping a coin biased 60–40 instead of actually throwing the ball. When a player makes a run of baskets pundits will eulogize him for playing well, and when he makes a run of misses he will be criticized for having an off day. Yet making or missing a basket in one shot has no effect on whether he will make or miss it on the next shot. Each shot is as random as the flip of a coin. Player B can be genuinely praised for having a 60–40 score ratio on average over many games, but praising him for any sequence of five baskets in a row is no different from praising the talent of a coin flipper who gets five consecutive heads. In both cases, they had a lucky streak. It is also possible – if not entirely probable – that player A, who is not as good overall at making baskets as player B, might have a longer run of successful shots in a match. This does not mean he is a better player. It is randomness giving A a lucky streak and B an unlucky one.

More recently, Simon Kuper and Stefan Szymanski looked at the 400 games the England football team has played since 1980. They write, in *Why England Lose*: 'England's win sequence ... is indistinguishable from a random series of coin tosses. There is no predictive value in the outcome of England's last game, or indeed in any combination of England's recent games. Whatever happened in one match appears to have no bearing on what will happen in the next one. The only thing you can predict is that over the medium to long term, England will win about half its games outright.'

The ups and downs of sporting performance are often explained by randomness. After a very big up you might get a call from *Sports Illustrated*. And you are almost guaranteed that your performance will slump.

LIFT
BROKEN
PLEASE
USE
STAIRS
ROOM

CHAPTER ELEVEN

The End of the Line

A few years ago Daina Taimina was reclining on the sofa at home in Ithaca, New York, where she teaches at Cornell University. A family member asked her what she was doing.

'I'm crocheting the hyperbolic plane,' she replied, referring to a concept that has mystified and fascinated mathematicians for almost two centuries.

'Have you ever seen a mathematician do *crochet*?' came the dismissive response.

The rebuff, however, made Daina even more determined to use handicraft in the course of scientific advancement. Which is just what she did, inventing what is known as 'hyperbolic crochet', a method of looping yarn that produces objects as intricate and beautiful as anything produced by the WI, and that has also contributed to an understanding of geometry in a way that mathematicians once never thought possible.

I'll come shortly to a detailed definition of *hyperbolic* and the insights gleaned by Daina's crochet models, but for the moment all you need to know is that hyperbolic geometry is an utterly counter-intuitive type of geometry that emerged in the early nineteenth century in which the set of rules that Euclid laid out so carefully in *The Elements* are taken as being false. 'Non-Euclidean' geometry was a watershed for mathematics in that it described a theory of physical space that totally contradicted our experience of the world, and therefore was hard to imagine, but nevertheless contained no mathematical contradictions, and so was as mathematically valid as the Euclidean system that came before.

Later that century an intellectual breakthrough of similar significance was made by Georg Cantor, who turned our intuitive understanding of the infinite on its head by proving that infinity comes in different sizes. Non-Euclidean geometry and Cantor's set

Hyperbolic crochet.

theory were gateways into two strange and wonderful worlds, and I'll visit them both in the following pages. Arguably, together they marked the beginning of modern mathematics.

The Elements, to recap from much earlier, is easily the most influential maths textbook of all time, having set out the basics of Greek geometry. It also established the *axiomatic method*, by which Euclid began with clear definitions of the terms to be used and the rules to be followed, and then built up his body of theorems from them. The rules, or *axioms*, of a system are the statements that are accepted without proof, so mathematicians always try to make them as simple and self-evident as possible.

Euclid proved all 465 theorems of *The Elements* with only five axioms, which are more commonly known as his five *postulates*:

1. There is a straight line from any point to any point.
2. A finite straight line can be produced in any straight line.
3. There is a circle with any centre and any radius.
4. All right angles are equal to one another.
5. If a straight line falling on two given straight lines makes the interior angles on the same side less than two right angles, the two given straight lines, if produced indefinitely, meet on that side on which the angles are less than the two right angles.

When we get to number 5, something does not feel right. The postulates start briskly enough. The first four are easy to state, easy to understand and easy to accept. Yet who invited the fifth to the party? It is long-winded, complicated and not especially, if at all, self-evident. And it is not even as clearly fundamental: the first time *The Elements* requires it is for Proposition 29.

Despite their love of Euclid's deductive method, mathematicians loathed his fifth postulate; not only did it go against their sense of aesthetics, they felt that it assumed too much to be an axiom. In fact, for 2000 years many great minds attempted to change the status of the fifth postulate by trying to deduce it from the other postulates so that it could be reclassified as a theorem instead of remaining as a postulate or axiom. But none succeeded. Perhaps the greatest evidence of Euclid's own genius was that he understood that the fifth postulate had to be accepted without proof.

Mathematicians had more success with restating the postulate in different terms. For example, the Englishman John Wallis in the seventeenth century realized that all *The Elements* could be proved by keeping the first four postulates as they were but by replacing the fifth postulate with the following alternative: *given any triangle, the triangle can be blown up or shrunk to any size so that the lengths of the sides stay in the same proportion to each other and the angles between the sides remain unaltered*. While it was quite an insight to realize that the fifth postulate could be rephrased as a statement about triangles rather than a statement about lines, it did not resolve mathematicians' concerns: Wallis's alternative postulate was perhaps more intuitive than the fifth postulate, though perhaps only marginally so, but it still wasn't as simple or self-evident as the first four. Other equivalents for the fifth postulate were also discovered;

Euclid's theorems still held true if the fifth postulate was substituted by the statement that the sum of angles in a triangle is 180 degrees, that Pythagoras's Theorem is true, or that for all circles the ratio of the circumference to diameter is pi. Extraordinary as it might sound, each of these statements is mathematically interchangeable. The equivalent that most conveniently expressed the essence of the fifth postulate, however, concerned the behaviour of parallel lines. From the eighteenth century mathematicians studying Euclid began to prefer using this version, which is known as the parallel postulate:

> *Given a line and a point not on that line, then there is at most one line that goes through the point and is parallel to the original line.*

It can be shown that the parallel postulate refers to the geometry of two distinct types of surface, hinging on the phrase 'at most one line' – which is mathspeak for 'either one line or no lines'. In the first case, illustrated by the diagram, for any line L and point P, there is *only one* line parallel to L (which is marked L') that goes through P. This version of the parallel postulate applies to the most obvious type of surface, a flat surface, such as a sheet of paper on your desk.

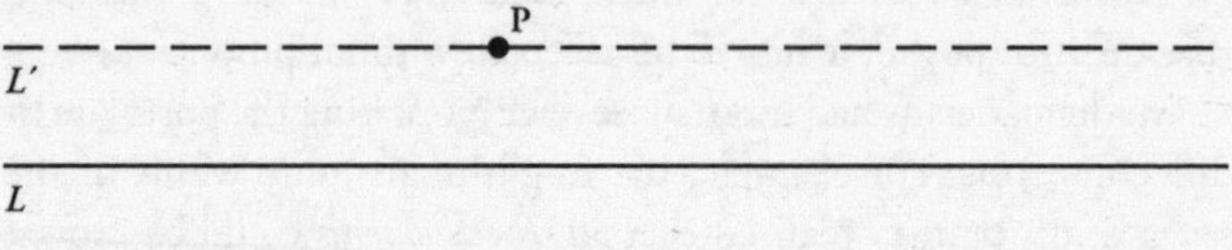

P

L'

L

The parallel postulate.

Now let's consider the second version of the postulate, in which for any line L and point P not on that line there are *no* lines through P and parallel to L. At first it is hard to think on what type of surface this might be the case. Where on Earth ...? On Earth is exactly where! Imagine, for example, that our line L is the equator, and imagine that point P is the North Pole. The only straight lines through the North Pole are the lines of longitude, such as the Greenwich Meridian, and all lines of longitude cross the equator. So, there are no straight lines through the North Pole that are parallel to the equator.

The parallel postulate provides us with a geometry for two types of surface: flat surfaces and spherical surfaces. *The Elements* was concerned with flat surfaces, so for 2000 years this was the main focus of mathematical enquiry. Spherical surfaces like the Earth were of less interest to theoreticians than they were to navigators and astronomers. It was only at the beginning of the nineteenth century that mathematicians found a wider theory that encompassed flat and spherical surfaces – and this happened only after they encountered a *third* kind of surface, the hyperbolic one.

One of the most determined aspirants in the quest to prove the parallel postulate from the first four postulates, and therefore show that it is not a postulate at all but a theorem, was Janós Bolyai, an engineering undergraduate from Transylvania. His mathematician father Farkas knew the scale of the challenge from his own failed attempts and implored his son to stop: 'For God's sake, I beseech you, give it up. Fear it no less than sensual passions because it too may take all your time and deprive you of your health, peace of mind and happiness in life.' But Janós stubbornly ignored his father's advice, and that was not his only rebellion: Janós dared to consider that the postulate might be false. *The Elements* was to mathematics what the Bible was to Christianity, a book of unchallengeable, sacred truths. While there was debate about whether the fifth postulate was an axiom or a theorem, no one had the temerity to suggest that it might actually not be true. As it turned out, doing so was the key to a new world.

The parallel postulate states that for any given line and a point not on that line there is *at most one* parallel line through that point. Janós's audacity was to suggest that for any given line and a point not on the line, *more than one* parallel line passes through that point. Even though it was not at all clear how to visualize a surface for which this statement was true, Janós realized that the geometry created by this statement, together with the first four postulates, was still mathematically consistent. It was a revolutionary discovery, and he recognized its momentousness. In 1823 he wrote to his father announcing that 'Out of nothing I have created a new universe'.

Janós was probably helped by the fact that he was working in isolation from any major mathematical institution, and so was less indoctrinated by traditional views. Even after making his discovery, he opted not to become a mathematician. On graduating, he joined the Austro-Hungarian army, where he was reportedly the best swordsman and dancer among his colleagues. He was also an outstanding musician, and it is said that he once challenged thirteen officers to duels on the condition that upon victory he could play the loser a piece on his violin.

Unbeknownst to János, another mathematician in an outpost even more distant from the hubs of European academia than Transylvania was making similar advances independently, but he had his work rejected by the mathematical establishment. In 1826 Nikolai Ivanovich Lobachevsky, a professor at Kazan University in Russia, submitted a paper that disputed the truth of the parallel postulate to the internationally renowned St Petersburg Academy of Sciences. It was turned down, so Lobachevsky then decided to submit it for publication in the local newspaper *Kazan Messenger*, and consequently no one took any notice.

The greatest irony about the toppling of Euclid's fifth postulate from the plinth of inviolable truth, however, is that several decades beforehand someone at the very heart of the mathematical establishment had indeed made the same discovery as János Bolyai and Nikolai Lobachevsky, yet this man had withheld his results from his peers. Quite why Carl Friedrich Gauss, the greatest mathematician of his day, decided to keep his work on the parallel postulate secret is not understood, although the received view is that he wanted to avoid getting embroiled in a feud about the primacy of Euclid with faculty members.

It was only on reading about János's results, which were published in 1831 as an appendix in a book by his father Farkas, that Gauss revealed to anyone that he had also considered the falsity of the parallel postulate. Gauss wrote a letter to Farkas, an old university classmate, in which he described János as a 'genius of the first order', yet added that he was unable to praise his breakthrough: 'For to praise it would be to praise myself. The entire content of the essay … coincide[s] with my own discoveries, some of which date back 30 to 35 years … I had intended to write all this down later so that at least it would not perish with me. It is therefore a pleasant surprise for me that I am spared this trouble, and I am especially glad that it is just the son of my old friend who takes precedence to me in this matter.' Janós was distressed when he learned that Gauss had got there first. And when, years later, Janós learned that Lobachevsky had also preceded him, he became haunted by the ludricrous notion that Lobachevsky was a fictional character invented by Gauss as a cunning ruse in order to deprive him of the credit for his work.

Gauss's final contribution to research on the fifth postulate came shortly before he died, when, already seriously ill, he set the title for the probationary lecture of one of his brightest students, 27-year-old Bernhard Riemann: 'On the hypotheses that lie at the foundations of geometry'. The cripplingly shy son of a Lutheran pastor, Riemann at first had some kind of breakdown struggling with what he would say, yet his solution to the problem would revolutionize maths. It would later revolutionize physics too, since his innovations were required by Einstein to formulate his general theory of relativity.

Riemann's lecture, given in 1854, consolidated the paradigm shift in our understanding of geometry resulting from the fall of the parallel postulate by establishing an all-embracing theory that included the Euclidean and non-Euclidean within it. The key concept behind Riemann's theory was the *curvature* of space. When a surface has zero curvature, it is *flat*, or Euclidean, and the results of *The Elements* all hold. When a surface has positive or negative curvature, it is *curved*, or non-Euclidean, and the results of *The Elements* do not hold.

The simplest way to understand curvature, continued Riemann, is by considering the behaviour of triangles. On a surface with zero curvature, the angles of a triangle add up to 180 degrees. On a surface with *positive* curvature, the angles of a triangle add up to *more than* 180 degrees. On a surface with negative curvature, the angles of a triangle add up to *less than* 180 degrees.

A sphere has positive curvature. We can see this by considering the sum of the angles of the triangle in the following diagram, which is made by the equator, the Greenwich Meridian and the line of longitude 73 degrees west of Greenwich (which goes through New York). Both angles where the longitude lines meet the equator are 90 degrees, so the sum of all three angles must be more than 180.

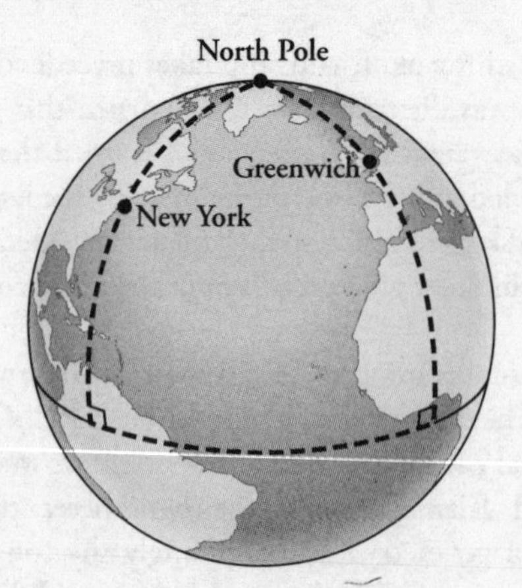

Triangle on a sphere: sum of angles greater than 180 degrees.

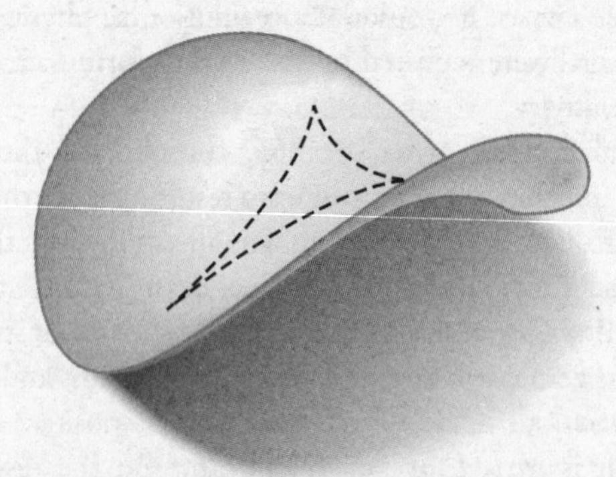

Triangle on a Pringle: sum of angles less than 180 degrees.

What type of surface has negative curvature? In other words, where are there triangles whose angles add up to less than 180 degrees? Pop open a pack of Pringles, and you'll see. Draw a triangle on the saddle part of the potato crisp (possibly with some fine French mustard) and the triangle looks 'sucked in' compared to the 'puffed out' triangle we see on a sphere. Its angles are clearly less than 180 degrees.

A surface with negative curvature is called *hyperbolic*. So, the surface of a Pringle is hyperbolic. The Pringle, however, is only an *hors d'oeuvre* in understanding hyperbolic geometry since it has an edge. Show a mathematician an edge and he or she will want to go over it.

Consider it this way. It is straightforward to imagine a surface with zero curvature and no edge: for example, this page, flattened on a desk, and extended infinitely in all directions. If we lived on such a surface and we started walking in a straight line in any direction, we would never reach an edge. Likewise, we have an obvious example of a surface with positive curvature and no edge: a sphere. If we lived on the surface of a sphere, we could walk for ever and ever in one direction and never reach an edge. (Of course, we do live on a rough approximation of a sphere. If the Earth were totally smooth, with no oceans or mountains to block our way, for example, and we started walking, we would return to our point of departure and continue going in circles.)

Now, what does a surface with negative curvature and no edge look like? It cannot look like a Pringle, since if we lived on an Earth-sized Pringle and we started walking in one direction, we would always eventually fall off it. Mathematicians have long wondered what an 'edgeless' hyperbolic surface might look like – one on which we could walk as far as we wished without coming to the end of it and without it losing its hyperbolic properties. We know it must be always curving like a Pringle, so what about sticking lots of Pringles together? Sadly, this wouldn't work since Pringles don't fit together neatly, and if we filled in the gaps with another surface these new areas would not be hyperbolic. In other words, the Pringle allows us to envisage only a local area with hyperbolic properties. What is incredibly difficult to envisage – and stretches even the most brilliant mathematical minds – is a hyperbolic surface that goes on for ever.

Spherical and hyperbolic surfaces are mathematical opposites, and here is a practical example that shows why. Cut a piece out of a spherical surface, such as a basketball. When we squash the piece on the ground to make it flat it will either stretch or rip, since there is not enough material to spread out in a flat way. Now imagine we had a rubber Pringle. When we try to flatten it, the Pringle would have *too much* material and some of it would fold on itself. Whereas the sphere closes in on itself, the hyperbolic surface expands.

Let's return to the parallel postulate, which provides us with a very concise way of classifying flat, spherical and hyperbolic surfaces.

For any given line and a point not on that line:

*On a **flat** surface there is **one and only one** parallel line through that point.*

*On a **spherical** surface there are **zero** parallel lines through that point.**

*On a **hyperbolic** surface there is **an infinite number** of parallel lines through that point.*

We can understand intuitively the behaviour of parallel lines on a flat or on a spherical surface, because we can easily visualize a flat surface that goes on for ever and we all know what a sphere is. It is much more challenging to understand the behaviour of parallel lines on a hyperbolic surface, since it is not at all clear what such a surface might look like as it spreads to infinity. Parallel lines in hyperbolic space get further and further apart from each other. They do not bend away from each other, since for two lines to be parallel they must also be straight, but they diverge because a hyperbolic surface is constantly curving away from itself, and as the surface curves away from itself it creates more and more space between any two parallel lines. Again, this idea is totally mind-boggling, and it's hardly surprising that, despite his genius, Riemann did not come up with a surface that had the properties he was describing.

The challenge of visualizing the hyperbolic plane galvanized many mathematicians in the final decades of the nineteenth century. One attempt, by Henri Poincaré, caught the imagination of M.C. Escher, whose famous *Circle Limit* series of woodcuts was inspired by the Frenchman's 'disc model' of a hyperbolic surface. In *Circle Limit IV*, a two-dimensional universe is contained on the circular disc in which angels and devils get progressively smaller the closer they get to the circumference. The angels and devils, however, are not aware that

* You might think that lines of latitude are parallel to the equator. This is not true because the lines of latitude (with the exception of the equator) are not straight lines, and only straight lines can be parallel to each other. A straight line is the shortest distance between two points, which is why a plane flying between New York and Madrid, which are both on the same line of latitude, does not fly along the line of latitude but instead has a path that looks curved when seen on a two-dimensional map.

Circle Limit IV.

they are getting smaller since as they shrink, so too do their measuring tools. As far as the inhabitants of the disc are concerned, they are all the same size and their universe goes on for ever.

The ingenuity of Poincaré's disc model is that it illustrates beautifully how parallel lines behave in hyperbolic space. First, we need to be clear what straight lines are in the disc. In the same way that straight lines on a sphere look curved when represented on a flat map (for example, flight paths are straight, but look curved on a map), lines that are straight in the discworld also look to us like they are curved. Poincaré defined a straight line in the disc as being a section of a circle that enters the disc at right angles. Figure 1 overleaf shows the straight line between A and B, which is made by finding the circle that goes through A and B and that enters the disc at right angles. The hyperbolic version of the parallel postulate states that for every straight line L and a point P not on that line, there is an infinite number of straight lines parallel to L that pass through P. This is shown in figure 2, where I have marked three straight lines – L′, L″ and L‴ – that pass through P but are all parallel to L. (Two lines are parallel if they are both straight and never meet.) Each of the lines L′, L″ and L‴ are parts of different circles that enter the disc at right angles. By looking at the disc we can now see how it must be the case that there is an infinite number of straight lines that are parallel to L and that pass through P, since we can draw an infinite number of circles that enter the disc at right angles and pass through P. Poincaré's model also helps us understand what it means for two parallel lines to diverge: L and L′ are parallel but become further and further apart the closer they get to the circumference of the disc.

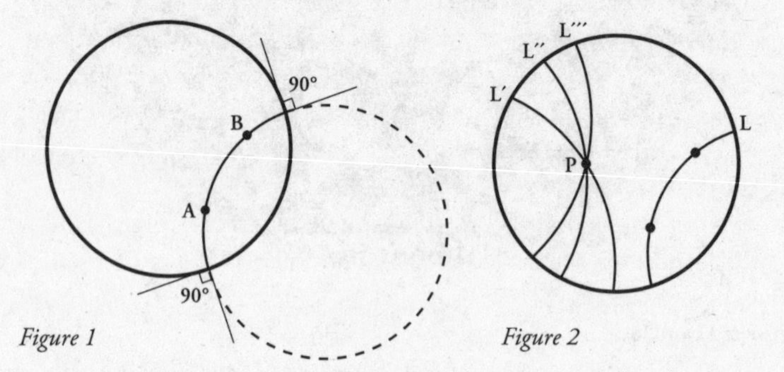

Figure 1

Figure 2

Poincaré's discworld is illuminating, but only up to a point. While it provides us with a conceptual model of hyperbolic space, distorted through a rather strange lens, it does not reveal what a hyperbolic surface looks like in our world. The search for more realistic hyperbolic models, which had looked promising in the last decades of the nineteenth century, was dealt a blow by the German mathematician David Hilbert in 1901, when he proved that it was impossible to describe a hyperbolic surface using a formula. Hilbert's proof was accepted by the mathematical community with resignation, since they concluded that if there was no way of describing such a surface with a formula, then such a surface must not actually exist. Interest in coming up with models for hyperbolic surfaces waned.

———

Which brings us back to Daina Taimina, whom I met at the South Bank, a riverside promenade of theatres, art galleries and cinemas in London. She gave me a brief summary of the history of hyperbolic space, which is a subject she has taught in her position as adjunct associate professor at Cornell. A consequence of Hilbert's proof that hyperbolic space cannot be generated by a formula, she said, is that computers are unable to create images of hyperbolic surfaces because computers can only create images based on formulas. In the 1970s, however, the geometer William Thurston found a low-tech approach to be much more fruitful. He had the idea that you didn't need a formula to make a hyperbolic model – all you needed was paper and scissors. Thurston, who in 1981 was awarded the Fields Medal (the supreme prize in maths) and who is now a colleague of Daina's at Cornell, came up with a model made from sticking together horseshoe-shaped slivers of paper.

Daina used a Thurston model with her students but it was so fragile that it always fell apart, and she had to make a new one each time. 'I hate gluing paper. It drives me crazy,' she said. Then she had a brainwave. What if it were possible to knit a model of the hyperbolic plane instead?

Her idea was simple: start with a line of stitches, and then for each subsequent line add a fixed amount relative to the number of stitches on the line before. For example, adding an extra stitch for every two stitches on the line before. In this case, if you started with a line of 20 stitches, the second line would have 30 stitches (adding 10), the third line 45 stitches (adding 15), and so on. (The fourth line should have an extra 22.5 stitches, but since you can't do half stitches you have to round up or down.) This, she hoped, would create a piece of fabric that became wider and wider – as if expanding out from itself hyperbolically. However, the knitting was too fiddly, since after any mistake she needed to unravel the whole line. So she swapped the knitting needles for a crochet hook. With crochet, there is no chance of unravelling, as you progress one stitch at a time. She got the knack pretty quickly. It helped that Daina was a demon at craftwork, a consequence of a childhood in 1960s Soviet Latvia.

For her first crochet model she added an extra stitch for every two stitches on the previous line. The result, however, was a piece of material with many tight ruffles. 'It was too curly,' she said.

'I couldn't see what was going on.' So for her next attempt, she changed the ratio, adding an extra stitch only for every five on the previous line. It worked better than she had expected. The material was now properly folding in on itself. She picked it up and followed straight lines in and out of the expanding flaps, and quickly realized that she could see parallel lines that diverged. 'It was the picture I always wanted to see,' she beamed. 'That was my excitement. It was also a quick thrill to make something with my hands that cannot be made by computer.'

Daina showed the hyperbolic crochet model to her husband – and he was as excited as she was. David Henderson is professor of geometry at Cornell. His specialism is topology, which Daina claims to know nothing about. He explained to her that topologists have long known that when an octagon is drawn on the hyperbolic plane it can be folded together in such a way as to resemble a pair of pants. 'We have to construct that octagon!' he told her, which is just what they did. 'No one had seen a hyperbolic pair of pants before!' Daina exclaimed, and opened a sports bag she had with her, took out a crocheted hyperbolic octagon and folded it to show me the model. It looked like a very cute pair of woollen toddler's shorts:

Word spread around the Cornell maths faculty about Daina's threaded creations. She told me she showed it to one colleague who is known for writing about hyperbolic planes. 'He looked at the model and started playing with it. Then his face lit up. "This is what a *horocycle* looks like!" he said,' having recognized a very complicated type of curve that he had never been able to picture before. 'He had been writing about them all his career,' added Daina, 'but they were all in his imagination.'

It is no exaggeration to say that Daina's hyperbolic models have given important new insight into a conceptually punishing area of maths. They give a visceral experience of the hyperbolic plane, allowing students to touch and feel a surface that was previously understood only in an abstract way. The models are not perfect, however. One problem is that the thickness of the stitches makes the crocheted models only a rough approximation of what in theory should be a smooth surface. Still, they are a great deal more versatile and accurate than a Pringle. If a piece of hyperbolic crochet had an infinite number of lines, it would theoretically be possible to live on that surface and walk for ever in one direction without ever coming to an edge.

One of the charms of Daina's models is that they are unexpectedly organic-looking for something conceived so formally. When the relative line-by-line increase in stitches is small, the models look like leaves of kale. When the increase is greater the material naturally folds itself into pieces that look like coral. In fact, the reason Daina came to London was for the opening of the Hyperbolic Crochet Coral Reef, an exhibition inspired by her models to promote awareness of marine destruction. Thanks to her mathematical innovation, she has unwittingly spawned a global movement of crochet activism.

Over the last decade, Daina has crocheted more than a hundred hyperbolic models. She brought her largest to London. It is pink, uses 5.5km of yarn, weighs 4.5kg and took her six months. Finishing it was an ordeal. 'As it got bigger it took a lot of energy to turn.' A remarkable property of the model is that it has an incredibly large surface area – 3.2 square metres, which is twice as much surface area as Daina herself. Hyperbolic surfaces maximize area with minimal volume, which is why they are favoured by some plants and marine organisms. When an organism needs a large surface area – to absorb nutrition, as is the case with coral – it grows in a hyperbolic way.

It is unlikely that Daina would ever have thought up the idea of hyperbolic crochet had she been born a man, which makes her inventions a noteworthy artefact in the cultural history of mathematics, where women have long been under-represented. Crochet, in fact, is just one example of a traditionally female craft inspiring mathematicians in recent years to explore new techniques. Together with mathematical knitting, quilt-making, embroidery and weaving, the academic discipline is now known as Math and the Fiber Arts.

When hyperbolic space was first conceived it appeared to go against any sense of reality, yet it has become accepted as equally 'real' as flat or spherical surfaces. Every surface has its own geometry, and we need to choose the one that best applies, or, as Henri Poincaré once said: 'One geometry cannot be more true than another; it can only be more convenient.' Euclidean geometry, for example, is the most appropriate for schoolchildren armed with rulers, compasses and flat pieces of paper, while spherical geometry is the most appropriate for airline pilots navigating flight paths.

Physicists are also interested in which geometry is most appropriate for their purposes. Riemann's ideas about the curvature of surfaces provided Einstein with the equipment to make one of his greatest breakthroughs. Newtonian physics assumed that space was Euclidean, or flat. Einstein's theory of general relativity, however, stated that the geometry of space-time (3-D space plus time considered as the fourth dimension) was not flat but curved. In 1919 a British scientific expedition in Sobral, a town in the northeast of Brazil, took images of the stars behind the sun during a solar eclipse and found that they had shifted slightly from their real positions. This was explained by Einstein's theory that the light from the stars was curving around the sun before it reached Earth. While the light appeared to bend around the sun when seen in three-dimensional space, which is the only way we can see things, it was actually following a straight line according to the curved geometry of space-time. The fact that Einstein's theory correctly predicted the position of the stars vindicated his general theory of relativity and it is what made him a global celebrity. The London *Times* headline blazoned: 'Revolution in Science, New Theory of the Universe, Newtonian Ideas Overthrown'.

Einstein was concerned with space-time, which he showed to be curved. What about the curvature of our universe without considering time as a dimension? In order to see which geometry best fits the behaviour of our three spatial dimensions on a large scale, we need to see how lines and shapes behave over extremely large distances. Scientists are hoping to discover this from the data that is being gathered by the Planck satellite, launched in May 2009, which is measuring cosmic background radiation – the so-called 'afterglow' of the Big Bang – to a higher resolution and sensitivity than ever before. Considered opinion is that the universe is either flat or spherical, although it is still possible that the universe might be hyperbolic. It is wonderfully ironic to think that a geometry originally thought to be nonsensical might actually reflect the way things really are.

At around the same time that mathematicians were exploring the counter-intuitive realm of non-Euclidean space, one man was turning upside-down our understanding of another mathematical notion: infinity. Georg Cantor was a lecturer at Halle University in Germany, where he developed a trail-blazing theory of numbers in which infinity could have more than one size. Cantor's ideas were so unorthodox that they initially provoked ridicule from many of his peers. Henri Poincaré, for example, described his work as 'a malady, a perverse illness from which some day mathematics would be cured', while Leopold Kronecker, Cantor's former teacher and professor of maths at Berlin University, dismissed him as a 'charlatan' and a 'corruptor of youth'.

This war of words probably contributed to Cantor's nervous breakdown in 1884, aged 39, the first of many mental-health episodes and hospitalizations. In his book on Cantor, *Everything and More*, David Foster Wallace writes: 'The Mentally Ill Mathematician seems now in some ways to be what the Knight Errant, Mortified Saint, Tortured Artist, and Mad Scientist have been for other eras: sort of our Prometheus, the one who goes to forbidden places and returns with gifts we can all use but he alone pays for.' Literature and film are guilty of romanticizing a link between maths and insanity. It's a cliché that suits the narrative requirements of a Hollywood script (exhibit A: *A Beautiful Mind*) but is, of course, an unfair generalization. The great mathematician, however, for whom the archetype could have been invented is Cantor. The stereotype fits him especially well since he was grappling with infinity, a concept that links mathematics, philosophy and religion. Not only was he challenging mathematical doctrine, but he was also setting out a brand-new theory of knowledge and, in his mind, of human understanding of God. No wonder he upset a few people along the way.

Infinity is one of the most brain-mangling concepts in maths. We saw earlier, in our discussion of Zeno's paradoxes, that envisaging an infinite number of ever-decreasing distances is full of mathematical and philosophical pitfalls. The Greeks tried to avoid infinity as much as they could. Euclid expressed ideas of infinity by making negative assertions. His proof that there is an infinite number of prime numbers, for instance, is actually a proof that there is no highest prime number. The ancients shied away from treating infinity as a self-contained concept, which is why the infinite series inherent in Zeno's paradoxes were so problematic for them.

By the seventeenth century, mathematicians were willing to accept operations involving an infinite number of steps. The work of John Wallis, who in 1655 introduced the symbol ∞ for infinity for the purpose of his work on infinitesimals (things that become infinitely small), paved the way for Isaac Newton's calculus. The discovery of useful equations that involved an infinite number of terms, such as $\frac{pi}{4} = 1 - \frac{1}{3} + \frac{1}{5} - \frac{1}{7} + \ldots$ showed that infinity was not an enemy, yet even so, it was still to be treated with care and suspicion. In 1831 Gauss stated the received wisdom when he said that infinity was 'merely a way of speaking' about a limit that one never reached, an idea that simply expressed the potential to carry on and on for ever. Cantor's heresy was to treat infinity as an entity in itself.

The reason why mathematicians pre-Cantor were nervous about treating infinity like any other number was that it contained many conundrums, the most famous of which Galileo wrote about in *Two New Sciences*, and is known as Galileo's paradox:

1. Some numbers are squares, such as 1, 4, 9 and 16, and some are not squares, such as 2, 3, 5, 6, 7, etc.

2. The totality of all numbers must be greater than the total of squares, since the totality of all numbers includes squares and non-squares.

3. Yet for every number, we can draw a one-to-one correspondence between numbers and their squares, for example:

$$\begin{array}{cccccccc} 1 & 2 & 3 & 4 & 5 & \ldots & n & \ldots \\ \downarrow & \downarrow & \downarrow & \downarrow & \downarrow & & \downarrow & \\ 1 & 4 & 9 & 16 & 25 & \ldots & n^2 & \ldots \end{array}$$

4. So, there are, in fact, as many squares as there are numbers. Which is a contradiction, since we have said, in point 2, that there are more numbers than squares.

Galileo's conclusion was that, when it comes to infinity, the numerical concepts 'more than', 'equal to' and 'less than' do not make sense. These terms may be understandable and coherent when discussing finite amounts, but not with infinite ones. It is meaningless to say there are more numbers than there are squares, or that there is an equal number of numbers and squares, since the totality of both numbers and squares is infinite.

Georg Cantor devised a new way to think about infinity that made Galileo's paradox redundant. Rather than thinking about individual numbers, Cantor considered collections of numbers, which he called 'sets'. The cardinality of any set is the number of members in the collection. So {1, 2, 3} is a set with a *cardinality* of three and {17, 29, 5, 14} is a set with cardinality four. Cantor's 'set theory' gets the pulse racing when considering sets with an infinite number of members. He introduced a new symbol for infinity, $\aleph_0$, (pronounced *aleph-null*), using the first letter of the Hebrew alphabet with the subscript 0, and said that this was the cardinality of the set of natural numbers, or {1, 2, 3, 4, 5 ...}. Every set whose members can be put in a one-to-one correspondence with the natural numbers also has cardinality $\aleph_0$. So, because there is a one-to-one correspondence between the natural numbers and their squares, the set of squares {1, 4, 9, 16, 25 …} has cardinality $\aleph_0$. Likewise, the set of odd numbers {1, 3, 5, 7, 9 …}, the set of prime numbers {2, 3, 5, 7, 11 …} and the set of numbers with 666 in them {666, 1666, 2666, 3666 …}, all have cardinality $\aleph_0$. If you have a set with an infinite number of members, and it is possible to count its members one by one so that eventually every member will be reached, then the cardinality of the set is $\aleph_0$. For this reason, $\aleph_0$ is also known as a 'countable infinity'. The reason this is exciting is because Cantor went on to show that we can go higher. Large though it may be, $\aleph_0$ is merely the baby in Cantor's family of infinities.

I'll introduce you to an infinity larger than $\aleph_0$ with the help of a story David Hilbert is said to have used in his lectures that concerns a hotel with a countably infinite, or $\aleph_0$, number of rooms. This well-known establishment, much loved by mathematicians, is sometimes called the Hilbert Hotel.

In the Hilbert Hotel there is an infinite number of rooms and they are numbered 1, 2, 3, 4 ... One day a traveller arrives at reception only to find that the hotel is full. He asks if there is any way a room can be found for him. The receptionist replies that of course there is! All the management needs to do is to reassign guests to different rooms in the following way: moving the guest in Room 1 to Room 2, moving the guest in Room 2 to Room 3, and so on, moving everyone in Room n to Room $n + 1$. If this is done, then every guest still has a room, and Room 1 is freed up for the new arrival. Perfect!

The following day a more complicated situation presents itself. A bus arrives, with all the passengers needing rooms. This bus has an infinite number of seats, numbered 1, 2, 3 and so on, all of which are occupied. Is there any way that a room can be found for each and every one of the passengers? In other words, even though the hotel is full, can the receptionist reshuffle the guests into different rooms in a way that leaves an infinite number of free rooms for the bus passengers? Easy peasy, comes the reply. All the management needs to do this time is to move every guest to the room numbered double the room he or she is already in, which takes care of Rooms 2, 4, 6, 8 … This leaves all the rooms with odd numbers empty, and the bus passengers can be given the keys for those. The passenger on the first seat gets Room 1, the first odd number, the passenger on the second seat gets Room 3, the second odd number, and so on.

On the third day, even more buses arrive at the Hilbert Hotel. In fact, an infinite number of buses arrives. The buses are lined up outside, with Bus 1 next to Bus 2, which is next to Bus 3, and so on. Each bus has an infinite number of passengers, like the bus that arrived the day before. And, of course, every passenger needs a room. Is there a way to find every passenger in every bus a room in the (already full) Hilbert Hotel?

No problem, says the receptionist. First, he needs to clear an infinite number of rooms. He does this by the trick he used the day before – move everyone to a room with twice the room number. This leaves all the odd-numbered rooms free. In order to fit in the infinite coach parties, all he needs to do is find a way of counting all the passengers, since once he has found a method, he can assign the first passenger to Room 1, the second to Room 3, the third to Room 5, and so on.

He does it like this: for each bus, list the passengers by seat, as in the table below. Each passenger is therefore represented by the form *m*/*n* where *m* is the number of the bus they are on and *n* is their seat number. If we start at the passenger in the first seat in the first bus (person 1/1), and then form the zigzag pattern below, by letting the second person be the passenger who is in the second seat of the first bus (1/2), and the third be the first passenger on the second (2/1), we will eventually count every single passenger.

	Seat 1	Seat 2	Seat 3	Seat 4	Seat 5	…
Bus 1	1/1	1/2	1/3	1/4	1/5	
Bus 2	2/1	2/2	2/3	2/4	2/5	
Bus 3	3/1	3/2	3/3	3/4	3/5	
Bus 4	4/1	4/2	4/3	4/4	4/5	
Bus 5	5/1	5/2	5/3	5/4	5/5	
…						

Now let's translate what we have learned with the Hilbert Hotel into some symbolic mathematics:

> When one person was found a room, this was the equivalent of showing that $1 + \aleph_0 = \aleph_0$
>
> When a countably infinite number of people could be found a room, we saw that $\aleph_0 + \aleph_0 = \aleph_0$
>
> When a countably infinite number of buses, each containing a countably infinite number of passengers could be found a room, this revealed that $\aleph_0 \times \aleph_0 = \aleph_0$

These rules are what we expect from infinity: add infinity to infinity and we get infinity, multiply infinity by infinity and we get infinity.

Let's stop here a second. We have already reached an amazing result. Look again at the table of seats and buses. Consider each person denoted *m*/*n* as the fraction $\frac{m}{n}$. The table, when extended infinitely, will cover every single positive fraction, since the positive fractions can also be defined as $\frac{m}{n}$ for all natural numbers m and n. For example, the fraction $\frac{5628}{785}$ will be covered on the 5628th row and 785th column. The zigzag counting method that counted every passenger in every bus can, therefore, also be used to count every positive fraction. In other words, the set of all positive fractions and the set of natural numbers have the same cardinality, which is $\aleph_0$. It seems intuitive that there should be more fractions than there are natural numbers, since between any two natural numbers there is an infinite number of fractions, yet Cantor showed that our intuition is wrong. There are as many positive fractions as there are natural numbers. (In fact, there are as many positive and negative fractions as there are natural numbers, since there are $\aleph_0$ positive fractions and $\aleph_0$ negative fractions and, as seen above, $\aleph_0 + \aleph_0 = \aleph_0$.)

We can appreciate how strange this result is by considering the number line, which is a way of understanding numbers by considering them as points on a line. Below is a number line starting at 0 and heading off towards infinity.

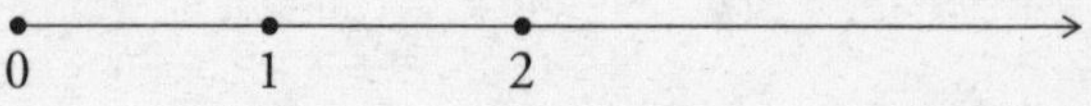

0 1 2

Every positive fraction can be considered as a point on this number line. From a previous chapter, we know that there is an infinite number of fractions between 0 and 1, as there is between 1 and 2, or between any two other numbers. Now imagine holding a microscope up to the line so that you can see between the points representing the fractions $\frac{1}{100}$ and $\frac{2}{100}$. As we also showed earlier, there is an infinite number of points representing fractions between these two points. In fact, wherever you place your microscope on the line, and however tiny the interval between two points that your microscope can see, there will always be infinitely many points representing fractions in this interval. Since there are infinite numbers of points representing fractions wherever you look, it comes as a bewildering surprise to realize that it is, in fact, possible to count them in an ordered list that will cover every single one without exception.

Now for the big event: proof that there is a cardinality larger than $\aleph_0$. Back we go to the Hilbert Hotel. On this occasion, the hotel is empty when an infinite number of people show up wanting rooms. But this time the travellers have not come in buses; they are in fact a rabble, with each wearing a T-shirt displaying a decimal expansion of a number between 0 and 1. No two people have the same decimal expansion on their chest and every single decimal expansion between 0 and 1 is covered. (Of course, the decimal expansions are infinitely long and so the T-shirts would need to be infinitely wide to display them, yet since we have suspended our disbelief in order to imagine a hotel with an infinite number of rooms, I figure it is not asking too much to envisage these T-shirts.)

A few of the arrivals charge into reception and ask if there is a way the hotel can accommodate them. For the receptionist to achieve this all he needs to do is find a way of listing every single decimal between 0 and 1, since once he has listed them, he can assign them rooms. This seems like a fair challenge, since, after all, he was able to find a way to list an infinite number of passengers from an infinite number of buses. This time, however, the task is impossible. There is no way to count every single decimal expansion between 0 and 1 in such a way that we can write them all down in an ordered list. To prove this I will show that for every infinite list of numbers between 0 and 1 there will always be a number between 0 and 1 that is not on the list.

This is how it's done. Let's imagine the first arrival has a T-shirt with the expansion 0.6429657…, the second has 0.0196012…, and the receptionist assigns them rooms 1 and 2. And say he carries on assigning rooms to the other arrivals, thus creating the infinite list that begins (remember, each of these expansions goes on for ever):

Room 1	0.6429657…
Room 2	0.0196012…
Room 3	0.9981562…
Room 4	0.7642178…
Room 5	0.6097856…
Room 6	0.5273611…
Room 7	0.3002981…
Room …	0…
…	…

Our aim, stated earlier, is to find a decimal expansion between 0 and 1 that is not on this list. We do this using the following method. First, construct the number that has the first decimal place of the number in Room 1, the second decimal place of the number in Room 2, the third decimal place of the number in Room 3 and so on. In other words, we are selecting the diagonal digits that are underlined here:

0.**6**429657…
0.0**1**96012…
0.99**8**1562…
0.764**2**178…
0.6097**8**56…
0.52736**1**1…
0.300298**1**…

This number is:
0.6182811…

We're almost there. We now need to do one final thing to construct our number that is not on the receptionist's list: we alter every digit in this number. Let's do this by adding 1 to every digit, so the 6 becomes a 7, the 1 becomes a 2, the 8 becomes a 9, and so on, to get this number: 0.7293922...

And now we have it. This decimal expansion is the exception that we were looking for. It cannot be on the receptionist's list because we have artificially constructed it so it cannot be. The number is not in Room 1, because its first digit is different from the first digit of the number in Room 1. The number is not in Room 2 because its second digit is different from the second digit of the number in Room 2, and we can continue this to see that the number cannot be in any Room *n* because its *n*th digit will always be different from the *n*th digit in the expansion of Room *n*. Our customized expansion 0.7293922... therefore cannot be equal to any expansion assigned to a room since it will always differ in at least one digit from the expansion assigned to that room. There may well be a number in the list whose first seven decimal places are 0.7293922, yet if this number is on the list then it will differ from our customized number by at least one digit further down the expansion. In other words, even if the receptionist carries on assigning rooms for ever and ever, he will be unable to find a room for the arrival with the T-shirt marked with the number we created beginning 0.7293922...

I chose a list starting with the arbitrary numbers 0.6429657... and 0.0196012... but equally I could have chosen a list starting with any numbers. For every list that it is possible to make, it will always be possible to create, using the 'diagonal' method opposite, a number that is not on the list. The Hilbert Hotel may have an infinite number of rooms, yet it *cannot* accommodate the infinite number of people defined by the decimals between 0 and 1. There will always be people left outside. The hotel is simply not big enough.

Cantor's discovery that there is an infinity *bigger* than the infinity of natural numbers was one of the greatest mathematical breakthroughs of the nineteenth century. It is a mind-blowing result, and part of its power is that the result really was quite straightforward to explain: some infinities are countable, and they have size $\aleph_0$, and some infinities are not countable, and hence bigger. These uncountable infinities come in many different sizes.

The easiest uncountable infinity to understand is called *c* and is the number of people who arrived at the Hilbert Hotel with T-shirts containing all the decimal expansions between 0 and 1. Again, it is instructive to interpret *c* by looking at what it means on the number line. Every person with a decimal expansion between 0 and 1 on his T-shirt can also be understood as a point on the line between 0 and 1. The initial *c* is used since it stands for the 'continuum' of points on a number line.

And here's where we come to another strange result. We know that there are *c* points between 0 and 1, and yet we know that there are $\aleph_0$ fractions on the totality of the number line. Since we have proved that *c* is bigger than $\aleph_0$, it must be the case that there are more points on a line between 0 and 1 than there are points that represent fractions on the entire number line.

Again, Cantor has led us to a very counter-intuitive world. Fractions, though they are infinite in number, are responsible for only a tiny, tiny part of the number line. They are much more lightly sprinkled along the line than the other type of number that makes up the number line, the numbers that cannot be expressed as fractions, which are our old friends the *irrational* numbers. It turns out that the irrational numbers are so densely packed that there are more of them in any finite interval on the number line than there are fractions on all of the number line.

We introduced *c* above as being the number of points on a number line between 0 and 1. How many points are there between 0 and 2, or between 0 and 100? Exactly *c* of them. In fact, between any two points on the number line there are exactly *c* points in between, no matter how far or close they are apart. What's even more amazing is that the totality of points on the entire number line is also *c*, and this is shown by the following proof, illustrated opposite. The idea is to show that there is a one-to-one correspondence between the points that lie between 0 and 1, and the points that lie on the entirety of the number line. This is done by pairing off every point on the number line with a point between 0 and 1. First, draw a semicircle suspended above 0 and 1. This semicircle acts like a matchmaker in that it fixes the couplings of the points between 0 and 1 and the points on the number line. Take any point on the number line, marked *a*, and draw a straight line from *a* to the centre of the circle. The line hits the semicircle at a point that is a unique distance between 0 and 1, marked *a′*, by drawing a line vertically down until it meets the number line. We can pair up every point marked *a* to a unique point *a′* in this way. As our chosen point *a* heads to plus infinity, the corresponding point between 0 and 1 closes in on 1, and as the chosen point heads to minus infinity, the corresponding point closes in on 0. If every point on the number line can be paired off with a unique point between 0 and 1, and vice versa, then the number of points on the number line must be equal to the number of points between 0 and 1.

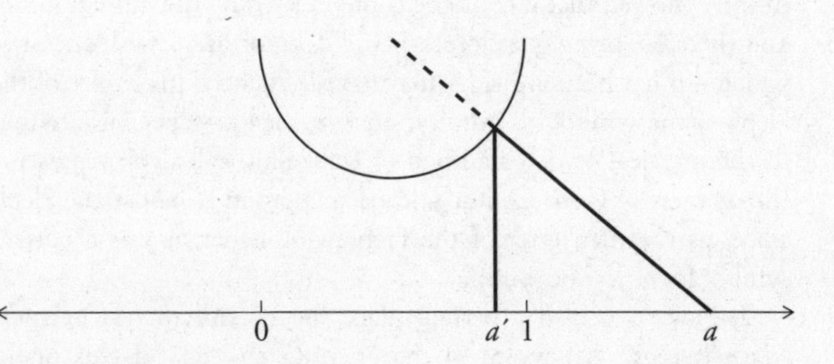

The difference between $\aleph_0$ and c is the difference between the number of points on the number line that are fractions and the total number of points, including fractions and irrationals. The leap between $\aleph_0$ and c, however, is so immense that were we to pick a point at random from the number line, we have 0 per cent probability of getting a fraction. There just aren't enough of them, compared to the uncountably infinite number of irrationals.

Difficult as Cantor's ideas were to accept at first, his invention of the aleph has been vindicated by history; not only is it now almost universally accepted into the numerical fold, but the zigzag and diagonal proofs are generally hailed as among the most dazzling in the whole of mathematics. David Hilbert said: 'From the paradise created for us by Cantor, no one will drive us out.'

Unfortunately for Cantor, this paradise came at the expense of his mental health. After he recovered from his first breakdown, he began to focus on other subjects, such as theology and Elizabethan history, becoming convinced that the scientist Francis Bacon wrote the plays of William Shakespeare. Proving Bacon's authorship became a personal crusade, and a focus for increasingly erratic behaviour. In 1911, giving a lecture at St Andrews University, where he had been invited to talk about mathematics, he instead discussed his views on Shakespeare, much to the embarrassment of his hosts. Cantor had several more breakdowns and was frequently hospitalized until his death in 1918.

A devout Lutheran, Cantor wrote many letters to clergymen about the significance of his results. He believed that his approach to infinity showed that it could be contemplated by the human mind, and therefore bring one closer to God. Cantor had Jewish ancestry, which – it has been argued – may have influenced his choice of the aleph as the symbol for infinity, since he may have been aware that in the mystical Jewish tradition of Kabbalah, the aleph represents the oneness of God. Cantor said he was proud he chose the aleph since, as the first letter of the Hebrew alphabet, it was a perfect symbol for a new beginning.

The aleph is also a perfect place for an end to our journey. Mathematics, as I wrote in the opening chapters of this book, emerged as part of man's desire to make sense of his own environment. By making notches on wood, or counting with fingers, our ancestors invented numbers. This was helpful for farming and trade, and ushered us into 'civilization'. Then, as mathematics developed, the subject became less about real things and more about abstract ones. The Greeks introduced concepts such as a point and a line, and the Indians invented zero, which opened the door to even more radical abstractions like negative numbers. While these concepts were at first counter-intuitive, they were assimilated quickly and we now use them on a daily basis. By the end of the nineteenth century, however, the umbilical cord linking mathematics to our own experience snapped once and for all. After Riemann and Cantor, maths lost its connection to any intuitive appreciation of the world.

After finding c Cantor kept going, proving that there are even bigger infinities. As we saw, c is the number of points on a line. It is also equal to the number of points on a two-dimensional surface.

(That's another surprising result, which you'll have to trust me on.) Let's call *d* the number of all possible lines, curves and squiggles that can be drawn on a two-dimensional surface. (These lines, curves and squiggles can be either continuous, as if drawn without taking your pen off the paper, or discontinuous, as if the pen has been lifted from the paper at least once, leaving a gap between different sections of the same line.) Using set theory, we can prove that *d* is bigger than *c*. And we can go one step further, showing that there must be an infinity larger than *d*. Yet no one has so far been able to come up with a set of naturally occurring things that has a cardinality larger than *d*.

Cantor led us beyond the imaginable. It is a rather wonderful place and one that is amusingly opposite to the situation of the Amazonian tribe I mentioned at the beginning of this book. The Munduruku have many things, but not enough numbers to count them. Cantor has provided us with as many numbers as we like, but there are no longer enough things to count.

Glossary

Algorithm: a set of rules or instructions designed to solve a problem.

Ambigram: a word (or set of words) written in such a way as to conceal other words, often the same word (or set of words) written upside-down.

Amicable number: two numbers are amicable if the factors of one add up to the other and vice versa.

Axiom: a statement that is accepted without proof, usually because it is self-evident, and used as a foundation of a logical system.

Base: in a number system, the base is the size of the number-grouping which, when using Arabic numerals, is equal to the number of different digits allowed in the system. Binary, using 0 and 1, is base two, while decimal, using 0 to 9, is base ten.

Cardinality: the size of a set.

Circumference: the perimeter of a circle.

Combinatorics: the study of combinations and permutations.

Constant: a fixed value.

Continuum: the points on a continuous line.

Convergent series: an infinite series that adds up to a finite number.

Correlation: a measure of the interdependence of two variables.

Countable infinity: an infinite set, the members of which can be put in a one-to-one correspondence with the natural numbers.

Cubic equation: an equation of the form $ax^3 + bx^2 + cx + d = 0$, where a, b, c and d are constants and a is non-zero.

Curvature: a property of space that can be determined by the behaviour of triangles or parallel lines.

Decimal fraction: a fraction when written with a decimal point, so 1.5 is the decimal fraction equal to $\frac{3}{2}$.

Denominator: the number beneath the line in a fraction.

Diameter: the width of a circle.

Distribution: a spread of possible outcomes and their likelihood of occurring.

Divergent series: an infinite series that does not add up to a finite number.

Divisor: a natural number that divides into another number with no remainder; so 5 is a divisor of 20.

Dyscalculia: a condition affecting one's ability to grasp numbers.

Edge: the chance of winning a gamble minus the chance of losing.

Egyptian triangle: a triangle whose sides have the ratios 3:4:5.

Expected value: the theoretical value of how much you can expect to win or lose in a bet.

Exponent: the power of a number, written as a raised symbol, such as the x in 3^x.

Factor: a divisor of a given number.

Factorize: to break a number down into its factors, usually just the ones that are prime numbers.

Fibonacci number: a number in the Fibonacci series, which begins 1, 1, 2, 3, 5, 8, 13…

Gambler's fallacy: the false idea that random outcomes are not random.

Gambler's ruin: the inevitability of bankruptcy if you gamble for long enough.

Gaussian distribution: the normal distribution.

Geometric progression: a sequence of numbers in which each new term is calculated by multiplying the previous term by a fixed number.

Hyperbolic plane: an infinitely large surface with negative curvature.

Hypotenuse: the side of a right-angled triangle opposite the right angle.

Infinite series: a series with an infinite number of terms.

Integer: a number that is one of the natural numbers, negative natural numbers or zero.

Inversion: the same thing as an ambigram.

Irrational number: a number that cannot be expressed as a fraction.

Latin square: a square grid where every element occurs only once in each row and column.

Law of large numbers: the rule that probability works out in the long run, in that the more examples of a random event you have (such as the flipping of a coin), the closer the actual results will be to the expected results.

Law of very large numbers: the rule that if the sample is large enough, then any result may occur no matter how unlikely.

Logarithm: if $a = 10^b$, then the logarithm of a is b, written $\log a = b$.

Magic square: a square grid containing consecutive numbers from 1 such that the sum of all rows, columns and diagonals are equal.

Mersenne prime: a prime number that can be expressed as $2^n - 1$.

Natural number: any whole number that can be reached counting upwards from 1.

Normal distribution: the most common type of distribution, which produces the bell curve.

Normal number: a number whose decimal digits consist of an equal number of 0s, 1s, 2s, 3s, 4s, 5s, 6s, 7s, 8s and 9s.

Number line: a visual representation of numbers as points on a continuous line.

Numerator: the number above the line in a fraction.

Order of magnitude: Most commonly, the scale of a number based on the positional value of the leftmost digit of that number. So, the order of magnitude of any number between 1 and 9 is one, between 10 and 99 is two, between 100 and 999 is three, and so on.

Parallel: two straight lines that never meet are parallel.

Perfect number: a number that is equal to the sum of its divisors (excluding itself).

Phi: the mathematical constant, the decimal expansion of which begins 1.618…, and is also known as the golden ratio, or the divine proportion.

Pi: the mathematical constant, the decimal expansion of which begins 3.14159265358979323846…, and that is equal to the ratio between the circumference and the diameter of a circle.

Platonic solid: the five solids whose sides are all identical regular polygons; in other words, the tetrahedron, cube, octahedron, icosahedron and dodecahedron.

Polygon: a two-dimensional closed shape made up of a finite number of straight lines.

Postulate: a statement that is assumed to be true, and used as an axiom.

Power: an operation that determines how many times a number is to be multiplied by itself, so if 10 is to be multiplied by itself four times, one writes 10^4 and calls this 'ten to the power of four'. Powers are not always natural numbers, but when one talks about the 'powers of *x*' it is assumed that one is referring only to the powers of *x* that are.

Prime number: a natural number that has only two divisors, itself and 1.

Probability: the chance of an event taking place, expressed as a fraction between 0 and 1.

Quadratic equation: an equation of the form $ax^2 + bx + c = 0$, where a, b, and c are constants and a is non-zero.

Radius: a straight line from the centre of a circle to the circumference.

Random walk: a visual interpretation of randomness, in which each random event is expressed as movement in a random direction.

Rational number: a number that can be expressed as a fraction.

Regression to the mean: the phenomenon that after extreme events less extreme ones are more likely.

Regular polygon: a polygon with sides of equal length and with equal internal angles.

Sequence: a list of numbers.

Series: the sum of the terms in a sequence.

Set: a collection of things.

Tessellation: an arrangement of tiles that fills a two-dimensional space completely with no overlaps.

Theorem: a statement that can be proved from other theorems and/or axioms.

Transcendental number: a number that cannot be expressed as a solution to a finite equation.

Uncountable infinity: an infinite set whose members cannot be put in a one-to-one correspondence with the natural numbers.

Unique solution: the situation when there is one and only one possible answer.

Variable: a quantity that can vary in value.

Vertex: where two lines meet to form an angle, or the angular points in a three-dimensional shape.

Appendix One

In order to see how Annairizi's tiled squares prove Pythagoras's Theorem, look at the marked triangle on p.88. All that we need to do is to rearrange the square of the hypotenuse into precisely the squares of the other two sides. The square of the hypotenuse is made up of five sections; three are light grey and two are dark grey. We can see by considering how the pattern repeats that the light grey sections make up exactly the square of one of the sides of the triangle, and that the dark grey sections make up the square of the other side.

For Leonardo's proof, first we need to show that the shaded sections in (i) and (ii) below are equal. We do this by rotating the section around point P. The two sections have the same side lengths and angles and therefore must be the same. Then we need to show that this section is equal to the section in (iii). This must be true since it is made up of identical parts.

With this information, we can complete the proof. The reflection of the first shaded section and its mirror image across the dotted line consists of two identical right-angled triangles and the squares on its two smaller sides. This area must be equal to the area covered by the sections shaded in (ii) and (iii) together, which consists of two identical right-angled triangles and the square of the hypotenuse. If we subtract the area of two triangles from both of these cases, the square of the hypotenuse must be equal to the square on the other two sides.

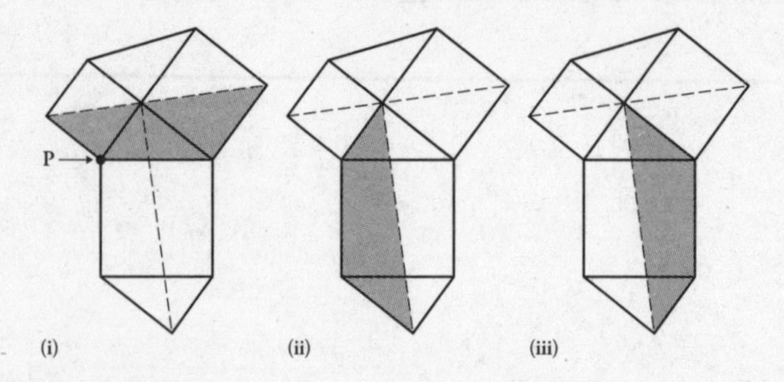

Appendix Two

In a unit square, the diagonal has length $\sqrt{2}$. To show that this is irrational I will use a *proof by contradiction* in which it is assumed that $\sqrt{2}$ is rational, and then I will show how this leads to a contradiction. If it is contradictory to say that $\sqrt{2}$ is rational, it must be irrational.

If $\sqrt{2}$ is rational, then there are natural numbers a and b, such that $\sqrt{2} = \frac{a}{b}$. Let's insist that this is the most reduced form of the fraction, so there is no way of rewriting $\frac{a}{b}$ as $\frac{m}{n}$ when m and n are natural numbers less than a and b.

If $\sqrt{2} = \frac{a}{b}$, then by squaring both sides of the equation, $2 = \frac{a^2}{b^2}$ which we can rewrite as $a^2 = 2b^2$.

Whatever the value of b^2, $2b^2$ must be even since multiplying any natural number by two makes an even number. If $2b^2$ is even, then a^2 is even. Now, since the square of an odd number is always an odd number, and the square of an even number is always even, this means that a must be even.

If a is even, then there is a number c less than a such that $a = 2c$, and therefore that $a^2 = (2c)^2 = 4c^2$.

By replacing a^2 with $4c^2$ in the equation above, we get $4c^2 = 2b^2$. Which reduces to $b^2 = 2c^2$. By the same reasoning, this means that b^2 is even, so b is even. If b is even, there is a number d less than b such that $b = 2d$.

Therefore $\frac{a}{b}$ can be rewritten $\frac{2c}{2d}$, or $\frac{c}{d}$ since the 2s cancel. We have our contradiction! From above we have stipulated that $\frac{a}{b}$ is the most reduced form of the fraction, which means that there are no values for c and d less than a and b such that $\frac{a}{b} = \frac{c}{d}$. Since we have arrived at a contradiction by assuming that $\sqrt{2}$ can be written as $\frac{a}{b}$ it must be the case that $\sqrt{2}$ cannot be written in such a way, so $\sqrt{2}$ is irrational.

Appendix Three

In Franklin's 16 × 16 magic square all the lines and columns add up to 2056. It is not a true magic square since the diagonals do not add up to 2056, but it is so rich in properties that Clifford A. Pickover writes that 'it is no exaggeration to say that one could spend a lifetime contemplating its wonderful structure'. For example, every 2 × 2 subsquare (and there are 225 of those) adds up to 514, which means that every 4 × 4 subsquare adds up to 2056. Many other symmetries and patterns are also contained in the square.

200	217	232	249	8	25	40	57	72	89	104	121	136	153	168	185
58	39	26	7	250	231	218	199	186	167	154	135	122	103	90	71
198	219	230	251	6	27	38	59	70	91	102	123	134	155	166	187
60	37	28	5	252	229	220	197	188	165	156	133	124	101	92	69
201	216	233	248	9	24	41	56	73	88	105	120	137	152	169	184
55	42	23	10	247	234	215	202	183	170	151	138	119	106	87	74
203	214	235	246	11	22	43	54	75	86	107	118	139	150	171	182
53	44	21	12	245	236	213	204	181	172	149	140	117	108	85	76
205	212	237	244	13	20	45	52	77	84	109	116	141	148	173	180
51	46	19	14	243	238	211	206	179	174	147	142	115	110	83	78
207	210	239	242	15	18	47	50	79	82	111	114	143	146	175	178
49	48	17	16	241	240	209	208	177	176	145	144	113	112	81	80
196	221	228	253	4	29	36	61	68	93	100	125	132	157	164	189
62	35	30	3	254	227	222	195	190	163	158	131	126	99	94	67
194	223	226	255	2	31	34	63	66	95	98	127	130	159	162	191
64	33	32	1	256	225	224	193	192	161	160	129	128	97	96	65

Appendix Four

The principle behind Gijswijt's sequence is to look for repeating blocks of numbers in the previous terms of the sequence. The 'block' has to be at the end of the sequence of previous terms, and the number of times it is repeated provides the next term.

Mathematically, the sequence is described as follows. Start with a 1, and then each subsequent term is the value k, when the previous terms are multiplied in order and written xy^k for the largest possible value of k.

The sequence is 1, 1, 2, 1, 1, 2, 2, 2, 3, 1, 1, 2, 1, 1, 2, 2, 2, 3, 1, 1 …

I think it is easiest to understand what is at work here by considering the first time a 3 appears, which is in position 9. The previous terms multiplied in order are $1 \times 1 \times 2 \times 1 \times 1 \times 2 \times 2 \times 2$. What Gijswijt requires us to do is to transform this sum into a term xy^k for the largest value of k. In this case, we get $(1 \times 1 \times 2 \times 1 \times 1) \times 2^3$. So, the following term is a 3. We are looking for the largest repeating block of numbers at the end of the sequence of previous terms, although in this case the block is a single number, 2, repeated three times.

But often the block will have several digits. Consider position 16. The previous terms multiplied together are $1 \times 1 \times 2 \times 1 \times 1 \times 2 \times 2 \times 2 \times 3 \times 1 \times 1 \times 2 \times 1 \times 1 \times 2$. This can be written $(1 \times 1 \times 2 \times 1 \times 1 \times 2 \times 2 \times 3) \times (1 \times 1 \times 2)^2$. So, the 16th term is a 2.

Going back to the beginning now, the second term is a 1 since the previous term 1 is not multiplied by anything. The third term is a 2 since the previous terms multiplied in order are $1 \times 1 = 1^2$, and the fourth term is 1 since the previous terms result in $(1 \times 1) \times 2$, where the final 2 is not multiplied by itself.

Appendix Five

We want to show that the harmonic series diverges, in other words that

$$1 + \frac{1}{2} + \frac{1}{3} + \frac{1}{4} + \frac{1}{5} + \ldots$$

will add up to infinity. This is done by showing that the harmonic series is larger than the following series, which does add up to infinity:

$$\frac{1}{2} + \frac{1}{2} + \frac{1}{2} + \frac{1}{2} + \frac{1}{2} + \ldots$$

Let's compare terms of the harmonic series in groups of two, four, eight and so on, starting from the third term. They are listed below. Because $\frac{1}{3}$ is bigger than $\frac{1}{4}$, $\frac{1}{3} + \frac{1}{4}$ must be bigger than $\frac{1}{4} + \frac{1}{4}$ which is $\frac{1}{2}$. Likewise, since $\frac{1}{5}$, $\frac{1}{6}$ and $\frac{1}{7}$ are all bigger than $\frac{1}{8}$, this means that $\frac{1}{5} + \frac{1}{6} + \frac{1}{7} + \frac{1}{8}$ is bigger than four eighths, which is also $\frac{1}{2}$. If we carry on, always considering double the number of terms, we can see that we will be able to add up these terms to make a value larger than $\frac{1}{2}$:

3rd and 4th terms	$\frac{1}{3} + \frac{1}{4}$	$>$	$\frac{1}{4} + \frac{1}{4} = \frac{1}{2}$
5th to 8th terms	$\frac{1}{5} + \frac{1}{6} + \frac{1}{7} + \frac{1}{8}$	$>$	$4(\frac{1}{8}) = \frac{1}{2}$
9th to 16th terms	$\frac{1}{9} + \ldots + \frac{1}{16}$	$>$	$8(\frac{1}{16}) = \frac{1}{2}$

…

The harmonic series, therefore, is bigger than $\frac{1}{2} + \frac{1}{2} + \frac{1}{2} + \frac{1}{2} + \frac{1}{2} + \ldots$, which is infinity times a half, which is infinity. So the harmonic series is bigger than infinity; in other words, it is infinite.

Appendix Six

The *continued fraction* is a strange type of fraction constructed by an infinite process of additions and divisions.

When phi is expressed as a continued fraction it looks like this:

$$\text{phi} = 1 + \cfrac{1}{1 + \cfrac{1}{1 + \cfrac{1}{1 + \ldots}}}$$

To understand how this works, let's take the fraction line by line and see that it closes in on phi:

$$1$$

$$1 + 1 = 2$$

$$1 + \frac{1}{1+1} = 1.5$$

$$1 + \frac{1}{1 + \frac{1}{1+1}} = 1 + \frac{2}{3} = 1.66\ldots$$

$$1 + \frac{1}{1 + \frac{1}{1 + \frac{1}{1+1}}} = 1.6$$

And so on.

Continued fractions provide mathematicians with a way of rating how irrational a number might be. Since the expression for phi contains only 1s, it is the 'purest' continued fraction that there is, and hence is considered the 'most irrational' number.

Notes on Chapters

During the writing of this book there were four tomes that were always on my desk, and whose contribution cannot be isolated to any individual chapter. Martin Gardner remains peerless in popular maths for his erudition, wit and clarity. Tobias Dantzig's *Number* is a classic about the cultural evolution of mathematics. Both the Ifrah and the Cajori are painstakingly well researched and endlessly fascinating.

Cajori, F., *A History of Mathematical Notations*, Dover, 1993 (facsimile of original by Open Court, Illinois, 1928/9)
Dantzig, T., *Number*, Plume, New York, 2007 (originally Macmillan, 1930)
Gardner, M., *Mathematical Games: The Entire Collection of His Scientific American Columns*, Mathematical Association of America, 2005
Ifrah, G., *The Universal History of Numbers*, John Wiley, New York, 2000

CHAPTER ZERO

This chapter is based on conversations in London with Brian Butterworth, and in Paris with Stanislas Dehaene and Pierre Pica. At University College London I was screened for dyscalculia by Teresa Iuculano and Marinella Cappelletti, with a computer program now used in schools in the UK. I'm not dyscalculic, which is probably no great surprise. If you would like to help support the Munduruku's protection of their traditional education and environment, donations can be sent to: The Munduruku Fund, The Arrow Rainforest Foundation, 5 Southridge Place, London SW20 8JQ, United Kingdom. More details can be found on:
www.thearrowrainforestfoundation.com

Butterworth, B., *The Mathematical Brain*, Macmillan, London, 1999
Dehaene, S., *The Number Sense*, Oxford University Press, Oxford, 1997
Matzusawa, T. (ed.), *Primate Origins of Human Cognition and Behavior*, Springer, Tokyo, 2001

Angier, N., 'Gut Instinct's Surprising Role in Math', *New York Times*, 2008
Dehaene, S., Izard, V., Spelke, E., and Pica, P., 'Log or Linear?', *Science*, 2008
Inoue, S., and Matsuzawa, T., 'Working memory of numerals in chimpanzees', *Current Biology*, 2007
Pica, P., Lerner, C., Izard, V., and Dehaene, S., 'Exact and Appropriate Arithmetic in an Amazonian Indigene Group', *Science*, 2004
Siegler, R.S., and Booth, J.L., 'Development of Numerical Estimation in Young Children', *Child Development*, 2004

CHAPTER ONE

Anyone wanting more information about the joys of base 12 can reach the Dozenal Society of America at contact@Dozenal.org, or 5106 Hampton Avenue Suite 205, Saint Louis, Missouri 63109-3115, USA. *Little Twelvetoes* is a classic of *Schoolhouse Rock!*, a series of musical cartoons about maths, science and grammar from the 1970s that can all be seen on the internet. My entry into the abacus world was only made possible by Kouzi Suzuki, a one-man *soroban* evangelist, who met me at a Tokyo rail station dressed up as Sherlock Holmes.

Andrews, F.E., *New Numbers*, Faber & Faber, London, 1936
Duodecimal Society of America, Inc., *Manual of the Dozen System*, Duodecimal Society of America, New York, 1960
Elbrow, Rear-Admiral G., *The New English System of Money, Weights and Measures and of Arithmetic*, P.S. King & Son, London, 1913
Essig, J., *Douze, notre dix future*, Dunod, Paris, 1955
Glaser, A., *History of Binary and Other Nondecimal Numeration*, Southampton, PA, 1971
Kawall Leal Ferreira, M. (ed.), *Idéias Matemáticas de Povos Culturalmente Distintos*, Global Editora, São Paulo, 2000
Suzuki, K., *Lectures on Soroban*, Institute for English Yomiagezan

Dowker, A., and Lloyd D., 'Linguistic influences on numeracy', *Education Transactions*, University of Bangor, 2005
Wassmann, J., and Dasen, P.R., 'Yupno Number System and Counting', *Cross-cultural Psychology Journal*, 1994
Hammarström, H., 'Rarities in Numeral Systems', 2007

CHAPTER TWO

Proofs Without Words is a gem, and was my source for the different Pythagoras proofs. Thanks to Tom Hull for much of the background about origami. The illustrations of how to make business-card tetrahedrons and cubes are inspired by his book. Another remarkable Japanese religio-geometric practice is *sangaku*, which didn't fit in the chapter but is too fascinating not to mention here. A *sangaku* is a wooden tablet hung at a Buddhist or Shinto shrine that has a proof of a geometric problem painted on it. Between the seventeenth and nineteenth centuries, thousands of *sangaku* were made by Japanese who had worked out geometrical problems but could not afford to publish them in books. Drawing the solutions on a tablet and hanging them at a shrine was a way of making a religious offering while also advertising their results.

Shortly before going to press, I learnt that Jerome Carter had died in a motorcycle accident in 2009.

Balliett, L.D., *The Philosophy of Numbers*, L.N. Fowler, 1908
Bell, E.T., *Numerology*, Century, 1933
Dudley, U., *Numerology*, Mathematical Association of America, 1997
du Sautoy, M., *Finding Moonshine*, Fourth Estate, London, 2008
Ferguson, K., *The Music of Pythagoras*, Walker, New York, 2008
Hull, T., *Project Origami*, A.K. Peters, Wellesley, MA, 2006
Kahn, C.H., *Pythagoras and the Pythagoreans, a Brief History*, Hackett, Indianapolis, IN, 2001
Loomis, E.S., *The Pythagorean Proposition*, Edwards Bros, Ann Arbor, MI, 1940
Maor, E., *The Pythagorean Theorem*, Princeton University Press, NJ, 2007
Mlodinow, L., *Euclid's Window*, Free Press, New York, 2001
Nelsen, R.B., *Proofs Without Words*, Mathematical Association of America, Washington DC, 1993
Riedwig, C., *Pythagoras, His Life, Teaching and Influence*, Cornell University Press, Ithaca, NY, 2002
Schimmel, A., *The Mystery of Numbers*, Oxford University Press, New York, 1993
Simoons, F.J., *Plants of Life, Plants of Death*, University of Wisconsin Press, Madison, WI, 1998
Sundara Rao, T., *Geometric Exercises in Paper Folding*, Open Court, Chicago, IL, 1901

Bolton, N.J., and MacLeod, D.N.G., 'The Geometry of the Sri Yantra', *Religion*, vol. 7, 1977
Burnyeat, M.F., 'Other Lives', *London Review of Books*, 2007

CHAPTER THREE

Even though the *Liber Abaci* was first published in 1202, its first English translation did not appear until its 800th anniversary, in 2002. Vedic mathematics is not the only type of speed arithmetic in the market. There are several 'systems' and many of them share the same tricks. The best known is the Trachtenberg System, devised by Jakow Trachtenberg while a political prisoner in a Nazi concentration camp. Self-styled 'mathemagician' Arthur Benjamin is an entertaining, recent purveyor of the speed arithmetician's art.

Fibonacci, L., *Fibonacci's Liber Abaci*, Springer, New York, 2002
Joseph, G.G., *Crest of the Peacock*, Penguin, London, 1992
Knott, K., *Hinduism: A Very Short Introduction*, Oxford University Press, 1998
Seife, C., *Zero*, Souvenir Press, London, 2000
Tirthaji, Jagadguru Swami S. B. K., *Vedic Mathematics*, Motilal Banarsidass, Delhi, 1992

Dani, S.G., 'Myths and reality: On "Vedic mathematics"'

CHAPTER FOUR

The least dweeby contestant in Leipzig was Rüdiger Gamm, a former bodybuilder who failed maths at school. After a career with exaggeratedly large biceps, he now has an exaggeratedly large brain. Gamm, whose calculation skills have made him a minor celebrity in Germany, told me that memory is his greatest asset: 'I think I have [stored] 200,000 to 300,000 numbers in my head.'

(I found this chapter a challenge because of having to restrain myself from the temptation to make terrible puns about pi. Mathematicians have a congenital propensity to overpun. When we see a word we can't help but break it down and rearrange it, which probably also explains why the world's top Scrabble players are maths and computer-science graduates, not linguists.)

Arndt, J., and Haenel, C., *Pi Unleashed*, Springer, London, 2002
Beckmann, P., *A History of Pi*, St Martin's Press, New York, 1971
Berggren L., Borwein J., and Borwein P., *Pi: A Source Book*, Springer, London, 2003
Bidder, G., *A short Account of George Bidder, the celebrated Mental Calculator: with a Variety of the most difficult Questions, Proposed to him at the principal Towns in the Kingdom, and his surprising rapid Answers!*, W.C. Pollard, 1821
Colburn, Z., *A memoir of Zerah Colburn, written by himself*, G. & C. Merriam, Springfield, MA, 1833
Rademacher, H., and Torplitz, O., *The Enjoyment of Mathematics*, Princeton University Press, NJ, 1957

Aitken, A.C., 'The Art of Mental Calculation; with Demonstrations', *Society of Engineers Journal and Transactions*, 1954
Preston, R., 'The Mountains of Pi', *New Yorker*, 1992

CHAPTER FIVE

Acheson, D., *1089 and all that*, Oxford University Press, Oxford, 2002
Berlinski, D., *Infinite Ascent*, The Modern Library, New York, 2005
Dale, R., *The Sinclair Story*, Duckworth, London, 1985
Derbyshire, J., *Unknown Quantity*, Atlantic Books, London, 2006
Hopp, P.M., *Slide Rules, Their History, Models and Makers*, Astragal Press, New Jersey, 1999
Maor, E., *e: The Story of a Number*, Princeton University Press, NJ, 1994
Rade, L., and Kaufman, B.A., *Adventures with Your Pocket Calculator*, Pelican, London, 1980
Schlossberg, E., and Brockman, J., *The Pocket Calculator Game Book*, William Morrow, New York, 1975
Vine, J., *Fun & Games with Your Electronic Calculator*, Babani Press,

London, 1977 (published in the US as *Boggle*, Price, Stern, Sloane Publishers, Los Angeles, CA, 1975)

CHAPTER SIX

In May 2010, a month after the first edition of this book was published, Martin Gardner died. He was 95, and still working. Two months later Tom Rokicki and his collaborators finally proved that God's number was 20, using 35 years of computer time donated by Google.

I found Dudeney's articles in *Strand Magazine* brilliantly well written, irrespective of the genius of the puzzles, and well worth a read. I am grateful to Angela Newing, the world expert on Henry Dudeney, for some of the biographical details, and to Jerry Slocum, for solving all my other puzzles about puzzles. If anyone wants an ambigram tattoo, check out Mark Palmer's creations at www.wowtattoos.com.

Bachet, C.G., *Amusing and Entertaining Problems that can be Had with Numbers (very useful for inquisitive people of all kinds who use arithmetic)*, Paris, 1612

Bodycombe, D.J., *The Riddles of the Sphinx*, Penguin, London, 2007

Danesi, M., *The Puzzle Instinct*, University of Indiana Press, Indianapolis, IN, 2002

Elffers, J., and Schuyt, M., *Tangram*, 1997

Gardner, M., *Mathematics, Magic and Mystery*, Dover, New York, 1956

Hardy, G.H., *A Mathematician's Apology*, Cambridge University Press, Cambridge, 1940

Hooper, W., *Rational Recreations, in which the principles of Numbers and Natural Philosophy are clearly and copiously elucidated by a series of easy, entertaining, interesting experiments, among which are all those commonly performed with the cards*, London, 1774

Loyd, S., *The 8th Book of Tan Part I*, 1903; new edition Dover, New York, 1968

Maor, E., *Trigonometric Delights*, Princeton University Press, NJ, 1998

Netz, R., and Noel, W., *The Archimedes Codex*, Weidenfeld & Nicolson, London, 2007

Pasles, P.C., *Benjamin Franklin's Numbers*, Princeton University Press, NJ, 2008

Pickover, C.A., *The Zen of Magic Squares, Circles and Stars*, Princeton University Press, NJ, 2002

Rouse Ball, W.W., *Mathematical Recreations and Problems*, Macmillan, London, 1892

Slocum, J., *The Tangram Book*, Sterling, New York, 2001

Slocum, J., and Sonneveld, D., *The 15 Puzzle*, Slocum Puzzle Foundation, California, 2006

Swetz, F.J., *Legacy of the Luoshu*, Open Court, Chicago, IL, 2002

Dudeney, H., 'Perplexities', column in *Strand Magazine*, London, 1910–30
Singmaster, D., 'The unreasonable utility of recreational mathematics', lecture at the First European Congress of Mathematics, Paris, July 1992

CHAPTER SEVEN

The *On-Line Encyclopedia of Integer Sequences* (www.research.att.com/~njas/sequences/) looks quite daunting at first to the non-specialist, but once you get the hang of it, is fascinating to surf. I found Chris Caldwell's online encyclopedia of primes, *The Prime Pages* (www.primes.utm.edu) an excellent resource.

Doxiadis, A., *Uncle Petros and Goldbach's Conjecture*, Faber & Faber, London, 2000
du Sautoy, M., *The Music of the Primes*, Fourth Estate, London, 2003
Reid, C., *From Zero to Infinity*, Thomas Y. Crowell, New York, 1955

Schmelzer, T., and Baillie, R., 'Summing a curious, slowly convergent series', *American Mathematical Monthly*, July 2008
Sloane, N.J.A., 'My Favorite Integer Sequences', 2000

CHAPTER EIGHT

It's a curious quirk that pi, phi and Fibonacci sound related when their etymologies are all completely different, although conspiracy theorists might not be convinced. Separating the cranks from the non-cranks when it comes to the golden ratio is not always easy. One definite non-crank is Ron Knott, whose website: www.computing.surrey.ac.uk/personal/ext/R.Knott/Fibonacci/ has all you ever wanted to know about 1.618…

Livio, M., *The Golden Ratio*, Review, London, 2002
Posamentier, A.S., and Lehmann, I., *The (Fabulous) Fibonacci Numbers*, Prometheus Books, New York, 2007

McManus, I.C., Cook, R., and Hunt, A., 'Beyond the Golden Section and normative aesthetics: why do individuals differ so much in their aesthetic preferences for rectangles?', *Perception*, vol. 36, 2007

CHAPTER NINE

The Kelly strategy is a lot more than just remembering the fraction $\frac{\text{edge}}{\text{odds}}$, since gambling situations are usually more complex than the very simple one I described. I apologize to Ed Thorp, who asked hopefully during our interview if I would be able to spell out Kelly in proper detail. Sorry, Ed, it's just too complicated for the scope of this book! William Poundstone's terrific book was a guiding light and I'm grateful he supplied me with data for the graph on p. 344.

Aczel, A.D., *Chance*, High Stakes, London, 2005
Bennett, D.J., *Randomness*, Harvard University Press, Cambridge, MA, 1998
Devlin, K., *The Unfinished Game*, Basic Books, New York, 2008
Haigh, J., *Taking Chances*, Oxford University Press, Oxford, 1999
Kaplan, M., and Kaplan, E., *Chances Are*, Penguin, New York, 2006
Mlodinow, L., *The Drunkard's Walk*, Allen Lane, London, 2008
Paulos, J.A., *Innumeracy*, Hill & Wang, New York, 1988
Poundstone, W., *Fortune's Formula*, Hill & Wang, New York, 2005
Rosenthal, J.S., *Struck by Lightning*, Joseph Henry Press, Washington DC, 2001
Thorp, E.O., *Beat the Dealer*, Vintage, New York, 1966
Tijms, H., *Understanding Probability*, Cambridge University Press, 2007
Venn, J., *The Logic of Chance*, Macmillan, London, 1888

CHAPTER TEN

Statistics is the one field of maths covered in this book that I never studied at school or college, so much of this was very new to me. Some mathematicians don't even consider statistics proper maths, occupied as it is with messy things like measurement. I enjoyed getting my hands dirty, although I'm not going back to Greggs for a very long time.

Blastland, M., and Dilnot, A., *The Tiger That Isn't*, Profile, London, 2007
Brookes, M., *Extreme Measures*, Bloomsbury, London, 2004
Cline Cohen, P., *A Calculating People: The Spread of Numeracy in Early America*, University of Chicago Press, IL, 1982
Cohen, I. B., *The Triumph of Numbers*, W. W. Norton, New York, 2005
Edwards, A.W.F., *Pascal's Arithmetical Triangle*, Johns Hopkins University Press, Baltimore, MD, 1987
Kuper S., and Szymanski S., *Why England Lose*, HarperCollins, London, 2009
Taleb, N.N., *The Black Swan*, Penguin, London, 2007

CHAPTER ELEVEN

While it is still an open question whether the universe is flat, spherical or hyperbolic, the universe is certainly pretty flat; if its curvature does indeed deviate from zero, it does so only very slightly. An irony of testing the universe for its curvature, however, is that it can never be conclusively proved that the universe is flat since there will always be measurement error. By contrast, it is theoretically possible to prove that the universe is curved, which would happen if the results produce a curvature, accounting for measurement error, that is non-zero.

The Hilbert Hotel sometimes goes by the name of Hotel Infinity, and the story has many different versions. The guests wearing T-shirts is my own adaptation.

Aczel, A.D., *The Mystery of the Aleph*, Washington Square Press, New York, 2000
Barrow, J.D., *The Infinite Book*, Jonathan Cape, London, 2005
Foster Wallace, D., *Everything and More*, W. W. Norton, New York, 2003
Kaplan, R., and Kaplan, E., *The Art of the Infinite*, Allen Lane, London, 2003
O'Shea, D., *The Poincaré Conjecture*, Walker, New York, 2007

Taimina, D., and Henderson, D.W., 'How to Use History to Clarify Common Confusions in Geometry', *Mathematical Association of America Notes*, 2005

INTERNET

It's impossible to research anything to do with maths without referring to Wikipedia and Wolfram MathWorld (www.mathworld.wolfram.com), which I conferred with on a daily basis.

GENERAL

The number of books I looked through is too long to list all of them here, but these ones directly contributed in one way or another to the material in this book. Anything by Keith Devlin, Clifford A. Pickover or Ian Stewart is always worth a read.

Bell, E.T., *Men of Mathematics*, Victor Gollancz, London, 1937
Bentley, P.J., *The Book of Numbers*, Cassell Illustrated, London, 2008
Darling, D., *The Universal Book of Mathematics*, Wiley, Hoboken, NJ, 2004
Devlin, K., *All the Math That's Fit to Print*, Mathematical Association of America, Washington DC, 1994
Dudley, U. (ed.), *Is Mathematics Inevitable?*, Mathematical Association of America, Washington DC, 2008
Eastaway, R., and Wyndham, J., *Why Do Buses Come in Threes?*, Robson Books, London, 1998
Eastaway, R., and Wyndham, J., *How Long is a Piece of String?*, Robson Books, London, 2002
Gowers, T., *Mathematics: A Very Short Introduction*, Oxford University Press, Oxford, 2002
Gullberg, J., *Mathematics*, W. W. Norton, New York, 1997
Hodges, A., *One to Nine*, Short Books, London, 2007
Hoffman, P., *The Man Who Loved Only Numbers: The Story of Paul Erdös and the Search for Mathematical Truth*, Fourth Estate, 1998
Hogben, L., *Mathematics for the Million*, Allen & Unwin, London, 1936
Mazur, J., *Euclid in the Rainforest*, Plume, New York, 2005
Newman, J. (ed.), *The World of Mathematics*, Dover, New York, 1956
Pickover, C.A., *A Passion for Mathematics*, Wiley, Hoboken, NJ, 2005
Singh, S., *Fermat's Last Theorem*, Fourth Estate, London, 1997

Acknowledgements

Firstly, thanks to Claire Paterson at Janklow & Nesbit, without whose encouragement this book would never have been written, and to my editors Richard Atkinson in London and Emily Loose in New York. I'm also very grateful to Andy Riley for his wonderful illustrations.

The success of my trips relied on the support of friends old and new: in Japan, Chieko Tsuneoka, Richard Lloyd Parry, Fiona Wilson, Kouzi Suzuki, Masao Uchibayashi, Tetsuro Matsuzawa, Chris Martin and Leo Lewis. In India, Gaurav Tekriwal, Dhananjay Vaidya and Kenneth Williams. In Germany, Ralf Laue. In the US, Colm Mulcahy, Tom Rodgers, Tom Hull, Neil Sloane, Jerry Slocum, David Chudnovsky, Gregory Chudnovsky, Tom Morgan, Michael de Vlieger, Jerome Carter, Anthony Baerlocher and Ed Thorp. In the UK, Brian Butterworth, Peter Hopp and Eddy Levin.

The manuscript is much improved thanks to comments from Robert Fountain, Colin Wright, Colm Mulcahy, Tony Mann, Alex Paseau, Pierre Pica, Stefanie Marsh, Matthew Kershaw, John Maingay, Morgan Ryan, Andreas Nieder, Daina Taimina, David Henderson, Stefan Mandel, Robert Lang, David Bellos and Ilona Morison. And thanks also to Natalie Hunt, Simon Veksner, Veronica Esaulova, Gavin Pretor-Pinney, Justin Leighton, Jeannine Mosely, Ravi Apte, Hugo de Klee, Maura O'Brien, Peter Dawson, Paul Palmer-Edwards, Elaine Leggett, Rebecca Folland, Kirsty Gordon, Tim Glister, Hugh Morison, Jonathan Cummings, Raphael Zarum, Mike Keith, Gareth Roberts, Gene Zirkel, Erik Demaine, Wayne Gould, Kirk Pearson, Angela Newing, Bill Eadington, Mike LeVan, Sheena Russell, Hartosh Bal, Ivan Moscovich, John Holden, Chris Ottewill, Mariana Kawall Leal Ferreira, Todd Rangiwhetu, William Poundstone, Frank Swetz and Amir Aczel. And lastly, Zara Bellos, my niece, who has promised to get an A star in maths if I mention her somewhere in these pages.

Picture Credits

p. 26 © Tetsuro Matsuzawa.
p. 48 Courtesy of Jürg Wassmann.
pp. 50–55 Dozenal fonts and DSA logo courtesy of Michael de Vlieger, Dozenal Society of America.
p. 59 From the collection of the Musée international d'horlogerie, La Chaux-de-Fonds, Switzerland.
p. 101 © Sir Roger Penrose.
p. 106 © Ravi Apte.
p. 189 Science Museum/Science & Society Picture Library.
p. 193 Science Museum/Science & Society Picture Library.
p. 210 © Justin Leighton.
p. 214 © Maki Kaji.
p. 220 © Jose/Fotolia.
p. 221 Courtesy of the Clendening History of Medicine Library, University of Kansas Medical Center.
p. 232 Courtesy of Jerry Slocum.
p. 235 Courtesy of Jerry Slocum.
pp. 238–9 Get Off The Earth is a registered trademark of the Sam Loyd Company and used by permission.
p. 244 Reproduced with permission of Scott Kim.
p. 245 Courtesy of Mark Palmer, Wow Tattoos.
p. 250 © Dániel Erdély, Walt van Ballegooijen and the Spidron Team, 2008. Artwork: Dániel Erdély.
p. 270 Courtesy of Paul Bateman.
p. 286 Used under license from Shutterstock.com.
p. 294 Kasia, 2009. Used under license from Shutterstock.com.
p. 301 © Alex Bellos.
p. 316 Scott W. Klette, courtesy of Nevada State Museum, Carson City, NV.
p. 344 Courtesy William Poundstone.
p. 384 © Daina Taimina.
p. 393 © The M.C. Escher Company, Holland, 2009. All rights reserved. www.mcescher.com.
p. 396 © Daina Taimina.

Index

Page numbers in *italic* refer to the illustrations

INDEX